Self-Assembly

Wiley Series on Surface and Interfacial Chemistry

Ponisseril Somasundaran and Nissim Garti, Series Editors

Multiple Emulsion: Technology and Applications
By A. Aserin
November 2007

Colloidal Nanoparticles in Biotechnology
Edited by Abdelhamid Elaissari
April 2008

Self-Assembled Supramolecular Architectures: Lyotropic Liquid Crystals
Edited by Nissim Garti, Ponisseril Somasundaran, and Raffaele Mezzenga
September 2012

Proteins in Solution and at Interfaces: Methods and Applications in Biotechnology and Materials Science
Edited by Juan M. Ruso and Ángel Piñeiro
March 2013

Ionic Liquid-Based Surfactant Science: Formulation, Characterization and Applications
Edited by Bidyut K. Paul and Satya P. Moulik
September 2015

Fluids, Colloids and Soft Materials: An Introduction to Soft Matter Physics
Edited by Alberto Fernandez-Nieves and Antonio Manuel Puertas
May 2016

Self-Assembly: From Surfactants to Nanoparticles
Edited by Ramanathan Nagarajan
January 2019

Self-Assembly

From Surfactants to Nanoparticles

Edited by

Dr. Ramanathan Nagarajan
Natick Soldier Research
Development & Engineering Center
Natick, USA

WILEY

This edition first published 2019

Registered Office
John Wiley & Sons, Inc., 111 River Street, Hoboken, NJ 07030, USA

Editorial Office
111 River Street, Hoboken, NJ 07030, USA

For details of our global editorial offices, customer services, and more information about Wiley products visit us at www.wiley.com.

Wiley also publishes its books in a variety of electronic formats and by print-on-demand. Some content that appears in standard print versions of this book may not be available in other formats.

Library of Congress Cataloging-in-Publication Data

Names: Nagarajan, R. (Ramanathan), editor.
Title: Self-assembly : from surfactants to nanoparticles / edited by Ramanathan Nagarajan.
Description: First edition. | Hoboken, NJ : Wiley, 2018. | Series: Wiley series on surface and interfacial chemistry | Includes bibliographical references and index. |
Identifiers: LCCN 2018028409 (print) | LCCN 2018028986 (ebook) | ISBN 9781119001393 (Adobe PDF) | ISBN 9781119001386 (ePub) | ISBN 9781119001362 (hardback)
Subjects: | MESH: Nanoparticles | Molecular Structure | Molecular Biology–methods
Classification: LCC QP801.P64 (ebook) | LCC QP801.P64 (print) | NLM QT 36.5 | DDC 572/.33–dc23
LC record available at https://lccn.loc.gov/2018028409

Cover Design: Wiley
Cover Images: Courtesy of Ramanathan Nagarajan

Set in 10/12pt WarnockPro by SPi Global, Chennai, India

Printed in United States of America

V10006486_112918

Contents

List of Contributors

Nicholas L. Abbott
Department of Chemical and Biological Engineering
University of Wisconsin-Madison
Madison
WI 53706

Anna C. Balazs
Department of Chemical Engineering
University of Pittsburgh
Pittsburgh
PA 15261

Stephen Z. D. Cheng
Department of Polymer Science
College of Polymer Science and Polymer Engineering
The University of Akron
Akron
OH 44325

Emily J. Crabb
Department of Chemical Engineering
University of Pittsburgh
Pittsburgh
PA 15261

Monica Olvera de la Cruz
Department of Materials Science and Engineering
Northwestern University
Evanston
IL 60208

Xue-Hui Dong
Department of Polymer Science
College of Polymer Science and Polymer Engineering
The University of Akron
Akron
OH 44325

Xuehui Dong
Department of Chemical Engineering
Massachusetts Institute of Technology
77 Massachusetts Avenue
Cambridge
MA 02139

Nathan C. Gianneschi
Department of Chemistry
Northwestern University
Evanston, IL 60208

Martin M. Hanczyc
Centre for Integrative Biology (*CIBIO*)
Università degli Studi di Trento
Via Sommarive, 9
Trento
Italy

Aaron Huang
Department of Chemical Engineering
Massachusetts Institute of Technology
77 Massachusetts Avenue
Cambridge
MA 02139

Mingjun Huang
Department of Polymer Science
College of Polymer Science and Polymer Engineering
The University of Akron
Akron
OH 44325

Olga Kuksenok
Department of Chemical Engineering
University of Pittsburgh
Pittsburgh
PA 15261

Lorraine Leon
Materials Science and Engineering
University of Central Florida
Orlando, FL32816

Ting Li
Department of Materials Science and Engineering
Northwestern University
Evanston
IL 60208

Yiwen Li
Department of Polymer Science
College of Polymer Science and Polymer Engineering
The University of Akron
Akron
OH 44325

Zhiwei Lin
Department of Polymer Science
College of Polymer Science and Polymer Engineering
The University of Akron
Akron
OH 44325

Hao Liu
Department of Polymer Science
College of Polymer Science and Polymer Engineering
The University of Akron
Akron
OH 44325

Gerald T. McFarlin IV
Department of Chemical Engineering
University of Pittsburgh
Pittsburgh
PA 15261

Rebecca J. McMurray
Department of Materials Science and Engineering
Northwestern University
Evanston
IL 60208

Nicholas M. Moellers
Department of Chemical Engineering
University of Pittsburgh
Pittsburgh
PA 15261

Pierre-Alain Monnard
Institute for Physics
Chemistry and Pharmacy
University of Southern Denmark
Campusvej, 55
4230 Odense M
Denmark

Ramanathan Nagarajan
Natick Soldier Research
Development and Engineering
Center
15 General Greene Avenue
Natick MA 01760

Allie Obermeyer
Department of Chemical Engineering
Massachusetts Institute of
Technology
77 Massachusetts Avenue
Cambridge
MA 02139

Bradley D. Olsen
Department of Chemical Engineering
Massachusetts Institute of
Technology
77 Massachusetts Avenue
Cambridge
MA 02139

Isaac Salib
Department of Chemical Engineering
University of Pittsburgh
Pittsburgh
PA 15261

Timothy J. Smith
Department of Chemical and
Biological Engineering
University of Wisconsin-Madison
Madison
WI 53706

Matthew P. Thompson
Department of Chemistry
Northwestern University
Evanston, IL 60208

Matthew Tirrell
Institute for Molecular Engineering
University of Chicago
Chicago
IL 60637

and

Argonne national laboratory
Argonne
IL 60439

Alexey I. Victorov
Institute of Chemistry
St. Petersburg State University
Universitetsky prospect 26
198504
St. Petersburg
Russia

Xin Yong
Department of Chemical Engineering
University of Pittsburgh
Pittsburgh
PA 15261

Xinfei Yu
Department of Polymer Science
College of Polymer Science and Polymer Engineering
The University of Akron
Akron
OH 44325

Kan Yue
Department of Polymer Science
College of Polymer Science and Polymer Engineering
The University of Akron
Akron
OH 44325

Wen-Bin Zhang
Key Laboratory of Polymer Chemistry and Physics of Ministry of Education
College of Chemistry and Molecular Engineering
Center for Soft Matter Science and Engineering
Peking University
Beijing 100871
China

Preface

Self-assembly was first recognized by James McBain in classical colloid science almost 100 years ago, with the discovery of spontaneous formation of multimolecular aggregates of soap molecules. For almost 80 years after the initial discovery, self-assembly studies were dominated by classical soap and surfactant molecules, and for the latter part of this period, studies on high-molecular-weight block copolymer systems were also prevalent. However, the term "self-assembly" did not appear in the literature until 1966, as revealed by a Web of Science search. In the following decade, the term began to appear in publications, but less than 10 times each year, and only to describe the self-assembly of protein or viral subunits. To the best of my knowledge, the first use of the term to describe amphiphilic systems was in the classic paper of Israelachvili, Mitchell, and Ninham "Theory of self-assembly of hydrocarbon amphiphiles into micelles and bilayers." Since then, there has been an explosion of studies in the literature, invoking the terminology of self-assembly.

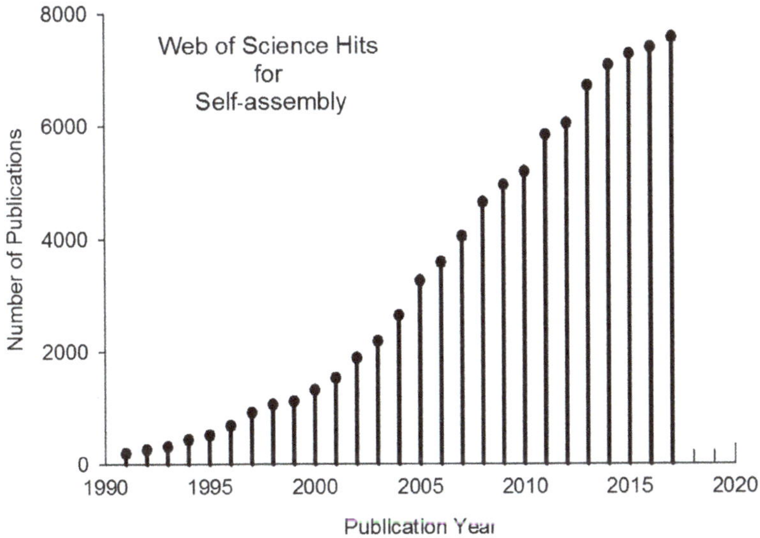

Over the last 20 years, the study of self-assembly has emerged as a distinct field, encompassing much larger and more complex molecular and nanoparticle systems. The study of self-assembly has extended far beyond surfactants and block copolymers and has been applied to peptide amphiphiles, DNA amphiphiles, protein–polymer conjugates, and nanoparticles. Self-assembly of molecules to create nanoparticles, self-assembly of nanoparticles to create new materials or devices, self-assembly in biological cell and its components contributing to essential life functions, self-assembly of proteins leading to neurodegenerative diseases, self-assembly of molecules/particles for nanomedicine applications of drug delivery, imaging, molecular diagnostics and theranostics, and self-assembly as the processing method to design materials to specification such as smart responsive materials and self-healing materials, have all made self-assembly a topic of great importance and have assured its continuing growth.

This book provides an effective entry for new researchers into this exciting field while also assessing state-of-the-art understanding of these diverse self-assembling systems. The book introduces the fundamentals and applications of self-assembled systems to academic and industrial scientists and engineers. Within its 10 chapters, the fundamental physical chemical principles that govern the formation and properties of self-assembled systems are considered. Important experimental techniques that can be used to characterize the properties of self-assembled systems, particularly the nature of molecular organization and structure at the nano-, meso-, or micro-scales, are reviewed. The synthesis and functionalization of self-assembled nanoparticles and the further self-assembly of the nanoparticles into one-, two-, or three-dimensional materials are discussed. Numerous potential applications of self-assembled structures are discussed. The book provides the first exhaustive accounting of self-assembly derived from various kinds of biomolecules including peptides, DNA, and proteins. Unifying as well as contrasting features of self-assembly, as we move from surfactant molecules to nanoparticles, are highlighted.

The first chapter discusses the essential similarity in the self-assembly behavior of low molecular weight surfactants and high molecular weight block copolymers from the point of view of the head-tail construct in amphiphilic systems. The emphasis on the head and neglect of the tail in surfactant free energy models is contrasted against the emphasis on the tail and minimal attention to the head in block copolymer free energy models. This difference, when resolved, allows for an unified treatment of self-assembly. The head–tail dependent free energy models are then suggested as a way to describe the self-assembly phenomena for a variety of non-classical amphiphilic systems involving dendrimers, DNA, peptides, proteins, and nanoparticles as critical head or tail components.

Chapter 2 is devoted to self-assembled systems of strongly growing and branching wormlike micelles that form reversible spatial networks in solutions.

Network reversibility and controllable viscosity make such systems very useful in numerous applications such as for drag reduction, paints, self-healing, and coatings. Relation of the observed viscoelasticity of a micellar solution to its structure is explained within the framework of the kinetic theories of breaking and recombining chainlike aggregates. The growth of non-ionic and ionic micelles, electrostatic rigidity, effects of branching, and scaling of the viscosity with the concentration of surfactant are all discussed in this chapter.

Chapter 3 reviews ways in which the self-assembly of redox-active and light-responsive surfactants have been used to achieve spatial and temporal control over interfacial and bulk properties of aqueous systems, including the interactions of surfactants with biomolecules. The switching of stimuli-responsive functional groups on the surfactants is shown to permit tuning of the surface tensions of aqueous systems, to induce surface tension gradient-driven flows, to change the state of aggregation of the surfactants in bulk solution, to permit temporal control over the transport of DNA across cell membranes and to achieve spatial control of surfactant-based microfluidic systems.

Chapter 4 highlights the importance of self-assembly to life processes. The knowledge about self-assembly of amphiphiles in aqueous environments is translated to the understanding of how lipids are uniquely connected to the formation of the cell, with cellular identity, with cellular functions, and also with cell death. The chapter discusses the idea that the spontaneous self-assembly of membranes may also be fundamental in the emergence of the first living cells in the context of an early Earth devoid of life. It describes how single-hydrocarbon-chain amphiphiles have been used to construct protocellular compartments in origin of life studies.

Chapter 5 shows how we can dynamically manipulate self-assembly. It develops the concept of programming the formation of synthetic assemblies using biomolecules, particularly peptides and nucleic acids. Biomolecules are utilized as recognition elements enabling the building of analytical probes or functional systems capable of performing sense-and-response processes in living systems. The focus is on the use of peptides and nucleic acids as the programming element. Examples are presented to highlight the ability of the programming element to control properties such as micelle formation, morphology, binding, reactivity, and spatial organization.

Protein analogous micelles (PAMs) resulting from the self-assembly of peptides conjugated to lipid tails or peptide amphiphiles are discussed in Chapter 6. Using the machinery of self-assembly, PAMs can be designed to include mixtures of different peptide amphiphiles leading to multifunctional, multivalent assemblies that can be stimuli responsive. This chapter discusses physicochemical aspects related to the design of PAMs including thermodynamic driving forces, the role of peptide secondary structure, micelle shape, amphiphile geometry, mixed micelles, and stimuli responsiveness. Based on these properties of PAMs, their applications for tissue engineering, diagnostics, and therapeutics are discussed, focusing on how the PAM structure dictates function.

Chapter 7 explores the approach to controlling the self-assembly of proteins into materials by incorporating them as one block in a block copolymer, creating the protein conjugate block copolymer. The folded conformation of the protein significantly impacts the nanostructure formation in the materials. This chapter focuses on the physics of self-assembly of the protein-conjugate block copolymers based on a categorization of the bioconjugates by the shape of the protein block: rod-like proteins, crystallizable proteins, cyclic proteins, coil-like proteins, and globular proteins. The thermodynamics of self-assembly for each shape is summarized, with an emphasis on general principles that guide the development of new materials.

Chapter 8 discusses the design and creation of novel materials with unique properties where DNA-nanoparticles self-assembly plays a critical role. The ability to independently alter individual components of the system, such as nanoparticle shape, size, and composition, as well as DNA length, sequence, and coating density results in a highly customizable system. The inherent self-assembly capability of the DNA-coated nanoparticles provides a unique platform for constructing complex crystalline structures. These nanoscale building blocks hold great potential for applications in medical diagnostics, plasmonics, catalysis, and photonics. This chapter emphasizes the recent progress in the field using multiscale modeling and simulation directed towards designing and predicting novel DNA-nanoparticle assemblies.

Chapter 9 addresses the intriguing phenomenon of using self-assembled lipid vesicles to controllably transport nanoparticles. It uses dissipative particle dynamics to model the interaction between fluid-driven lipid vesicles and Janus nanoparticles in order to establish design rules for the vesicle-mediated particle transport. The transport is enabled by adaptive behavior of the vesicle, shedding lipids to cover the Janus particle and undergoing a self-healing process after the particle deposition, so that the vesicle can be used in successive particle pick-up and delivery events.

Chapter 10 describes "giant surfactants," which are analogs of classical surfactants but with one or more molecular nanoparticles as headgroups. The combination of the molecular nanoparticle heads having diverse symmetries and surface functionalities with the tails possessing variable compositions and architectures is shown to generate a large family of giant surfactants. These novel giant amphiphiles self-assemble into a great variety of ordered supramolecular structures in solution. The universal principles that govern their self-assemblies are explored in this chapter in order to provide guidance to the rational design and manipulation of new functional materials for technologically relevant applications.

It is my hope that this book will most efficiently introduce the reader to the field of self-assembly, providing from basic to advanced information, on each of the multiple topics covered. There are no textbooks or courses (or even professional short courses) covering all of these topics in any one place. The contributors are pioneers in their respective topical areas of research. I hope the book stimulates both entrant and experienced researchers to become active participants in this field of research.

Ramanathan Nagarajan

Acknowledgments

This book project was initiated in March 2014. The first chapter was delivered in December 2014, but it has taken this long to bring it to a conclusion. I first want to thank all the chaper authors for their contributions and infinite patience. The active interest from Mr. Jonathan Rose, Wiley senior editor, and Ms. Aruna Pragasam, content editor, was critical in the final stages of this book project to get it completed. My work on this book would not have been possible without the support from the Natick Soldier Research, Development and Engineering Center, my colleagues there, and its director, Mr. Douglas Tamilio. I very much appreciate their understanding that while Soldiers have many immediate capability needs that researchers should work on addressing, it is very important to also take a long-term view and explore new scientific ideas that can qualitatively transform the technological capabilities that can be made available to them. Finally, the support from my family always remains the foundation for my work.

1

Self-Assembly from Surfactants to Nanoparticles – Head vs. Tail

Ramanathan Nagarajan

Natick Soldier Research, Development & Engineering Center,
15 General Greene Avenue, Natick MA 01760, USA

1.1 Introduction

Surfactant molecules are amphiphilic, composed of a polar headgroup that likes water and a nonpolar tail that dislikes water, thus contributing to an intrinsic duality in their molecular characteristics. Despite their mutual antipathy, the headgroup and tail of the surfactant cannot leave one another because they are covalently connected. The dilemma of mutual antipathy and forced coexistence faced by these molecules is resolved in nature by the intriguing phenomenon of molecular self-assembly, wherein the surfactant molecules self-assemble into three-dimensional structures with distinct and separate regions composed of the nonpolar parts and the polar parts, having minimal contact with one another. Block copolymers are an important class of high molecular weight polymer molecules that share great molecular similarity with the surfactants. A diblock copolymer is made up of repeating units A and B, with the repeating units occurring as blocks, covalently connected to one another. If one block (B) is hydrophilic or solvophilic (head) while the other block (A) is hydrophobic or solvophobic (tail), the block copolymer becomes a high molecular weight analog of the low molecular weight surfactant. Surfactants and block copolymers display characteristic molecular self-assembly behavior in solutions, at interfaces as well as in bulk, generating nanoscale structures of different shapes. These nanoscale features determine many macroscopic properties of these amphiphile systems, relevant for their practical applications.

The ability to generate desired nanoscale morphologies by synthesizing novel amphiphiles so that the amphiphilic systems can be tailored for specific applications as well as the ability to manipulate the morphologies using chemical and physical stimuli remain active goals of research in this field. Critical to

Self-Assembly: From Surfactants to Nanoparticles, First Edition. Edited by Ramanathan Nagarajan.

achieving these goals is an understanding of the link between the molecular structure of the amphiphiles and their self-assembly behavior. Studies on surfactants have a long history, starting with the pioneering recognition of the existence of aggregates of soap molecules by McBain [1], over 100 years ago. He coined the term micelles to describe these aggregates and visualized a lamellar morphology for these aggregates. The proposal of a spherical micelle structure was made by Hartley [2] who suggested that the "aggregates are essentially liquid and since they will tend to present the minimum surface to the water, they will presumably be roughly spherical and of the largest radius consistent with none of the heads being submerged in the paraffin interior." Theoretical developments relating to surfactants have a rich history of 70 years starting with a pioneering model proposed by Debye [3]. The historical developments in the evolution of theories applied to surfactants leading up to current state of the art have recently been summarized [4].

The first synthesis of block copolymers was reported by Dunn and Melville [5] who synthesized and characterized a diblock copolymer of poly(methyl methacrylate)-b-poly(styrene). The observation that solvents can cause segregation of polymer blocks was first made by Merrett [6] based on studies of graft copolymers with rubber as backbone and grafted blocks of poly(methyl methacrylate) interacting with solvent mixtures involving benzene and methanol, benzene being a good solvent for rubber and methanol a good solvent for poly(methyl methacrylate). Merrett also used the term micelle to describe the graft copolymer aggregates. The first report of block copolymer aggregates in solution was from Climie and White [7] who studied poly(methyl methacrylate)-b-poly (acrylonitrile) in a mixed solvent of dimethyl formamide, which is a good solvent for both blocks and benzene, which is a nonsolvent for the poly(acrylonitrile) block. The clear demonstration of block copolymer micelles formed in dilute solutions was reported by Krause [8]. In this study using poly(styrene)-b-poly(methyl methacrylate) block copolymer, micelles were obtained in acetone, a non-solvent for poly(styrene), and in triethylbenzene, a nonsolvent for poly(methyl methacrylate). Block copolymer aggregates thus have a rich history of nearly 60 years and these early developments along with a comprehensive discussion of the synthesis and solution properties of block copolymers have been reviewed some years ago by Riess [9].

In this chapter, we focus mainly on the theoretical ideas based on which quantitative models to describe the self-assembly of surfactants and block copolymers have been developed. Tanford [10] and Israelachvili et al. [11] pioneered two of the most important models that currently dominate our understanding of surfactant self-assembly. Tanford proposed the concept of opposing forces to formulate a quantitative expression for the standard free energy change when a singly dispersed surfactant molecule in solution becomes part of a multimolecular surfactant aggregate, also in the solution. In his model, the formation of the equilibrium aggregate resulted from balancing the interfacial free energy at the micelle-water interface against the repulsions between the surfactant headgroups also located at the interface. Using this free energy expression, with the controlling role played by the headgroup, he was able to explain why surfactant aggregates form in aqueous solutions, why they grow, and why they do not keep growing but remain finite in size. Israelachvili, Mitchell, and Ninham proposed the concept of a molecular packing parameter P defined as $P = v_o/a_e\ell_o$, where v_o and ℓ_o are the volume and length of the surfactant tail and a_e is the equilibrium area per molecule of the surfactant aggregate at the hydrophobic core-water interface. They demonstrated how the size and the shape of the aggregate at equilibrium can be predicted from the magnitude of P in accordance with simple geometry driven molecular packing considerations. In the context of the Tanford model, P is mainly dependent on the headgroup interactions and therefore the aggregate size and shape predicted by the packing parameter are clearly dominated by only the headgroup interactions. Remarkably, in this model, the surfactant tail has no explicit role in influencing the aggregate shapes.

Theoretical studies of self-assembly of block copolymers evolved without any obvious contact with the surfactant self-assembly literature. Theoretical understanding of how pure block copolymers organize into microdomains was advanced through the work of Meier [12, 13] and Helfand [14, 15]. Theoretical treatments of block copolymer micelles in selective solvents or in homopolymers have been pioneered by de Gennes [16], Leibler et al. [17], Noolandi and Hong [18], and Whitmore and Noolandi [19]. de Gennes [16] analyzed the formation of a diblock copolymer micelle in selective solvents by minimizing the free energy per molecule of an isolated micelle with respect to the aggregation number or core radius. The micelle core was assumed fully segregated and devoid of any solvent. In this model framework, the free energy of formation of the core-corona interface and the elastic free energy of stretching of the core blocks compete to control the micellization behavior. Leibler et al. [17] treated the problem of micelle formation of a symmetric diblock copolymer in a homopolymer solvent. In their study, as in de Gennes' work, the interface was taken to be sharp. Noolandi and Hong [18] and Whitmore and Noolandi [19] formulated mean field models taking into account the possibility of a diffuse interface between the core and corona regions. In all treatments of block copolymer self-assembly, the elastic deformation of the core forming block (the tail) played the central role. Notably, in these models, the solvophilic block (the analog of surfactant headgroup) had no explicit role. In Section 1.2 we outline the models for surfactants and block copolymers and then show how the mutually exclusive emphasis on headgroup or tail has now given place to consideration of both head and tail, thereby improving their quantitative predictive abilities. In Section 1.3 we show that the headgroup vs. tail-based free energy concepts can rationalize self-assembly behavior observed in many non-classical amphiphile systems involving dendrimers, DNA, peptides, proteins, and nanoparticles as head or tail components.

1.2 Classical Surfactants and Block Copolymers

1.2.1 Tanford Model for Surfactant Micelles

Surfactant molecules self-assemble into spherical, globular or cylindrical micelles, or spherical bilayers, also known as vesicles (Figure 1.1). In these aggregates, the hydrophobic domain is made up of the surfactant tails and the hydrophilic headgroups are crowded at the aggregate core-water interface.

If the density in the hydrophobic domain is considered equal to that in similar liquid hydrocarbons, the surfactant tails must entirely fill the space in these domains. As a result, irrespective of the shape of the aggregate, no point within the aggregate can be farther than ℓ_0 from the aggregate-water interface, where ℓ_0 is the extended length of the surfactant tail. Therefore, at least one dimension of the surfactant aggregates should be smaller than or at most equal to $2\ell_0$. This is purely a geometrical or packing constraint that the surfactant aggregate will have to satisfy [10, 11]. The geometrical relations governing these different aggregate shapes are summarized in Table 1.1.

The size and shape of the aggregates are dependent on the surfactant molecular structure as well as the solution conditions. The Gibbs equilibrium condition

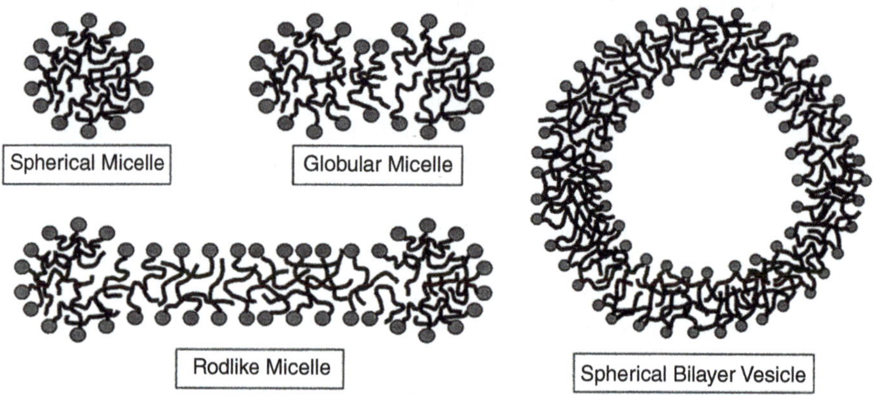

Figure 1.1 Schematic representation of surfactant aggregates in dilute aqueous solutions. The structures include spherical micelles, globular micelles not much larger than the spherical micelles, large cylindrical micelles with globular endcaps, and spherical bilayer vesicles. The length of the surfactant tail constrains one characteristic dimension in each of these aggregates (radius for the micelles and half-bilayer thickness for the vesicles). More complex aggregates involving branching of cylindrical micelles and network formation among them are discussed in detail by Victorov in Chapter 2 of this book.

Table 1.1 Geometrical relations for spherical and cylindrical micelles and bilayers.[a)]

Variable	Sphere	Cylinder	Planar bilayer
Volume of core $V = g\,v_o$	$4\pi\, R^3/3$	$\pi\, R^2$	$2R$
Surface area of core $A = g\,a$	$4\pi\, R^2$	$2\pi\, R$	2
Area per molecule a	$3v_o/R$	$2v_o/R$	v_o/R
Aggregation number g	$36\pi v_o^2/a^3$	$4\pi v_o/a^2$	$2/a$

a) R is the core radius in case of spherical, globular or cylindrical micelles and in the case of bilayer, it represents the half-bilayer thickness. g denotes the number of surfactant molecules in the aggregate. v_0 is the volume and ℓ_0 is the extended length of the surfactant tail. The variables V, A, and g refer to the entire aggregate in the case of a sphere, unit length in the case of a cylinder or unit area in the case of a bilayer.

stipulates the equality of chemical potential of a surfactant molecule present in the singly dispersed state and that incorporated within a multimolecular aggregate of any given size and shape [10]. This chemical potential equality allows one to relate the concentration X_g of surfactant aggregates of size g and any shape to the concentration X_1 of the singly dispersed surfactant through the equation

$$X_g = X_1^g \exp(-g\,\Delta\mu_g^o/kT), \quad \Delta\mu_g^o = \frac{\mu_g^o}{g} - \mu_1^o \tag{1.1}$$

where, μ_g^o is the standard state chemical potential of an isolated micelle of aggregation number g while μ_1^o is the standard state chemical potential of a singly dispersed surfactant, both in bulk solvent. k is the Boltzmann constant and T is the temperature. Based on Eq. (1.1), the fundamental quantity of interest that will allow one to determine all aggregation properties is the standard state chemical potential difference $\Delta\mu_g^o$, also referred to as the standard free energy change on aggregation. Tanford [10, 20] formulated a quantitative expression for $\Delta\mu_g^o$ on phenomenological grounds, invoking the concept of opposing forces. The standard free energy change is composed of three contributions:

$$\left(\frac{\Delta\mu_g^o}{kT}\right) = \left(\frac{\Delta\mu_g^o}{kT}\right)_{Tail} + \left(\frac{\Delta\mu_g^o}{kT}\right)_{Int} + \left(\frac{\Delta\mu_g^o}{kT}\right)_{Head} \tag{1.2}$$

The first term $(\Delta\mu_g^o/kT)_{Tail}$ is a negative free energy contribution arising from the transfer of the surfactant tail from its unfavorable contact with water to the favorable hydrocarbon-like environment of the aggregate core. This transfer free energy contribution depends on the surfactant tail but not on the aggregate shape or size. The second term $(\Delta\mu_g^o/kT)_{Int}$ is a positive free energy contribution that accounts for the residual contact between the surfactant tails and water at the aggregate core-water interface. This is represented as the product of a contact free energy per unit area σ (or an interfacial free energy per unit area, or interfacial tension) and the surface area per molecule of the aggregate core, a. The third term $(\Delta\mu_g^o/kT)_{Head}$ is another positive free energy contribution that accounts for the repulsive interactions between the headgroups because they crowd at the aggregate surface. The repulsions may be due to steric interactions (for any type of headgroup) and also electrostatic interactions (dipole–dipole interactions for zwitterionic headgroups and ion–ion repulsions for ionic headgroups). Since the repulsions would increase if the headgroups come close to one another, Tanford proposed an expression for this free energy contribution with an inverse dependence on a. Thus, the standard free energy change per molecule on aggregation proposed by Tanford has the form:

$$\left(\frac{\Delta\mu_g^o}{kT}\right) = \left(\frac{\Delta\mu_g^o}{kT}\right)_{Tail} + \left(\frac{\sigma}{kT}\right)a + \left(\frac{\alpha}{kT}\right)\frac{1}{a} \tag{1.3}$$

where α represents the headgroup repulsion parameter. One may note that this phenomenological parameter α has to be connected to the surfactant headgroup features and solution conditions if one wants to carry out any predictive computations of aggregation.

From the free energy model of Tanford, the equilibrium aggregation behavior can be examined either by treating the surfactant solution as consisting of aggregates with a distribution of sizes (as represented by Eq. (1.1)) or by treating the aggregate as constituting a pseudophase. If the aggregate is viewed as a pseudophase, in the sense of small systems thermodynamics, the equilibrium condition corresponds to a minimum in the standard free energy change per molecule, $\Delta\mu_g^o/kT$. The minimization can be done with respect to either the aggregation number g or the core surface area per molecule a, since they are dependent on one another through the geometrical relations given in Table 1.1. One obtains, in this manner, the equilibrium core surface area a_e per molecule characterizing the aggregate:

$$\frac{\partial}{\partial a}\left(\frac{\Delta\mu_g^o}{kT}\right) = \left(\frac{\sigma}{kT}\right) - \left(\frac{\alpha}{kT}\right)\frac{1}{a^2} = 0, at\ a = a_e \quad \Rightarrow \quad a_e = \left(\frac{\alpha}{\sigma}\right)^{1/2} \tag{1.4}$$

The critical micelle concentration (cmc, denoted as X_C in mole fraction units), in the pseudophase approximation, is obtained from the relation,

$$\ln X_C = \left(\frac{\Delta\mu_g^o}{kT}\right)_{Tail} + \left(\frac{\sigma}{kT}\right)a_e + \left(\frac{\alpha}{kT}\right)\frac{1}{a_e} = \left(\frac{\Delta\mu_g^o}{kT}\right)_{Tail} + \left(\frac{2\sigma^{1/2}\alpha^{1/2}}{kT}\right) \tag{1.5}$$

The principal outcomes from the free energy model are the equilibrium area per molecule and the cmc. From the equilibrium area per molecule, one can determine the shape of the aggregate based on purely geometrical considerations. For example, if we consider a spherical micelle of core radius R, made up of g molecules, the volume of the core $V = g\, v_o = 4\pi\, R^3/3$, the surface area of the core $A = g\, a = 4\pi\, R^2$, and hence $R = 3\, v_o/a$. Introducing the constraint that the core radius R cannot exceed the extended length ℓ_o of the tail, one obtains the limit, $0 \leq v_o/a\ell_o \leq 1/3$, for spherical micelles. This dimensionless group $v_o/a\ell_o$ is well known as the molecular packing parameter P, first explicitly introduced by Israelachvili et al. [11]. The application of this idea to other aggregate shapes using the geometrical relations given in Table 1.1, leads to the well-known [11] connection between the molecular packing parameter and the aggregate shape: $0 \leq v_o/a\ell_o \leq 1/3$ for sphere, $1/3 \leq v_o/a\ell_o \leq 1/2$ for cylinder, and $1/2 \leq v_o/a\ell_o \leq 1$ for bilayer. The ratio v_o/ℓ_o is a property of the surfactant tail and since both the volume and length are proportional to the number of carbon atoms in an aliphatic tail, the ratio v_o/ℓ_o is a constant, practically independent of the tail length, equal to 0.21 nm^2 for single-chain surfactants and 0.42 nm^2 for double-chain surfactants. Therefore, once the equilibrium area is determined from the Tanford free energy model, one can identify the size and shape of the aggregate and the magnitude of the cmc.

In the Tanford free energy expression (Eq. (1.3)), only the first contribution, the tail transfer free energy, is negative, and this allows the free energy change on aggregation to be negative. Hence, this contribution is solely responsible for the aggregation to occur. However, it affects only the cmc (as shown by Eq. (1.5)) but not the equilibrium area a_e (as shown by Eq. (1.4)), which determines the size and shape of the aggregate. The second contribution, the free energy of residual contact between the aggregate core and water, is positive and decreases in magnitude as the area a decreases. A decrease in the area a corresponds to an increase in the aggregation number g for all aggregate shapes, as shown on Table 1.1. Hence, this contribution promotes the growth of the aggregate. The third contribution, the free energy due to headgroup repulsions, is also positive and increases in magnitude if the area a decreases or the aggregation number g increases. Hence, this contribution is responsible for limiting the growth of aggregates to a finite size. Thus, the Tanford model clearly identifies why aggregates form, why they grow, and why they do not keep growing but remain finite in size.

The Tanford model explicitly attributes a central role to the surfactant headgroup in controlling self-assembly. The magnitude of headgroup repulsions determines the equilibrium area per molecule, and hence the size and shape of the surfactant aggregate. In contrast, the surfactant tail does not affect the equilibrium area per molecule but influences only the magnitude of the cmc. We have shown earlier [21, 22] that although the surfactant tail does not affect the equilibrium area in the framework of Tanford model, there are other implicit ways by which the tail does exert an influence on the aggregate size and shape. We describe four situations that demonstrate the implicit role of the tail in determining the aggregate size and shape, as follows

(i) *Spherical micelles from nonionic or zwitterionic surfactants.* Consider the formation of spherical micelles from a family of nonionic or zwitterionic surfactants with the same headgroup but differing tail lengths. The headgroup repulsion parameter α is a constant for the homologous family of molecules. Therefore, the equilibrium area a_e calculated from Eq. (1.4) is independent of the tail length, and the packing parameter $P = v_o/a_e\ell_o$ is also independent of the tail length (since v_o/ℓ_o does not change with tail length). Using the geometrical relations in Table 1.1, the aggregation number of the spherical micelles is given by

$$g = \frac{36\,\pi\,v_o^2}{a_e^3} \tag{1.6}$$

One may note that because of the proportionality of v_o to the number of carbon atoms n_C in an aliphatic tail and the tail length independence of a_e, the aggregation number g is proportional to n_C^2 or the square of the surfactant tail length. The tail does determine the actual size or aggregation number of the spherical micelle.

(ii) *Spherical micelles from ionic surfactants.* Consider a family of ionic surfactants, with the same headgroup but differing tail lengths. The implicit influence of the tail arises as follows. The equilibrium area a_e depends on the headgroup repulsion parameter α, which in this case accounts for electrostatic interactions between the ionic headgroups. The electrostatic interactions are influenced by the ionic strength of the solution through the inverse Debye length κ, which for uni-univalent ionic systems is given by

$$\kappa = \left[\frac{8\,\pi\,e^2 n_o}{\varepsilon\,kT}\right]^{1/2} = \left[\frac{8\,\pi\,e^2}{\varepsilon\,kT}\,\frac{cmc}{1000}\,N_{Avo}\right]^{1/2} \tag{1.7}$$

In Eq. (1.7), n_o is the number of counterions in solution per cm^3. In the absence of any added electrolyte, the ionic strength is determined primarily by the concentration of the singly dispersed surfactant in solution, which is practically equivalent to the cmc. The second equality in Eq. (1.7) represents this condition where n_o is related to the cmc in moles/liter, N_{Avo} is the Avogadro number and the factor 1000 converts concentration per liter to that per cm^3 of solution. The cmc is dependent on the transfer free energy of the tail, which directly depends on the length of the tail. Thus, the tail length implicitly influences the headgroup interaction parameter α, hence, the equilibrium area a_e, and the size of the micelles, for ionic surfactants. In the presence of significant amount of added electrolyte, this tail length effect on the Debye length and the headgroup interactions diminishes and even disappears.

(iii) *Formation of cylindrical micelles.* Consider the formation of large spherocylindrical micelles of size g containing g_{cap} molecules in the two spherical endcaps and (g-g_{cap}) molecules in the middle cylindrical part. Eq. (1.1) for the size distribution of micelles can be rearranged to explicitly represent the molecules in the middle cylindrical domain and those in the endcap spherical domains. Based on this, the size distribution of aggregates and the average aggregation numbers are given [11, 21] by the expressions:

$$X_g = \frac{1}{K} Y^g, \quad Y = X_1 \exp\left(-\frac{\Delta\mu^o_{cyl}}{kT}\right), \quad \ln K = g_{cap}\left[\frac{\Delta\mu^o_{cap} - \Delta\mu^o_{cyl}}{kT}\right]$$

$$g_n = g_{cap} + [K(X_{tot} - X_1)]^{1/2}, \quad g_w = g_{cap} + 2[K(X_{tot} - X_1)]^{1/2} \quad (1.8)$$

In Eq. (1.8), $\Delta\mu^o_{cyl}$ and $\Delta\mu^o_{cap}$ are the changes in the standard chemical potentials when singly dispersed surfactant molecule becomes incorporated in the cylindrical middle and spherical endcaps, respectively, of the spherocylindrical aggregate. The variable K is a measure of the free energy cost of having g_{cap} molecules in the spherical endcaps compared to the energetically favored cylindrical middle. The parameter Y indicates the possibility of occurrence of cylindrical aggregates at a given concentration of the singly dispersed surfactant molecules. g_n and g_w denote the number and weight average aggregation numbers, respectively, calculated as a function of the total surfactant concentration X_{tot}.

The tail length of the surfactant determines g_{cap} since for increasing tail lengths, larger number of molecules could be accommodated in the endcaps. For single-tail amphiphiles, g_{cap} varies from 27 for C_8, 55 for C_{12} and 115 for C_{18} tails. Therefore, for a homologous family of surfactants, even when the equilibrium area per molecule remains independent of tail length, the parameter K would increase dramatically with increasing tail length. Therefore, the tail length significantly influences the size of the cylindrical micelles even for a constant equilibrium area per molecule. This also implies that even when the equilibrium area per molecule favors the formation of cylindrical micelles, we may observe only spherical or globular micelles when the tail length is short if K is not large enough.

(iv) *Formation of spherical bilayer vesicles.* Consider solution conditions that favor the formation of spherical bilayer vesicles. When bilayer vesicles form with an enclosed aqueous cavity, the headgroups of the surfactants in the inner layer of the bilayer are located within this aqueous cavity. Therefore, for bilayer vesicles to form, at a minimum, the volume per surfactant molecule of this inner aqueous cavity must be larger than the volume of the headgroup. Furthermore, the inner radius of the vesicle will have to be larger than the length of the headgroup (in the direction normal to the hydrophobic surface).

Consider a vesicle whose hydrophobic domain has an outer radius R_o and the half-bilayer thickness is R. R is usually smaller than the extended length ℓ_o of the tail. The inner radius of the hydrophobic domain is thus $(R_o - 2R)$. Israelachvili et al. [11] have shown that R_o can be calculated from

$$R_o = \frac{R\left[3 + \sqrt{3\left(\frac{4v_o}{a_e R} - 1\right)}\right]}{6\left(1 - \frac{v_o}{a_e R}\right)} \tag{1.9}$$

If the number of surfactant molecules in the inner layer of the bilayer is g_i, then the volume per molecule of the aqueous cavity V_{cav} is given by

$$g_i = \frac{4\pi (R_o - 2R)^2}{a_e}, \quad V_{cav} = \frac{4\pi (R_o - 2R)^3}{3g_i} = \frac{(R_o - 2R)\, a_e}{3} \tag{1.10}$$

It may be noted that in writing Eq. (1.10), we have assumed for simplicity that the equilibrium areas per molecule and the thicknesses of the layers are the same for the inner and the outer layers of the spherical bilayer vesicle. This assumption can be relaxed, but for the purposes of demonstrating the tail length effect, those additional details are not necessary. From Eq. (1.9) we note that for a homologous family of surfactants, with the equilibrium area a_e remaining the same, the outer vesicle radius R_o is roughly proportional to the tail length. From Eq. (1.10) we see that with decreasing chain length, the aqueous cavity volume per surfactant molecule (in the inner layer of the bilayer) decreases. Some illustrative calculations are presented in Table 1.2 for different values of the packing parameter P. In performing the calculations, for simplicity, we have assumed that the half bilayer thickness R of the vesicle is equal to the extended length ℓ_o of the surfactant tail. The calculations have been performed for a surfactant with two alkyl chains in the tail, taking $v_o/\ell_o = 0.42\ \text{nm}^2$.

For each packing parameter, the corresponding equilibrium area per molecule calculated as $a_e = (v_o/\ell_o)/P$, is also listed in Table 1.2. For alkyl chains having 8–18 carbon atoms, (n_C), the outer radius R_o of the vesicle is calculated using Eq. (1.9) and the aqueous cavity volume per molecule, V_{cav} is calculated using Eq. (1.10) and the results are summarized in Table 1.2. Therefore, for a homologous family of double-tailed surfactants with the unchanging equilibrium area per molecule (or constant packing parameter), the surfactants with longer tails may form vesicles, but as the tail length decreases, the cavity volume can become smaller than the volume necessary to accommodate the surfactant headgroup, thereby preventing the formation of the bilayer vesicle. Indeed, even when the packing parameter is above 1/2

Table 1.2 Tail length dependence of vesicle interior volume per molecule.[a)]

		$P = 0.75$ $a_e = 0.56\,nm^2$		$P = 0.80$ $a_e = 0.53\,nm^2$		$P = 0.90$ $a_e = 0.47\,nm^2$		$P = 0.95$ $a_e = 0.44\,nm^2$	
n_C	ℓ_o (nm)	R_o (nm)	V_{cav} (nm^3)	R_o (nm)	V_{cav} (nm^3)	R_o (nm)	V_{cav} (nm^3)	R_o (nm)	V_{cav} (nm^3)
8	1.15	4.17	0.348	5.3	0.531	11.1	1.368	22.6	2.992
10	1.4	5.08	0.423	6.5	0.647	13.5	1.666	27.5	3.643
12	1.65	5.99	0.499	7.7	0.762	15.9	1.963	32.4	4.293
14	1.9	6.90	0.575	8.8	0.878	18.3	2.261	37.3	4.944
16	2.15	7.80	0.650	10.0	0.943	20.7	2.558	42.2	5.594
18	2.4	8.71	0.726	11.1	1.109	23.2	2.856	47.2	6.245

a) Detailed computations of free energy minimizations allowing the half bilayer thicknesses of the inner and outer layers of the vesicle to be different and the areas per molecule at the inner and the outer surface to be different will give different predictions for the radius R_o and the cavity volume V_{cav}. But the qualitative result that the cavity volume decreases with decreasing tail length for any given packing parameter remains valid.

required for bilayers to form, bilayers may not form for surfactants of shorter tail lengths unless the packing parameter is large enough to accommodate the headgroups in the vesicle interior.

1.2.2 de Gennes Model for Block Copolymer Micelles

In contrast to the surfactant aggregates discussed above that are formed in water as the solvent, block copolymer micelles can be generated in any solvent as long as the solvent is nonselective to one block (solvophobic block denoted as A) and selective to the other block (solvophilic block denoted as B). The solvophobic block forms the core of the aggregate and the solvophilic block forms the shell or the corona of the aggregate (Figure 1.2).

de Gennes [16] analyzed the formation of AB diblock copolymer micelles in a selective solvent by minimizing the free energy per molecule of an isolated micelle ($\Delta\mu_g^o$). The micelle core was assumed to be fully segregated, devoid of any solvent, and the interface is treated as sharp. In the de Gennes model, the free energy of formation of the micellar core–corona interface and the free energy of stretching of the solvophobic block constituting the micelle core control the micellization behavior.

The free energy model of de Gennes can be written as

$$(\Delta\mu_g^o) = (\Delta\mu_g^o)_{A,Tr} + (\Delta\mu_g^o)_{A,def} + (\Delta\mu_g^o)_{Int}$$

$$(\Delta\mu_g^o) = (\Delta\mu_g^o)_{A,Tr} + \frac{R^2}{N_A} + \sigma\frac{N_A}{R} \tag{1.11}$$

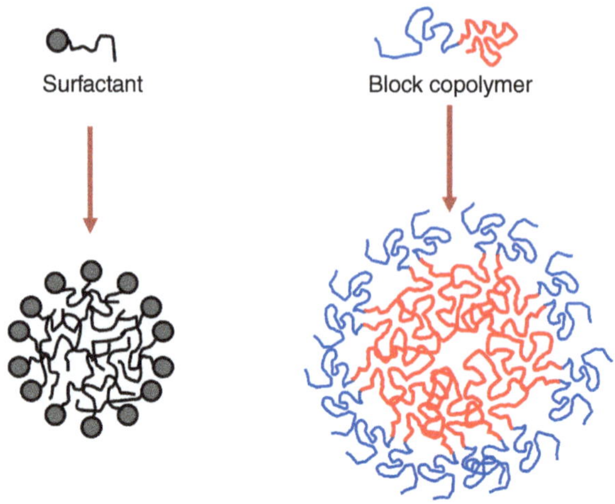

Figure 1.2 The formation of block copolymer micelle in a solvent selective to one of the blocks is analogous to the formation of the surfactant micelle. The solvophobic block of the block copolymer is equivalent to the surfactant tail while the solvophilic block of the block copolymer plays the role of surfactant headgroup.

The first term is the transfer of the solvophobic block A from the incompatible solvent to the micelle core resembling a melt of polymer block A. This term was not explicitly written in the original de Gennes paper but was obviously implied as it is the negative free energy contribution promoting self-assembly to occur. This is entirely analogous to the tail transfer free energy considered for surfactants. The second term is the stretching or elastic deformation of the solvophobic block A within the micelle core. The third term is the free energy of formation of the interface between the micelle core and the corona. The core block stretching is related to the core dimension R and the number of segments N_A in the A block while the interfacial free energy is related to the area per molecule a, which is inversely related to the core radius R. Explicit expressions for each of the contributions are also shown in Eq. (1.11) in the spirit of scaling models, excluding any numerical coefficients and treating all variables as nondimensional. In the last term accounting for the interfacial free energy, the relation between the area per molecule and core radius given in Table 1.1, $a \sim N_A/R$, has been used, noting that the volume of block A is proportional to the number of segments N_A in the block. From the minimization of the free energy, the de Gennes model predicts that the micelle core radius R and the aggregation number g are related to the size N_A of the solvophobic block as

$$R \sim N_A^{2/3}\sigma^{1/3}, \quad g \sim R^3/N_A \sim N_A\sigma \tag{1.12}$$

In the de Gennes free energy model, the solvophilic block B is considered to have no influence on the micelle characteristics and therefore the aggregate size has no dependence on the number of segments N_B in the B block. Note that the solvophilic block interactions in block copolymer micelles are analogous to the headgroup interactions in surfactant micelles. This offers a marked contrast to the Tanford model for surfactants, where the headgroup repulsions were the principal contribution controlling the finite size of surfactant aggregates and the influence of the tail was not explicitly recognized.

1.2.3 Surfactant Self-Assembly Model Incorporating Tail Effects

The block copolymer micelle model treats the elastic stretching of the solvophobic block when micelle core comes into existence as a key contribution to the micelle free energy, but this is not considered in the Tanford model for surfactants. Therefore, we have introduced this as a new contribution in the free energy model for surfactant micelle [21, 22] by adopting the treatment pioneered by Semenov [23] for block copolymer chain stretching in confined geometries. The derivation of this expression takes into account the fact that the tail has to deform nonuniformly along its length to fill the aggregate core with uniform density.

The free energy model for surfactants, incorporating the contribution from tail packing (or elastic deformation), can be written as

$$\left(\frac{\Delta\mu_g^o}{kT}\right) = \left(\frac{\Delta\mu_g^o}{kT}\right)_{Tail} + \left(\frac{\Delta\mu_g^o}{kT}\right)_{Int} + \left(\frac{\Delta\mu_g^o}{kT}\right)_{Head} + \left(\frac{\Delta\mu_g^o}{kT}\right)_{Pack}$$
$$\left(\frac{\Delta\mu_g^o}{kT}\right)_{Pack} = \frac{3\pi^2}{80}\frac{R^2}{NL^2}, \quad \frac{5\pi^2}{80}\frac{R^2}{NL^2}, \quad \frac{10\pi^2}{80}\frac{R^2}{NL^2} \tag{1.13}$$

The first three free energy contributions come from the original Tanford model while the fourth contribution for packing is based on the tail stretching contribution viewed as a dominant factor in the block copolymer theory. The three expressions shown for packing free energy contributions correspond to spheres, cylinders, and planar bilayers, respectively. In the packing free energy contribution, L is a characteristic segment length that is taken to be 0.46 nm (see Ref. [21] for details) and N is the number of segments in a tail such that $N L = \ell_o$ and $N L^3 = v_o$. It should be mentioned that the treatment of Semenov focused on the arrangement of polymer segments in domains formed by high molecular weight block copolymers. In such a case, the dimension R would be typically much larger than the unperturbed end to end distance $N^{1/2}L$ of the polymer block, with N being in many 10s or 100s. Therefore, in developing the free energy model [23], the microscopic elastic energy was written considering only chain stretching. For classical single chain surfactants with 8–18 carbon atoms in the tail, the number of segments N is typically two to five and the direct application of the result from block copolymer analysis need not be quantitatively accurate. Nevertheless, because of the simplicity of this analytical expression and the alternative of computationally intensive estimation methods for this packing free energy contribution, we have applied this analytical equation for all of our predictive computations [21].

From geometrical relations for the aggregates, we have $R = 3\,v_o/a$, $2\,v_o/a$, and v_o/a for the three geometries (Table 1.1). Introducing these in Eq. (1.13), the packing free energy contribution can be rewritten as

$$\left(\frac{\Delta\mu_g^o}{kT}\right)_{Pack} = \frac{Q}{a^2}, \quad Q_{sph} = \frac{27}{8}v_o L, \quad Q_{cyl} = \frac{20}{8}v_o L, \quad Q_{bilayer} = \frac{10}{8}v_o L, \tag{1.14}$$

where the symbol Q is used to denote the coefficient of $1/a^2$ in the free energy expression and it stands for Q_{sph}, Q_{cyl}, or $Q_{bilayer}$ depending upon the aggregate shape. The equilibrium area a_e given before by Eq. (1.4) is now obtained from the modified relation

$$\frac{\partial}{\partial a}\left(\frac{\Delta\mu_g^o}{kT}\right) = \left(\frac{\sigma}{kT}\right) - \left(\frac{\alpha}{kT}\right)\frac{1}{a^2} - \frac{2Q}{a^3} = 0, \quad at \quad a = a_e$$

$$a_e = \left(\frac{\alpha}{\sigma} + \frac{2Q/a_e}{\sigma/kT}\right)^{1/2} \tag{1.15}$$

Since the variable Q is dependent on the tail, the tail has direct influence over the equilibrium area a_e and the packing parameter P. From Eq. (1.15), we can see that the incorporation of the tail-packing effect causes an increase in the equilibrium area per molecule and correspondingly a decrease in the packing parameter, compared to the free energy model that neglects the tail-packing effect. The consequence is the direct control exerted by the tail over the size and shape of the equilibrium aggregate.

Illustrative numerical calculations have been carried out taking into account this packing free energy contribution. For this purpose, we have chosen a value of $(\alpha/kT) = 5\,\text{nm}^2$ for the headgroup interaction parameter and $(\sigma/kT) = 12\,\text{nm}^{-2}$, consistent with σ being around $50\,\text{mN}\,\text{m}^{-1}$ for aliphatic hydrocarbons-water interface. Q depends on the tail length as well as the shape of the aggregate within which the tails have to pack, as shown in Eq. (1.14). A summary of the calculated results for various tail lengths of single tail surfactant micelles is given in Table 1.3. Shown on the table are the equilibrium area a_e calculated from the Tanford model (Eq. (1.4)) and the model that explicitly considers tail packing (Eq. (1.15)) and takes into account the elastic deformation of the surfactant tail.

Since the Tanford model does not attribute any role for the surfactant tail in controlling the micelle size or shape, the calculated equilibrium area per molecule is independent of the tail length. In contrast, in the model that includes the hydrophobic tail stretching in the micelle core, the area per molecule shows a tail length dependence and increases as the tail length increases. All surfactants listed on the table form only spherical micelles. One can observe that the tail length influence is appreciable enough when it comes to considering the aggregation number of the spherical micelles.

Table 1.3 Influence of tail-packing free energy contribution on the size of spherical surfactant micelles.

		a_e (nm^2)	g	Q_{sph} (nm^4)	a_e (nm^2)	g
n_C	v_o (nm^3)	Based on Eq. (1.4)		Based on Eqs. (1.14) and (1.15)		
8	0.243	0.645	25	0.377	0.711	19
10	0.297	0.645	37	0.461	0.723	26
12	0.351	0.645	52	0.545	0.735	35
14	0.405	0.645	69	0.629	0.747	45
16	0.459	0.645	89	0.713	0.758	55

1.2.4 Star Polymer Model of Block Copolymer Self-Assembly Incorporating Headgroup Effects

The star polymer model for block copolymer micelles explicitly recognizes the role of the solvophilic block B. The star polymer model for micelles is built on the analogy between the conformation of star-shaped polymers and micelles. Star polymers resemble micelles, but for the important difference that a chemical link exists at the center between the various branches of the star polymer (Figure 1.3). Daoud and Cotton [24] studied the conformation of a uniform star polymer, with g branches joining at the origin, present in a good solvent. They identified three distinct regions in this conformation – a swollen region, an unswollen region, and a uniform density region. For each of these three regions, they developed expressions for the radial concentration profiles utilizing the concept of blobs employed in theories of semi-dilute solutions of linear polymers. They also derived an expression for the spatial extension of the branches as a function of the number of branches in the star polymer.

The results from the Daoud-Cotton model are the basis of the scaling analysis of spherical and cylindrical micelles in good and theta solvents, pioneered by Zhulina and Birshtein [25] and the star model for spherical micelles in good solvents independently formulated by Halperin [26]. We have applied the Daoud-Cotton results to formulate a star polymer model for both micellization as well as solubilization [27] in block copolymer solutions. The development of various free energy contributions and the predictions for different cases of block composition, block size, and solvent-block interactions are detailed in that study. Here, we will present only the final results from the star polymer model, focusing on the role of the soluble block. Therefore, the results are given only for the case when the selective solvent is a good or theta solvent for the solvophilic block B, since this is the condition where we expect

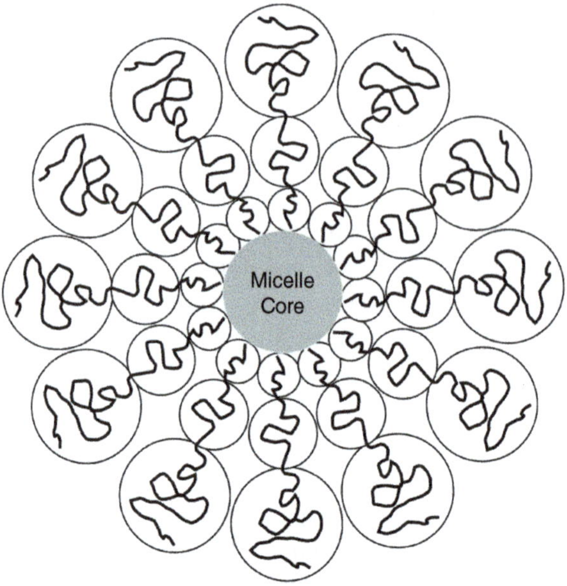

Figure 1.3 Schematic representation of the star polymer model of block copolymer micelle. The star is made up of multiple branches all connected to the surface of the micelle core. Each branch is made of a succession of blobs whose size increases radially outwards. The free energy is calculated as the number of blobs.

the solvophilic block interactions to be most important. As is conventional in scaling analysis, all numerical coefficients are ignored while writing various equations. All spatial variables such as the core radius R, the corona thickness D, and the core area per molecule a, are expressed as dimensionless quantities and the free energy contributions are expressed in units of kT.

Based on the application of the star polymer concept, the free energy per molecule of the block copolymer micelle is written as

$$(\Delta\mu_g^o) = (\Delta\mu_g^o)_{A,Tr} + (\Delta\mu_g^o)_{A,def} + (\Delta\mu_g^o)_{Int} + (\Delta\mu_g^o)_B$$

$$(\Delta\mu_g^o) = (\Delta\mu_g^o)_{A,Tr} + \frac{R^2}{N_A} + \sigma\frac{N_A}{R} + g^{1/2}K_B, \quad K_B = \ln\left(1 + \frac{D}{R}\right) \qquad (1.16)$$

Note that the free energy of transfer of the solvophobic block A from the incompatible solvent to the micelle core is usually not explicitly written but is obviously implied. This term provides the negative free energy contribution driving aggregate formation but is independent of the size and shape of the aggregate and therefore is not needed for any aggregate size and shape calculations. The core block elastic deformation term and the interfacial energy term are as in the de Gennes model. The only new addition is the B block dependent term, written here based on the blob picture of the star polymer model. We carry out free energy minimization for the limiting case when $(\Delta\mu_g^o)_B$ is dominant compared to $(\Delta\mu_g^o)_A$. When N_B is much larger than N_A, the shell block free energy contribution $(\Delta\mu_g^o)_B$ can be dominant compared to the core block free energy contribution $(\Delta\mu_g^o)_A$. One may note that when $N_B \gg N_A$, we have correspondingly $D/R \gg 1$. In this case, if the solvent is a good solvent for the B block, we get

$$R \sim N_A^{3/5}\sigma^{2/5}K_B^{-2/5}, \quad g \sim N_A^{4/5}\sigma^{6/5}K_B^{-6/5}, \quad D \sim N_A^{4/25}N_B^{3/5}\sigma^{6/25}\omega_B^{1/5}K_B^{-6/25}$$

$$K_B = \ln\left(1+\frac{D}{R}\right) \approx \ln(D/R) \approx \ln\left(N_A^{-\frac{11}{25}}\, N_B^{\frac{3}{5}}\, \sigma^{-\frac{4}{25}}\, \omega_B^{\frac{1}{5}} K_B^{\frac{4}{25}}\right) \approx \ln\left(N_A^{-\frac{11}{25}}\, N_B^{\frac{3}{5}}\right) \tag{1.17}$$

where $\omega_B = (1/2 - \chi_{BS})$ and χ_{BS} is the block B – solvent S Flory interaction parameter. If the solvent S is a theta solvent for the B block, we get:

$$R \sim N_A^{3/5}\sigma^{2/5}K_B^{-2/5}, \quad g \sim N_A^{4/5}\sigma^{6/5}K_B^{-6/5}, \quad D \sim N_A^{1/5}N_B^{1/2}\sigma^{3/10}3K_B^{-3/10}$$

$$K_B = \ln\left(1+\frac{D}{R}\right) \approx \ln(D/R) \approx \ln\left(N_A^{-\frac{2}{5}}\, N_B^{\frac{1}{2}}\, \sigma^{-\frac{1}{10}}\, K_B^{\frac{4}{25}}\right) \approx \ln\left(N_A^{-\frac{2}{5}}\, N_B^{\frac{1}{2}}\right) \tag{1.18}$$

Note that we have the same relations for the core radius R and the micelle aggregation number g as in Eq. (1.17) but a different dependence for the corona thickness D depending on whether the solvent is a good solvent or a theta solvent for the B block. In both cases, we find the micelle core radius and aggregation number are only weekly influenced by the solvophilic block, with a logarithmic dependence on N_B (appearing through K_B). Certainly this is much weaker than the case of surfactant micelles, where the headgroup interactions are paramount.

1.2.5 Mean Field Model of Block Copolymer Self-Assembly Incorporating Headgroup Effects

The surfactant micelle model treats the repulsive interactions between the hydrophilic headgroups as a central contribution to the micelle free energy. Indeed, without this contribution finite micelles cannot be realized. In block copolymers, an analogous free energy contribution should arise from the difference in state of the hydrophilic block B when the singly dispersed copolymer becomes part of the micelle. However, this contribution is not considered in the free energy model of de Gennes. While the star polymer model considered the B block dependent free energy, it did not yield a significant dependence of the micelle core radius and aggregation number on the size N_B of the B block but only a weak logarithmic dependence because the core does not appear as a significant factor in the asymptotic star model that focuses mainly on the corona. Therefore, we have developed mean field models [28–31] of block copolymer micelles, explicitly accounting for the coronal region's contribution to the free energy of micellization. We have considered two alternate options.

In one option, we treat the corona to be of uniform solvent concentration and correspondingly the corona B block is nonuniformly stretched as described by Semenov [23]. Alternately, we treat the corona B block to be uniformly stretched as in the Flory model [32], and correspondingly the corona region has a radially varying solvent concentration profile. The free energy model of de Gennes (Eq. (1.5)) is now modified to include the osmotic (dilution term) and elastic (deformation term) free energy contributions associated with the B block, equivalent to the headgroup repulsions for surfactants.

The free energy model based on uniform concentration and nonuniform chain stretching in the corona region has the form

$$(\Delta\mu_g^o) = (\Delta\mu_g^o)_{A,Tr} + (\Delta\mu_g^o)_{A,def} + (\Delta\mu_g^o)_{Int} + (\Delta\mu_g^o)_{B,dil} + (\Delta\mu_g^o)_{B,def}$$

$$\begin{aligned}
\left(\frac{\Delta\mu_g^o}{kT}\right)_{A,def} &= q\,\frac{p\pi^2}{80}\,\frac{R^2}{(N_A/q)L^2} \\
\left(\frac{\Delta\mu_g^o}{kT}\right)_{Int} &= \frac{\sigma}{kT}\,a, \quad \sigma = \frac{kT}{L^2}\left(\frac{\chi_{AS}}{6}\right)^{1/2} \\
\left(\frac{\Delta\mu_g^o}{kT}\right)_{B,dil} &= N_B\left[\frac{v_B}{v_S}\frac{1-\varphi_B}{\varphi_B}\ln(1-\varphi_B) + \frac{v_B}{v_S}(1-\varphi_B)\,\chi_{BS}\right] \\
\left(\frac{\Delta\mu_g^o}{kT}\right)_{B,def} &= \frac{3}{2}\,\frac{L\,R}{\left(\frac{a}{q}\right)\varphi_B}\,P
\end{aligned} \tag{1.19}$$

Eq. (1.19) is written to be valid for AB diblock and symmetric BAB triblock copolymers, with $q = 1$ for AB diblock and $q = 2$ for BAB triblock copolymers. The first term is the free energy of transfer of the solvophobic block A from the incompatible solvent to the micelle core and it provides the negative free energy driving micelle formation to occur. The second term is the A block elastic deformation energy contribution with $p = 3$ for spheres, $p = 5$ for cylinders, and $p = 10$ for lamella. The A block deformation term is written based on the Semenov treatment that we had applied also to surfactant tails, as shown in Eq. (1.13). The third term is the interfacial energy term where the interfacial tension σ can be an experimentally measured value or it can be calculated based on the expression provided by Helfand [14], as shown. Here, the variable χ_{AS} denotes the Flory interaction parameter of the A block with solvent S and L is the characteristic segment size. The fourth term is the B block osmotic or dilution term. It is written using the formulation of Flory polymer solution thermodynamics. Here, χ_{BS} denotes the Flory interaction parameter of the B block with solvent S, L is the characteristic segment size, and φ_B is the volume fraction of polymer B segments in the corona region. The last term is the B block deformation term, also based on the approach developed by Semenov [23]. In this expression, $P = (D/R)/(1 + D/R)$ for spheres, $P = \ln(1 + D/R)$ for cylinders, and

Table 1.4 Geometrical relations for block copolymer aggregates.

Property	Sphere	Cylinder	Lamella
Core volume, V_C	$4\pi R^3/3$	πR^2	$2R$
Corona/Core volume ratio, V_S/V_C	$[(1+D/R)^3-1]$	$[(1+D/R)^2-1]$	$[(1+D/R)-1]$
Aggregation number, g		$V_C/(N_A v_A)$	
Core surface area/molecule, a	$3\,N_A v_A/R$	$2\,N_A v_A/R$	$N_A v_A/R$
Volume fraction of B in corona, φ_B		$(g\,N_B v_B)/V_S = (N_B v_B/N_A v_A)(V_C/V_S)$	

The symbols v_A and v_B refer to the molecular volumes of A and B segments, v_S is the molecular volume of the solvent, N_A and N_B are the number of segments of blocks A and B for both AB diblock and symmetric BAB triblock copolymers. The number of block copolymer molecules g in an aggregate, the aggregate core volume V_C, and the corona volume V_S all refer to the total quantities in the case of spherical aggregates, quantities per unit length in the case of cylindrical aggregates and quantities per unit area in the case of lamellar aggregates.

$P = (D/R)$ for lamellae [29, 30]. The Semenov treatment accounts for the fact that the chain deformation has to be nonuniform in order to maintain uniform concentrations of segments in the corona region. We note in passing that in our first model for block copolymer micelles [28], we had used the Flory model for chain elasticity that assumes uniform chain deformation to calculate the deformation free energy contribution, while also assuming the corona region to be uniform in concentration.

This free energy model is applicable to aggregates having spherical, cylindrical, and lamellar morphologies. All the geometrical properties of the aggregates appearing in the free energy expressions are summarized in Table 1.4. Note that in our detailed treatment of block copolymer micelles [28–30], we have also included a free energy contribution to account for the localization of the AB joint to a narrow volume of the aggregate and also a contribution that accounts for backfolding in the case of a BAB triblock copolymer. Both of these contributions are practically independent of the aggregate size and shape and therefore not shown in Eq. (1.19).

For illustrative purposes, calculations have been carried out for the PEO–PPO diblock copolymers aggregating in water, with water being a selective solvent for the PEO block and a poor solvent for the PPO block. One can expect the B block dependent free energy contribution to gain greater importance when the selective solvent changes from being a theta solvent to a good solvent for the B block. Our calculations for PEO–PPO block copolymer micelles considering uniform concentration in the corona and nonuniform chain stretching for the coronal B block (Semenov approach) yielded results summarized in Table 1.5. The predicted aggregate core radius R, corona thickness D, area per molecule at the core–corona interface a_e and the

Table 1.5 Model predictions of spherical micelles of PEO–PPO in water at 25 °C.

Block copolymer	R (nm)	D (nm)	a_e (nm^2)	g
$E_{34}P_{60}$	8.29	3.69	2.09	413
$E_{40}P_{70}$	9.02	4.21	2.25	455
$E_{26}P_{30}$	5.07	2.91	1.71	189
$E_{38}P_{43}$	6.16	3.95	2.02	235
$E_{54}P_{61}$	7.43	5.24	2.38	292
$E_{38}P_{29}$	4.48	3.81	1.88	134
$E_{48}P_{35}$	4.89	4.56	2.07	145
$E_{52}P_{40}$	5.33	4.89	2.17	164
$E_{74}P_{56}$	6.33	6.43	2.56	196
$E_{104}P_{35}$	3.93	7.27	2.58	75
$E_{122}P_{40}$	4.18	8.14	2.77	79
$E_{200}P_{64}$	5.27	11.6	3.52	99
$E_{154}P_{29}$	3.04	8.51	2.76	42
$E_{208}P_{39}$	3.53	10.6	3.20	49
$E_{236}P_{45}$	3.81	11.6	3.42	53
$E_{266}P_{50}$	4.00	12.6	3.62	56

aggregation number g for spherical micelles are listed for each of the diblock copolymer considered.

By correlating the predicted results, we find the following scaling relations for the PEO–PPO-water system:

$$R \sim N_A^{0.73} N_B^{-0.29}, \qquad g \sim N_A^{1.19} N_B^{-0.87}, \qquad D \sim N_A^{0.06} N_B^{0.60} \tag{1.20}$$

The core radius and the aggregation number show dependence on the solvophilic block B that was entirely absent in the case of de Gennes model and was much weaker in the case of the star polymer model. Indeed, the free energy contributions analogous to surfactant headgroup repulsions are very important for block copolymers and the contribution leads to stronger dependencies on the B block when the selective solvent is a very good solvent for the B block.

1.2.6 Tail Effects on Shape Transitions in Surfactant Aggregates

The rationalization of why surfactant molecules choose to self-assemble as spherical micelles, cylindrical micelles, or spherical bilayer vesicles and how one can cause a transition in shapes is a fundamental problem of interest.

In surfactant aggregates, since the hydrophobic domain is made up of the surfactant tails, irrespective of the shape of the aggregate, no point within the aggregate can be farther than ℓ_0 from the aggregate-water interface, where ℓ_0 is the extended length of the surfactant tail. In view of this constraint, and considering the geometrical properties of aggregates of different shapes, we have already referred to the well-known [11] connection between the molecular packing parameter and the aggregate shape: $0 \leq v_0/a\ell_0 \leq 1/3$ for sphere, $1/3 \leq v_0/a\ell_0 \leq 1/2$ for cylinder, and $1/2 \leq v_0/a\ell_0 \leq 1$ for bilayer. Note that the free energy plays a fundamental role, as it should, in determining the magnitude of the packing parameter, because the equilibrium area per molecule a_e is obtained from the condition of minimum free energy. We have already discussed how tail packing contributes to the free energy, as described in Eq. (1.13). To assess the consequences of this contribution to determining the shape of the aggregate, we can compare the equilibrium areas predicted by the Tanford model (Eqs. (1.3) and (1.4)), which ignores this tail contribution against the modified model (Eqs. (1.13) and (1.15)) for illustrative cases where shape transitions may occur. The calculations have been performed for a surfactant with a dodecyl alkane tail [22] and the results are summarized in Table 1.6. As mentioned earlier, $(\sigma/kT) = 12\,\text{nm}^{-2}$, consistent with σ being around 50 mN/m for the aliphatic hydrocarbon-water interface. The headgroup interaction parameter α/kT is taken as a variable in the range 5.0–0.6 nm^2 to reflect the condition where the headgroup repulsions are decreased by adding salt to an ionic surfactant solution.

The inclusion of the tail-packing free energy (Eq. (1.15)) results in the equilibrium area a_e for surfactant aggregates being larger than that estimated from Eq. (1.3). Since in all cases the area per molecule is increased, the predicted packing parameter is always decreased when the tail-packing contribution is considered. Therefore, the predictions of equilibrium aggregate shapes are altered by the consideration of the tail elastic stretching. The calculated results show that when compared to the predictions of the Tanford model, accounting for the tail deformation results in larger spheres becoming smaller spheres at

Table 1.6 Predicted aggregate shape transitions for surfactant with dodecyl tail.

	a_e (nm^2)	$v_0/a_e\ell_0$	Shape	a_e (nm^2)	$v_0/a_e\ell_0$	Shape
α/kT (nm^2)	Based on Eqs. (1.3) and (1.4)			Based on Eqs. (1.13) and (1.15)		
5.0	0.645	0.326	Sphere	0.735	0.286	Sphere
3.0	0.500	0.42	Cylinder	0.629	0.333	Sphere
2.4	0.447	0.47	Large cylinder	0.565	0.372	Small cylinder
1.2	0.316	0.664	Bilayer	0.488	0.430	Cylinder
0.6	0.224	0.938	Large vesicle	0.374	0.561	Small vesicle

$\alpha/kT = 5.0\,\text{nm}^2$, spheres being preferred over cylinders at $\alpha/kT = 3.0\,\text{nm}^2$ and cylinders being preferred over bilayer for $\alpha/kT = 1.2\,\text{nm}^2$. It should be noted that in applying Eq. (1.15) for spherical bilayer vesicles, it would be necessary to allow for the inner and the outer layers of the vesicle to differ in their layer thicknesses and also the area per molecule at the inner and the outer surfaces to be different. Such refinements are essential for any predictive computations but for the limited purposes of demonstrating the influence of the tail, they are left out in Eq. (1.15) and the resulting computed results summarized in Table 1.6.

1.2.7 Headgroup Effects on Shape Transitions in Block Copolymer Aggregates

We have already seen how the solvophilic block free energy contributions (equivalent to the surfactant headgroup repulsions) significantly influences the size of spherical block copolymer micelles. The block copolymer aggregates also exist as cylinders or bilayers just as surfactant aggregates. The shape transitions have been examined for PEO–PPO diblock copolymers using the mean field model (Eq. (1.19)). The results obtained for spherical micelles was shown in Table 1.5 and the results for cylinders and lamella are given in Table 1.7.

From Tables 1.5 and 1.7, we note that for block copolymers that have similar hydrophobic block lengths ($EO_{16}PO_{50}$ and $EO_{266}PO_{50}$, or $EO_{12}PO_{35}$, $EO_{18}PO_{32}$, and $EO_{45}PO_{35}$, for example) aggregate shapes change from spheres at large PEO block size to cylinders and lamella as the block size of PEO decreases, clearly reflecting the solvophilic block (headgroup) effect on the equilibrium aggregate shapes. Further, for the block copolymers that have similar hydrophobic block lengths mentioned above, the core radius of the sphere is larger than the core radius of the cylinder, which, in turn, is larger than the half-bilayer

Table 1.7 Shapes of PEO–PPO diblock copolymer aggregates in water at 25 °C.[a)]

Block copolymer	*R* (nm)	*D* (nm)	a_e (nm^2)	*g*	Shape
$E_{12}P_{35}$	2.62	1.73	1.29	1.55	Lamella
$E_{12}P_{38}$	2.80	1.72	1.31	1.53	Lamella
$E_{16}P_{50}$	3.26	2.19	1.48	1.35	Lamella
$E_{22}P_{69}$	3.90	2.85	1.71	1.17	Lamella
$E_{18}P_{32}$	4.02	2.27	1.54	16.5	Cylinder

a) The aggregation number *g* is defined as the number of molecules per nm length in the case of cylinder and the number of molecules per nm^2 area in the case of lamella.

Table 1.8 Predicted aggregate shapes for PEO–PPO–PPO triblock copolymer.[a)]

Trade name	Block copolymer	*R* (nm)	*D* (nm)	a_e (nm^2)	*g*	Aggregate shape
L62	$E_6P_{35}E_6$	1.75	0.96	1.93	1.04	Lamella
L72	$E_6P_{38}E_6$	1.87	0.96	1.96	1.02	Lamella
L92	$E_8P_{50}E_8$	2.19	1.22	2.20	0.91	Lamella
L122	$E_{11}P_{69}E_{11}$	2.65	1.60	2.52	0.80	Lamella
L63	$E_9P_{32}E_9$	1.45	1.39	2.13	0.94	Lamella
P103	$E_{17}P_{60}E_{17}$	3.87	2.17	2.99	8.1	Cylinder
P123	$E_{20}P_{70}E_{20}$	4.21	2.48	3.21	8.3	Cylinder
L64	$E_{13}P_{30}E_{13}$	3.42	1.64	2.54	58	Sphere
P84	$E_{19}P_{43}E_{19}$	4.19	2.24	2.97	74	Sphere
P104	$E_{27}P_{61}E_{27}$	5.09	2.99	3.47	94	Sphere
P65	$E_{19}P_{29}E_{19}$	3.06	2.18	2.74	43	Sphere
P75	$E_{24}P_{35}E_{24}$	3.36	2.63	3.02	47	Sphere
P85	$E_{26}P_{40}E_{26}$	3.66	2.82	3.16	53	Sphere
P105	$E_{37}P_{56}E_{37}$	4.38	3.73	3.70	65	Sphere
F77	$E_{52}P_{35}E_{52}$	2.75	4.31	3.69	26	Sphere
F87	$E_{61}P_{40}E_{61}$	2.92	4.84	3.96	27	Sphere
F127	$E_{100}P_{64}E_{100}$	3.70	6.98	5.00	35	Sphere
F68	$E_{77}P_{29}E_{77}$	2.13	5.13	3.94	15	Sphere
F88	$E_{104}P_{39}E_{104}$	2.48	6.39	4.55	17	Sphere
F98	$E_{118}P_{45}E_{118}$	2.67	7.03	4.86	18	Sphere
F108	$E_{133}P_{50}E_{133}$	2.83	7.66	5.14	20	Sphere

a) The aggregation number *g* is defined as the number of molecules per nm length in the case of cylinder and the number of molecules per nm^2 area in the case of lamella.

thickness of the bilayer. This prediction is the direct result of the chain elastic deformation free energy contribution. The shape transitions have also been explored for the commercially available triblock copolymers PEO–PPO–PEO of various molecular weights and block compositions and the results are summarized in Table 1.8.

The shape transition of block copolymer aggregates as a function of the block copolymer composition shown on Table 1.8 follows the behavior exhibited by surfactant aggregates as a function of headgroup repulsions. For surfactants, when the headgroup repulsions are weak, lamellar aggregates are favored while for strong headgroup repulsions, spherical micelles are formed. Cylindrical micelles result for intermediate values of headgroup repulsions. For the block copolymer aggregates, the free energy contributions associated with the hydrophilic block B are analogous to the headgroup repulsions in surfactant micelles. Therefore, the pattern of aggregation observed for increasing headgroup repulsions in surfactants is exactly reproduced with increasing size of the hydrophilic block B in the case of block copolymers.

For surfactants, the packing parameter $P = v_o/a_e\ell_o$ provided the bounds for various aggregate shapes. Although the free energy determines the equilibrium area a_e, the constraint that the radius of spherical or cylindrical micelle or half bilayer thickness cannot exceed ℓ_o plays a critical role in influencing the shape transitions. For example, a surfactant with a dodecyl tail can accommodate only 54 or fewer molecules if the aggregate shape is to remain spherical. However, in the case of the block copolymer, there is no equivalent constraint imposed by the extended length of the hydrophobic block since typically the block sizes are large. For example, if we look at the family of diblock copolymers in Table 1.5 with approximately 35 PO units, the extended length of the hydrophobic PPO block can be estimated roughly to be 16.1 nm, taking the segment length to be about 0.46 nm [33]. Corresponding to this extended length, spherical aggregates having a PO_{35} core block can accommodate as many as 5175 molecules. However the calculations show that we form spherical micelles of core radius 4.9 nm and aggregation number 145 for $EO_{48}PO_{35}$. When the PEO chain length is decreased, keeping the same hydrophobic block, even though there are no packing constraints that would prevent the sphere from growing from 145 molecules all the way up to 5175 molecules, the micelle does not grow retaining spherical shape but changes to a cylinder for $EO_{18}PO_{32}$ and bilayer for $EO_{12}PO_{35}$ (Table 1.7). This is a consequence of the free energy control over shape transitions. The predicted core radii are only 2.6, 4.0, and 4.9 nm for $EO_{12}PO_{35}$, $EO_{18}PO_{32}$, and $EO_{48}PO_{35}$ block copolymers forming lamellar, cylindrical, and spherical aggregates, much smaller than the extended length of 16.1 nm for the PPO block.

1.3 Self-Assembly of Nonclassical Amphiphiles Based on Head–Tail Competition

Over the last decade, there have been extensive studies on nonclassical amphiphiles that combine biological or inorganic elements or nanoparticles into classical surfactant, lipid, or polymer structures [34–48]. This book presents the most recent advances in these areas by pioneering scientists in these fields. Here, we present a few examples where the authors have invoked the application of existing theoretical models and the packing parameter concept to rationalize the observed self-assembly behavior. In all cases, the interactions involving the headgroups and the tail groups and the head-tail competition can be used to explain the patterns of aggregation observed.

1.3.1 Dendritic Amphiphiles

Dendritic polymers have a high degree of branching and therefore multiple number of end groups. These end groups can be readily modified by different functional groups such as reactive ligands, targeting moieties, etc., allowing for the possibility of multivalent binding interactions and enabling the practical applications of these molecules including for nanomedicine. This has generated significant interest in the design of amphiphiles built with dendritic polymers as the headgroups. An example is the series of dendritic amphiphiles synthesized in the Haag lab [34] involving glycerol-based generation 2 (G2) and generation 3 (G3) dendrons as hydrophilic headgroups and dialkyl chains as the tailgroup (Figure 1.4). They created six distinct dendritic amphiphiles by varying the generation of the headgroup and the length and geometry of the hydrophobic tailgroup.

The size distribution of the self-assembled structures were measured using dynamic light scattering (DLS) and the aggregate morphologies were observed

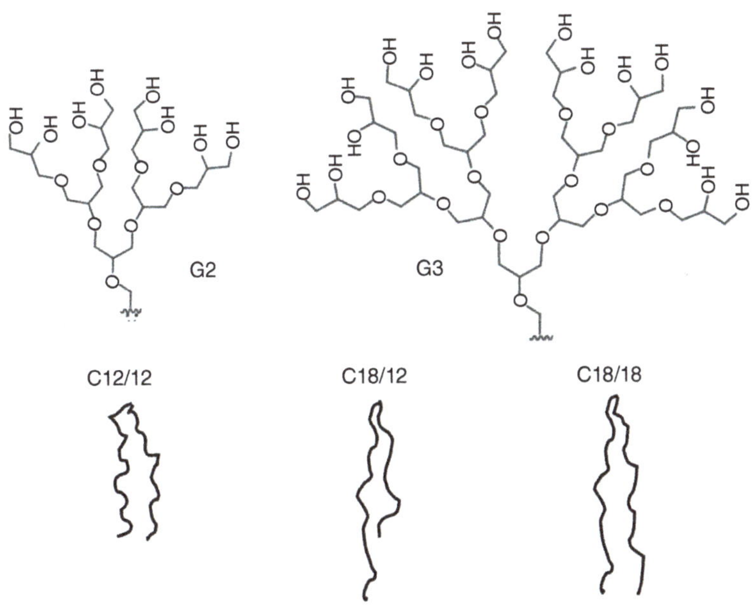

Figure 1.4 Headgroups and tailgroups of dendritic amphiphiles. The headgroups G2 and G3 are second- and third-generation dendrimers with hydroxyl surface groups. The tails are dialkyl chains with symmetric lengths of 12 and 18 carbon atoms for two molecules and asymmetric lengths of 12 and 18 carbon atoms for the third molecule, as shown in the figure. By combining these headgroups and tailgroups, six amphiphile structures were generated by the Haag lab [34]. They are depicted as G2C12/12, G2C12/18, G2C18/18, G3C12/12, G3C12/18, and G3C18/18.

using cryo-transmission electron microscopy (cryo-TEM). The DLS data predicted the existence of small micelles for all G3 amphiphiles (G3C12/12, G3C12/18, and G3C18/18) and the narrow intensity distributions indicated that the micelles were monodisperse. In contrast, the G2 amphiphiles revealed cylindrical aggregates of a markedly larger size by DLS, with the size distributions being wider and the average sizes considerably different for the three molecules considered. Cryo-TEM showed that for the symmetric G2 amphiphiles (G2C12/12 and G2C18/18), increasing the alkyl chain length from 12 to 18 led to a morphological transition from wormlike micelles to vesicles. However, the dissymmetric amphiphile G2C12/18 formed coexisting spherical as well as wormlike micelles reflected in the bimodal distribution obtained from light scattering.

The free energy model for classical surfactants (Eq. (1.13)) can be applied here. Because the dendritic amphiphiles considered are nonionic, the headgroup interactions are due only to steric repulsions. Based on the volumes and lengths of aliphatic hydrocarbon chains [21], the ratio v_0/ℓ_0 is equal to 0.42 nm^2 for the C12/12 and C18/18 tails and is 0.35 nm^2 for the C12/18 tail. For the G2 dendritic amphiphile with 8 hydroxyl groups on the surface, the headgroup volume can be roughly estimated using group contributions to be about half of that for the G3 headgroup with 16 hydroxyl groups, implying that the headgroup interactions should be larger in magnitude for the G3 amphiphiles compared to the G2 amphiphiles. For illustrative purposes, taking the headgroup repulsion parameter $\alpha/kT = 5$ and 10 nm^2, to represent the G2 and G3 headgroups, the areas per molecule are calculated using Eqs. (1.14) and (1.15) for each of the three aggregate shapes. In calculating the area corresponding to a given shape, the chain packing free energy corresponding to that shape is used. If the results show that the packing parameter calculated assuming the spherical shape is above 1/3, then that solution is invalid. Similarly, if the packing parameter calculated assuming the cylindrical shape is greater than 1/2, then that solution is invalid. The calculated results are summarized in Table 1.9.

For the G2 amphiphiles, the packing parameter corresponds to a bilayer for both G2C12/12 and G2C18/18. Fort G2C18/18, the tail length is long enough and the aqueous cavity volume of the vesicle is large enough to accommodate the headgroups and vesicles can form. However, because the tail length is shorter for G2C12/12, the aqueous cavity volume is small to accommodate the headgroups and vesicle formation is prevented. In this case cylindrical micelles corresponding to $P = 1/2$ will form. For G2C12/18, cylindrical micelles are predicted because of the smaller v_0/ℓ_0. For the G3 amphiphiles, because the headgroup repulsion parameter is large, the area per molecule is larger and the packing parameter is smaller. In all three cases the packing parameter is larger than but close enough to 1/3 and therefore, globular micelles not much large than the largest spherical micelles are predicted. The calculations based on the free energy expression (Eq. (1.13)) is only illustrative and detailed predictive

Table 1.9 Aggregation behavior of dendritic amphiphiles.

	a_e (nm^2) from Eqs. (1.14) and (1.15)			$P = v_o/a_e\ell_o$		
Molecule	Sphere	Cylinder	Bilayer	Sphere	Cylinder	Bilayer
G2C12/12	0.803	0.769	0.715	0.523 invalid	0.546 invalid	0.587
G2C12/18	0.828	0.791	0.728	0.423 invalid	0.442	0.481 invalid
G2C18/18	0.855	0.812	0.741	0.491 invalid	0.517 invalid	0.567
G3C12/12	1.006	0.985	0.951	0.417 invalid	0.426	0.441 invalid
G3C12/18	1.025	0.997	0.959	0.341 invalid	0.351	0.365 invalid
G3C18/18	1.043	1.015	0.966	0.403 invalid	0.414	0.435 invalid

calculations would require quantitative modeling of the headgroup repulsions as we have done in our earlier work [21] and not the use of the phenomenological parameter α appearing in the Tanford free energy expression.

1.3.2 DNA Amphiphiles

DNA can be attached to hydrophobic polymers or lipids to provide DNA-polymer or DNA-lipid conjugates with amphiphilic properties [35–37]. The DNA not only serves as the hydrophilic portion of the amphiphile but can also be used to manipulate the self-assembly taking advantage of the DNA hybridization properties. Here, we discuss an example of DNA amphiphiles based on the work conducted in the Gianneschi laboratory [35] where they showed that short oligonucleotide conjugated lipids spontaneously assemble into unilamellar vesicles that can undergo reversible, vesicle-to-micelle transitions in response to specific DNA manipulations.

The system demonstrated by Thompson et al. [35] involves a DNA amphiphile that consist of two C_{18} alkyl hydrophobic tails, linked covalently to the 5′-termini of a 9-mer single-stranded ssDNA oligonucleotide (Lipid-DNA1 in Figure 1.5). This DNA amphiphile was found to self-assemble into uniformly shaped, spherical vesicles approximately 500 nm in diameter as characterized by multiple techniques. The unilamellar bilayer architecture of these vesicle assemblies was confirmed by cryoTEM with bilayer thicknesses of 8–9 nm, roughly equivalent to double the extended length of the DNA amphiphile. The vesicles were mixed with partially complementary 19 base ssDNA sequences

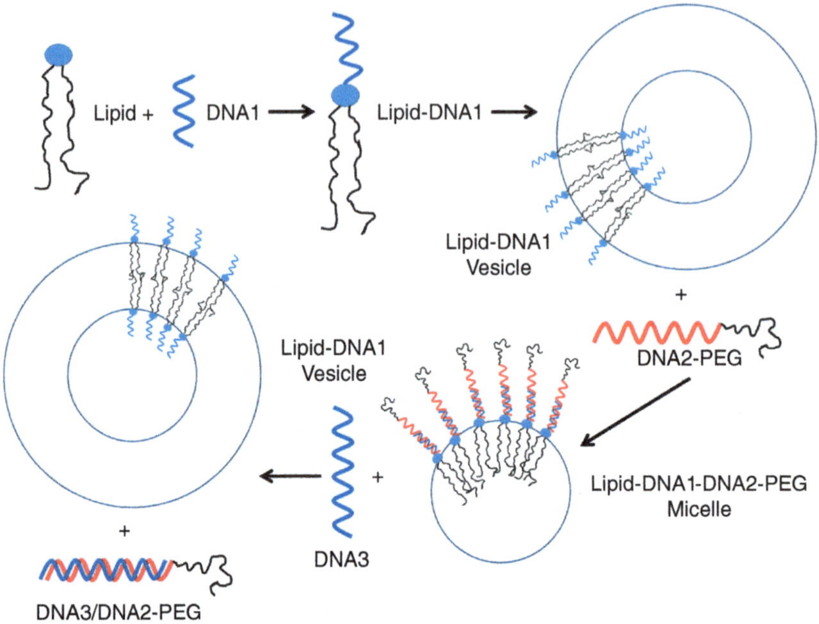

Figure 1.5 Use of DNA-programming to manipulate aggregate structures formed DNA amphiphiles [35]. The DNA1-lipid self assembles to form spherical unilamellar vesicles. The vesicles are then mixed with DNA2-PEG, which is partially complementary to DNA1. The complexation of DNA2-PEG with the DNA1 headgroups effectively creates a new headgroup, introducing additional steric and electrostatic repulsions between headgroups. This makes the vesical structure less optimal compared to spherical micelles and therefore the vesicle transform into micelles. These micelles are now mixed with DNA3, which is perfectly complementary to DNA2. Since DNA2-DNA3 binding is stronger than the DNA1-DNA2 binding, the DNA2-PEG is released from the micelles to form DNA3-DNA2-PEG hybrids. The headgroup of the amphiphile now reverts to DNA1 and the corresponding reduction in the headgroup repulsions makes the vesicle structure to be optimal allowing vesicles to reversibly reform from the spherical micelles. Extensive discussions on DNA containing amphiphiles are presented by Thompson and Gianneschi in Chapter 5 of this book.

modified at their 5′-termini with two 18-member ethylene glycol phosphoramidites (DNA2-PEG in Figure 1.5). The complexation of DNA2-PEG with DNA1 creates a new headgroup, which gives rise to a change in the structure of the aggregates from vesicles to micelles of about 20–25 nm in size. The spherical micelles were then combined with DNA3, which is a 19-base ssDNA strand perfectly complementary to DNA2. The hybridization of DNA2 with DNA3 being much stronger than the partial hybridization of DNA1 with DNA2, the DNA2-PEG was completely removed from the micelles, resulting in the reversible formation of vesicles of DNA1-lipid along with the formation of the stable 19-base pair duplex, DNA2-DNA3.

Considering DNA1-lipid, the DNA1 headgroup has negative charges from its phosphate groups but the net charge is always smaller than the number of phosphate groups. Therefore, the headgroup repulsions involve both electrostatic and steric interactions. The solution experiments were conducted in 20 or 50 mM Tris buffer and also in the presence of various amounts of $MgCl_2$. The presence of these electrolytes would decrease the extent of electrostatic interactions. Further, since the charges on the phosphate ions are located at some significant distance away from the hydrophobic core surface, the magnitude of electrostatic interactions would be further smaller. As a result, the headgroup repulsion parameter α/kT can be taken to be a small value. Taking, for example, $\alpha/kT = 3.0\,\text{nm}^2$, the equilibrium area calculated from Eq. (1.15) will be $0.63\,\text{nm}^2$ and the corresponding packing parameter P is 0.67, denoting the formation of a vesicle. When PEG-DNA2 complexes with the DNA1-lipid, the new headgroup is the DNA1-PEG-DNA2 complex. For this new headgroup, there would be some small increase in the electrostatic interactions and significant increase in the steric repulsions because of the large PEG group present. Taking, for example, a larger value of $\alpha/kT = 12.0\,\text{nm}^2$, the equilibrium area calculated from Eq. (1.15) will be $1.11\,\text{nm}^2$ and the corresponding packing parameter P is 0.378, implying a globular micelle, not much larger than the largest spherical micelle. Subsequently, on contact with DNA3 if all of the DNA2-PEG is removed by complete hybridization, we are left with DNA1 as the headgroup of the amphiphile, resulting in the reversible formation of vesicles. As mentioned previously, quantitative predictions would require more detailed modeling of the vesicle, allowing for the inner and the outer surfaces to be different and also modeling more precisely for the nature of headgroup repulsions.

1.3.3 Peptide Amphiphiles

Another class of nonclassical amphiphiles is that incorporating peptides such as for example, a molecule with hydrophobic polymer as the tail and a hydrophilic peptide with one or more hydrophilic amino acids as the headgroup [38–40]. Alternately, the hydrophobic tail could be an alkyl chain or a peptide itself composed of hydrophobic amino acids. Such peptide amphiphiles have been investigated in view of potential applications in nanomedicine. For example, the peptides can be selected to be substrates for cancer-associated proteins and thus amenable to proteolysis. Each peptide in the headgroup can be chemically modified by proteolysis by a specific enzyme, allowing for an enzyme-responsive switching of the morphology of the micelles used as drug delivery vehicles or as *in vivo* sensors [40]. Tirrell et al. investigated [38] peptide amphiphiles that form wormlike micelles resembling nanofibers, which are viewed as potential synthetic extracellular matrix materials for tissue engineering and regenerative medicine. The amphiphilic peptides are also being evaluated for materials design applications involving biomimetic mineralization. The self-assembled peptide amphiphiles act as templates to induce nucleation, growth, and alignment of nanocrystals, thereby resulting in the formation of metallic nanowires and other highly ordered superstructures of minerals or composites.

Here we consider as an example studies on a series of amphiphilic peptides that were synthesized by Xu et al. [39]. In these amphiphiles, one lysine residue at the C terminus acts as the hydrophilic head and a chain of three to nine alanine residues serves as the hydrophobic tail. The C terminus was amidated, and the N terminus was acetylated, thus generating one positive charge per molecule (from the lysine residue side chain) at neutral pH. The molecular structures of the three amphiphiles denoted as A_3K, A_6K, and A_9K are shown on Figure 1.6.

The self-assembly of these peptide amphiphiles in aqueous solution has been examined by atomic force microscopy (AFM) and transmission electron microscopy (TEM). These studies have revealed that bilayer sheets were formed from A_3K, long wormlike micelles were formed from A_6K, and short nanorods were formed from A_9K. All of these peptide nanostructures were produced in solutions at a pH of about 6.0 when the lysine can have one positive charge. When the solution pH was changed to above 11, no characteristic supramolecular structures were formed, but instead irregular aggregates were

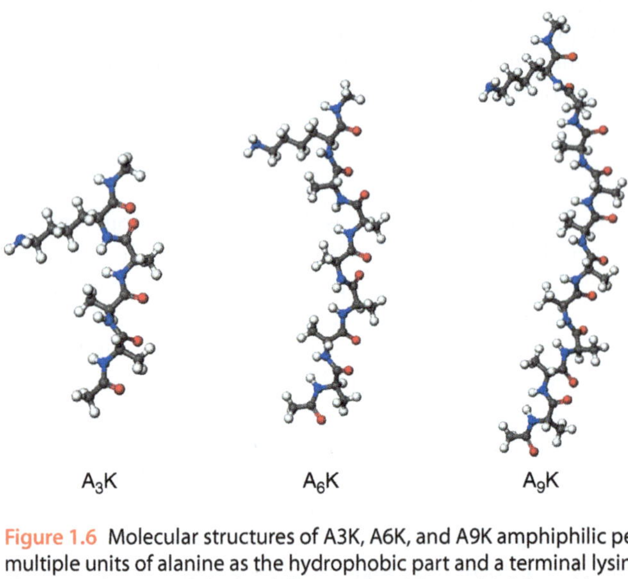

Figure 1.6 Molecular structures of A3K, A6K, and A9K amphiphilic peptides composed of multiple units of alanine as the hydrophobic part and a terminal lysine with one positive charge as the cationic hydrophilic part. Peptide amphiphiles with peptides as headgroups connected to a hydrophobic alkyl chain are extensively discussed by Leon and Tirrell in Chapter 6 of this book, focusing on unique peptide–peptide interactions and various aggregate structures that can result from it.

observed by TEM. Since the side chain amino group of lysine has a pKa of around 10, at pH 11, the headgroups became deprotonated, losing their charge and thereby changing the amphiphilic peptide to a hydrophobic molecule that simply precipitates in solution.

In the framework of the free energy model given by Eqs. (1.13)–(1.15), let us assume that for the appropriate headgroup parameter α/kT, the equilibrium area and the resulting packing parameter for A_3K correspond to a bilayer. Based on the free energy model, if the hydrophobic tail length is increased, the area per molecule would increase and the packing parameter would decrease in magnitude. This would qualitatively rationalize why for A_6K and A_9K cylindrical micelles are favored. Again, for A_6K the packing parameter would be larger than that for the A_9K amphiphile, implying that the A_6K can form long worm-like micelles while A_9K forms somewhat smaller cylindrical micelles. It should be noted, however, that the chain length effect modeled in Eqs. (1.13)–(1.15) is based on polymers with segments as basic units. The molecular structure and chain conformation for the peptides are significantly different from those of a polymer, and one would have to rederive an appropriate packing free energy contribution for these cases before any quantitative predictions can be made.

1.3.4 Protein–Polymer Conjugates

Protein–polymer conjugates combine natural biological molecules like polypeptides or proteins with synthetic polymer molecules to generate synergistic benefits from the presence of both components [41]. Many applications of polymer–protein conjugates are found in the area of therapeutics using the conjugates to construct targeted drug delivery vehicles. Because of conjugation with the polymer, the effective size of the protein is increased and its renal excretion rate decreased. The polymer chains also increase the half-life of the protein in the circulation system by protecting the protein from enzymatic degradation or removal by receptor recognition or complexation with antibodies. The attachment of the polymer to the protein also decreases the propensity of the protein to aggregate. One can also take advantage of the responsive characteristics of the polymer to generate stimuli-responsive protein function. In Figure 1.7, a number of polymer–protein conjugate types are schematically depicted incorporating homopolymers or block copolymers.

The polymer–protein conjugates can self-assemble into aggregates similar to the behavior exhibited by surfactants and block copolymers. Schematic of aggregates that can form from the polymer–protein conjugates are illustrated in Figure 1.8 assuming spherical shapes. Obviously, other shapes are also possible based on which of the shape-dependent free energies is the smallest and also the geometrical packing constraints applicable to the systems. Figure 1.7a depicts a hydrophobic homopolymer connected to the protein. In this case, self-assembly is possible if the protein is hydrophilic and plays the

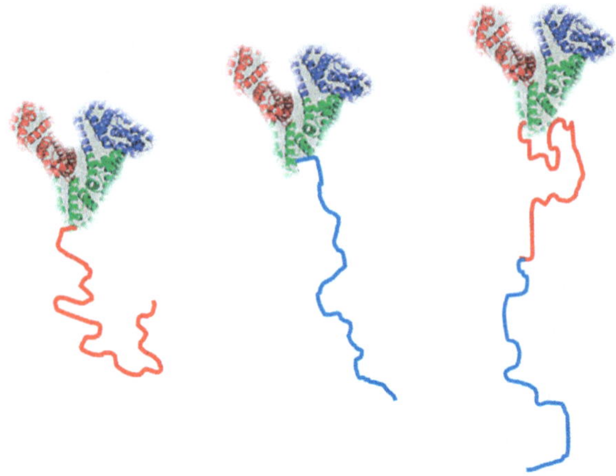

(a) (b) (c) (d)

Figure 1.7 Schematic representations of protein–polymer conjugates. In (a), the protein is conjugated to a hydrophobic polymer chain while in (b), The protein is conjugated to a hydrophilic polymer chain. In (c) the protein is conjugated to the hydrophobic block of a diblock copolymer while in (d), the protein is conjugated to the hydrophilic block of a diblock copolymer.

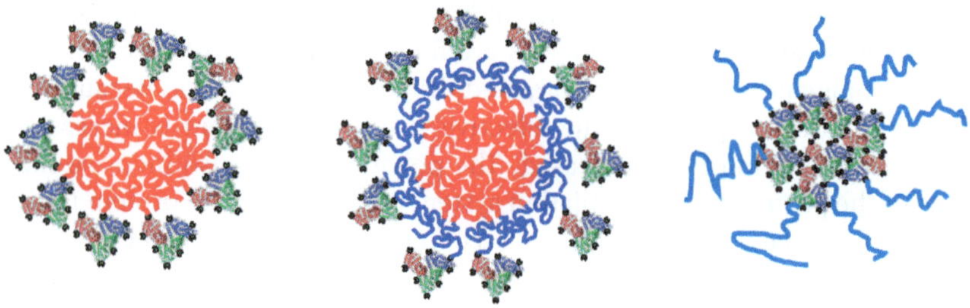

(a) (b) (c)

Figure 1.8 Schematic of spherical micelles anticipated from different protein–polymer conjugates. In (a), the hydrophobic polymer forms the core of the micelle while the hydrophilic protein forms the corona of the micelle. In (b), the hydrophobic block of the block copolymer controls micellization and the hydrophilic block and the hydrophilic protein conjugated to it are in the coronal region. If the protein is hydrophobic, other aggregation patterns could result for this molecule. In (c), the protein is hydrophobic and forms a core of protein molecules surrounded by the hydrophilic polymer conjugated to it. Such hydrophobic protein–polymer conjugates are discussed in detail by Olsen in Chapter 7 of this book.

role of the headgroup. Possible spherical micelle resulting from this conjugate is shown on Figure 1.8a. Figure 1.7b shows a hydrophilic homopolymer conjugated to a protein. In this case, self-assembly is possible provided the protein is hydrophobic. The resulting aggregate in spherical shape is shown on Figure 1.8c. Figures 1.7c,d show the protein conjugated to a diblock copolymer, either at the hydrophobic block end or at the hydrophilic block end. In this case, the block copolymers themselves are capable of forming aggregates, the presence of the protein can play the role of strengthening or weakening the aggregate depending on its hydrophilicity/hydrophobicity characteristics.

Depending on the nature of the protein, additional interactions should be considered while developing an expression for the free energy of aggregation. For example, in Figure 1.7a, the protein is connected to a hydrophobic homopolymer. The structure of a spherical micelle formed from such protein–polymer conjugate is depicted in Figure 1.8a. One can calculate the free energy of formation of aggregates for this system using Eq. (1.19) by introducing both steric and ionic repulsions between the protein molecules in the corona region of the micelle to represent the headgroup repulsions.

$$\left(\frac{\Delta\mu_g^o}{kT}\right) = \left(\frac{\Delta\mu_g^o}{kT}\right)_{A,Tr} + \left(\frac{\Delta\mu_g^o}{kT}\right)_{A,def} + \left(\frac{\Delta\mu_g^o}{kT}\right)_{Int} + \left(\frac{\Delta\mu_g^o}{kT}\right)_{Head}$$

$$\left(\frac{\Delta\mu_g^o}{kT}\right)_{Head} = \left(\frac{\Delta\mu_g^o}{kT}\right)_{Steric} + \left(\frac{\Delta\mu_g^o}{kT}\right)_{Ionic}$$

$$\left(\frac{\Delta\mu_g^o}{kT}\right)_{Head} = -\ln\left(1 - \frac{v_H}{a\delta}\right) + \frac{2\pi Z^2 e^2}{\varepsilon a \kappa kT} \tag{1.21}$$

Equation (1.21) presents the simplest representation of the headgroup repulsions. The headgroup repulsion term α/a introduced by Tanford is now replaced by the first term denoting steric interactions represented by the van der Waals excluded volume model and the second term denoting ionic interactions represented by the Debye model [21]. Here v_H is the volume of the protein headgroup present in the unit volume $a\delta$ of the corona region, a being the area per molecule and δ denoting the thickness of the corona. Z is the net charge on the protein, e is the electronic charge, and κ is the inverse Debye length that accounts for any counterions and added electrolytes.

For modeling the self-assembly of the hydrophobic protein-hydrophilic homopolymer conjugate shown on Figure 1.7b, one would have to introduce attractive interactions between the proteins. One can adapt Eq. (1.19) for this case recognizing that the proteins act as the equivalent of the tail group A.

$$(\Delta\mu_g^o) = (\Delta\mu_g^o)_A + (\Delta\mu_g^o)_{Int} + (\Delta\mu_g^o)_{B,dil} + (\Delta\mu_g^o)_{B,def}$$

$$\left(\frac{\Delta\mu_g^o}{kT}\right)_A = -S_H \frac{\sigma}{kT}$$

$$\left(\frac{\Delta\mu_g^o}{kT}\right)_{int} = \frac{\sigma}{kT} a \tag{1.22}$$
$$\left(\frac{\Delta\mu_g^o}{kT}\right)_{B,dil} = N_B \left[\frac{v_B}{v_S}\frac{1-\varphi_B}{\varphi_B}\ln(1-\varphi_B) + \frac{v_B}{v_S}(1-\varphi_B)\,\chi_{BS}\right]$$
$$\left(\frac{\Delta\mu_g^o}{kT}\right)_{B,def} = \frac{3}{2}\frac{LR}{a\,\varphi_B}P$$

In Eq. (1.22), S_H is the solvent accessible surface area of the protein molecule and the attractive protein interactions term is simply represented as the product of this surface area and a characteristic interfacial tension σ between the hydrophobic protein and the solvent water. The negative sign denotes that the interactions are attractive and provide the driving force for the self-assembly. All the remaining terms are as defined in Eq. (1.19) for the block copolymer self-assembly.

For the protein-block copolymer conjugates shown in Figures 1.7c,d, we have various possible aggregation patterns depending on whether the protein is hydrophilic or hydrophobic. If the protein is hydrophilic, the conjugate in Figure 1.7c can be simply treated like a triblock copolymer with a hydrophobic middle block and hydrophilic end blocks. The corona region in this case would consist of both the hydrophilic block of the copolymer and the protein. The conjugate in Figure 1.7d can be treated as a simple diblock copolymer by adding the additional headgroup repulsions term appearing in Eq. (1.21). On the other hand, if the protein is hydrophobic, then the conjugate in Figure 1.7c can be treated as a diblock copolymer, simply by adding the hydrophobic proteins contribution as represented in Eq. (1.22). The conjugate in Figure 1.7d can be represented as a triblock copolymer with end hydrophobic blocks and a middle hydrophilic block. The core now consists of a mixture of the hydrophobic block of the copolymer and the protein. Detailed quantitative models are not presented here but can be developed as described in this chapter and compared against aggregate structural data from systematic experiments with various polymers and proteins if they become available.

1.3.5 Amphiphilic Nanoparticles

An important class of nonclassical amphiphiles that is generating significant interest is that incorporating nanoparticles as a part of the amphiphile [42–48]. One example is an amphiphile containing a purely inorganic multinuclear headgroup of the polytungstate type R-$[PW_{11}O_{39}]_3$ [42], where R is the hydrophobic tail. Such polyoxometalate (POM) amphiphiles self-assemble into micelles and at higher concentrations into lyotropic liquid crystalline phases. These amphiphiles play the role of classical surfactants such as for emulsification but because of their nanoparticle headgroups, also act as catalysts for chemical reactions.

Cheng and collaborators [45–48] have introduced the terminology of "molecular nanoparticles" to suggest that the nanoparticles can be viewed as molecules and the terminology of "giant surfactants" to represent the amphiphiles composed of compact and rigid nanoparticles as the headgroup and flexible polymer chains as tail (Figure 1.9a,b). Examples of such giant amphiphiles studies in their laboratory include those based on functionalized fullerene, polyhedral oligomeric silsesquioxanes (POSS), and POM derivatives, with variable surface functionalities. They have also constructed amphiphiles where both the head and tail groups are nanoparticles by combining fullerene as the hydrophobic tail with functionalized fullerene as the hydrophilic head (Figure 1.9c,d).

Yu et al. [45] found that AC_{60}-PS_n (polystyrene with block size n) amphiphiles (Figure 1.9a) formed spherical micelles at low concentrations while at higher concentrations they transformed into cylindrical micelles and vesicles depending on the chain length of the hydrophobic polystyrene block. The concentration effect was attributed to the increasing concentration of protons in solution leading to reduction in the carboxylic acid dissociation and the net charge on the headgroup. The AC_{60}-$2PS_n$ amphiphiles (Figure 1.9b) aggregated into vesicles. These results can be readily explained qualitatively by the free energy model given by Eq. (1.21).

Lin et al. [48] found that the AC_{60}-C_{60} and AC_{60}-$2C_{60}$ amphiphiles (Figures 1.9c,d) made up of nanoparticles as both the hydrophobic head and the hydrophilic tail also self-assembled into various morphologies when studied in either tetrahydrofuran (THF) or dimethyl formamide (DMF)-water mixture as solvents. They found that in THF, vesicles were formed for AC_{60}–C_{60} and AC_{60}–$2C_{60}$, but with different inter layer packing arrangements. In the DMF/water system, spherical, and cylindrical micelles were formed for AC_{60}–C_{60} and AC_{60}–$2C_{60}$, respectively. The authors have suggested that in THF as the solvent, the dissociation of carboxylic acid is small and therefore the electrostatic headgroup repulsions are small. As a result, the area per molecule will be small, the packing parameter will be large, and vesicles are expected. In contrast, in DMF/water, the dissociation of the carboxylic acid will be larger and therefore the headgroup repulsions will be of larger magnitude. As a consequence, the area per molecule will be larger, the packing parameter will be smaller, and cylindrical and spherical micelles are thus formed. These interpretations are consistent with the application of the free energy models developed for the classical amphiphilic systems, with accounting made for the geometrical properties of the nanoparticle head and tail groups.

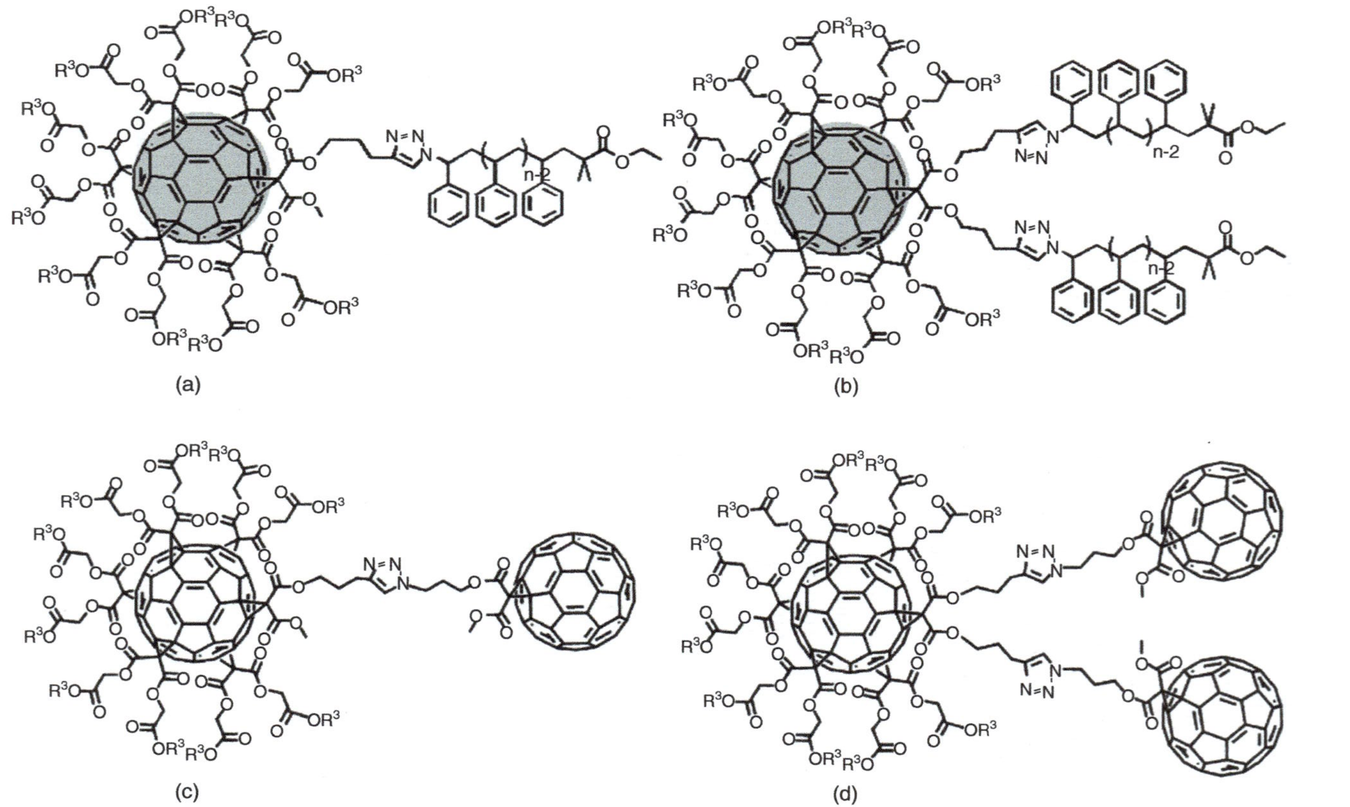

Figure 1.9 Structures of some of the nanoparticle amphiphiles synthesized and studied in Cheng laboratory. Shown are amphiphiles with hydrophilic AC_{60} headgroup and with one (a) or two (b) hydrophobic polystyrene tails. A new class of giant molecules have also been created with hydrophilic AC_{60} and hydrophobic C_{60}. Shown are such giant molecules with one (c) hydrophobic C_{60} or two (d) hydrophobic C_{60}. The notation AC_{60} represents C_{60} Fullerene functionalized with 10 carboxylic acid groups and A denotes the carboxylic acid functionality. Cheng et al. discuss in detail such nanoparticle amphiphiles they have created in Chapter 10 of this book.

1.4 Conclusions

Studies of self-assembly in solutions over almost a century have focused on surfactants and block copolymer molecules forming aggregates of various shapes. Although surfactants and block copolymers exhibit analogous self-assembly behavior, the two fields have evolved independent of one another, with the theoretical models emphasizing mutually exclusive molecular features. In the Tanford model for surfactant micelle, the balancing of surfactant headgroup repulsions against the interfacial energy determines the equilibrium properties of the aggregate. Noticeably absent was any influence attributed to the hydrophobic tail over the aggregate size and shape. In the de Gennes model for block copolymer micelle, the micelle characteristics are determined by the balancing of elastic deformation of the solvophobic block against the interfacial energy. The solvophilic block interactions (which are the analog of the surfactant headgroup interactions) play no role in determining the micelle characteristics. By incorporating the mutually neglected free energy contributions between these two theoretical models, we can have a unitary theory applicable to both surfactants and block copolymers. The incorporation of surfactant tail deformation free energy by borrowing from the block copolymer model has led to more accurate predictions of the critical micelle concentration and micelle characteristics for many classical surfactants. Similarly, the incorporation of solvophilic block interactions by borrowing from the surfactant model provides predictions about block copolymer micelles that are consistent with experiments that demonstrate the significant influence of the solvophilic block.

The free energy models of surfactants and block copolymers can be adapted and extended to account for interactional terms connected to novel headgroups such as peptides, oligonucleotides, POSS, POM, etc., and also to account for novel hydrophobic tails such as hydrophobic peptides, fullerenes, etc., in order to quantitatively describe the wide pattern of aggregation observed in these nonclassical amphiphilic systems. The design of such novel amphiphiles incorporating responsiveness and programmability and the ability to combine organic, inorganic, and biological moieties within a single molecular structure makes this area a very interesting and fertile field of research for years to come. The novel amphiphiles make possible entirely new areas of applications in materials science and nanomedicine, far beyond the traditional colloidal applications of surfactants and block copolymers.

Acknowledgments

Support from Natick Soldier Research, Development and Engineering Center, is acknowledged.

References

1 McBain, J.W. (1913). General discussion on colloids and their viscosity. *Trans. Faraday Soc.* 9: 99–101.
2 Hartley, G.S. (1936). *Aqueous Solutions of Paraffin Chain Salts*. Paris: Hermann et Cie.
3 Debye, P. (1949). Light scattering in soap solutions. *J. Phys. Colloid Chem.* 53: 1–8.
4 Nagarajan, R. (2014). One hundred years of micelles: evolution of the theory of micellization. Chapter 1. In: *Surfactant Science and Technology: Retrospects and Prospects* (ed. L.S. Romsted), 3–52. Boca Raton, FL: Taylor and Francis.
5 Dunn, A.S. and Melville, H.W. (1952). Synthesis of 'block' copolymers. *Nature* 169: 699–700.
6 Merrett, F.M. (1957). Graft polymers with preset molecular configurations. *J. Polym. Sci.* 24: 467–477.
7 Climie, I.E. and White, E.F.T. (1960). The aggregation of random and block copolymers containing acrylonitrile in mixed solvents. *J. Polym. Sci.* 47: 149–156.
8 Krause, S. (1964). Dilute solution properties of a styrene-methyl methacrylate block copolymer. *J. Phys. Chem.* 68: 1948–1955.
9 Riess, G. (2003). Micellization of block copolymers. *Prog. Polym. Sci.* 28: 1107–1170.
10 Tanford, C. (1973). *The Hydrophobic Effect*. New York: Wiley.
11 Israelachvili, J.N., Mitchell, J.D., and Ninham, B.W. (1976). Theory of self-assembly of hydrocarbon amphiphiles into micelles and bilayers. *J. Chem. Soc., Faraday Trans. 2* 72: 1525–1568.
12 Meier, D.J. (1969). Theory of block copolymers. I. Domain formation in A-B block copolymers. *J. Polym. Sci. C* 26: 81–98.
13 Meier, D.J. (1970). A theory of the morphology of block copolymers. *Polym. Prepr.* 11: 400–405.
14 Helfand, E. and Tagami, Y. (1971). Theory of the interface between immiscible polymers. *J. Polym. Sci., Part B: Polym. Lett.* 9: 741–746.
15 Helfand, E. and Sapse, A.M. (1975). Theory of unsymmetric polymer-polymer interfaces. *J. Chem. Phys.* 62: 1327–1331.
16 de Gennes, P.G. (1978. . Suppl. 14). Macromolecules and liquid crystals: reflections on certain lines of research. In: *Solid State Physics* (ed. J. Liebert), 1–18. Academic Press.
17 Leibler, L., Orland, H., and Wheeler, J.C. (1983). Theory of critical micelle concentration for solutions of block copolymers. *J. Chem. Phys.* 79: 3550–3557.
18 Noolandi, J. and Hong, K.M. (1983). Theory of block copolymer micelles in solution. *Macromolecules* 16: 1443–1448.

19 Whitmore, M.D. and Noolandi, J. (1985). Theory of micelle formation in block copolymer- homopolymer blends. *Macromolecules* 18: 657–665.
20 Tanford, C. (1974). Theory of micelle formation in aqueous solutions. *J. Phys. Chem.* 78: 2469–2479.
21 Nagarajan, R. and Ruckenstein, E. (1991). Theory of surfactant self-assembly: a predictive molecular thermodynamic approach. *Langmuir* 7: 2934–2969.
22 Nagarajan, R. (2002). Molecular packing parameter and surfactant self-assembly: the neglected role of the surfactant tail. *Langmuir* 18: 31–38.
23 Semenov, A.N. (1985). Contribution to the theory of microphase layering in block-copolymer melts. *Sov. Phys. JETP* 61: 733–742.
24 Daoud, M. and Cotton, J.P. (1982). Star shaped polymers – a model for the conformation and its concentration-dependence. *J. Phys.* 43: 531–538.
25 Zhulina, Y.B. and Birshtein, T.M. (1985). Conformations of molecules of block copolymers in selective solvents (Micellar structures). *Polym. Sci. USSR* 27: 570–576. (Original article in Russian published in Vysokomol. Soyed. A27: No. 3, 511–517, 1985).
26 Halperin, A. (1987). Polymeric micelles – a star model. *Macromolecules* 20: 2943–2946.
27 Nagarajan, R. and Ganesh, K. (1993). Solubilization in spherical block copolymer micelles: scaling analysis based on star model. *J. Chem. Phys.* 98: 7440–7450.
28 Flory, P.J. (1962). *Principles of Polymer Chemistry*. Ithaca, NY: Cornell University Press.
29 Nagarajan, R. and Ganesh, K. (1989). Block copolymer self-assembly in selective solvents – spherical micelles with segregated cores. *J. Chem. Phys.* 90: 5843–5856.
30 Nagarajan, R. and Ganesh, K. (1989). Block copolymer self-assembly in selective solvents – theory of solubilization in spherical micelles. *Macromolecules* 22: 4312–4325.
31 Nagarajan, R. and Ganesh, K. (1996). Comparison of solubilization of hydrocarbons in (PEO–PPO) diblock versus (PEO–PPO–PEO) triblock copolymer micelles. *J. Colloid Interface Sci.* 184: 489–499.
32 Nagarajan, R. (1996). Solubilization of hydrophobic substances by block copolymer micelles in aqueous solutions. In: *Solvents and Self-Organization of Polymers*, NATO ASI Series, vol. 327 (ed. S.E. Webber, P. Munk and Z. Tuzar), 121–165. Kluwer Publishers.
33 Lettow, J.S., Lancaster, T.M., Glinka, C.J. et al. (2005). Small-angle neutron scattering and theoretical investigation of poly(ethylene oxide)-poly(propylene oxide)-poly(ethylene oxide) stabilized oil-in-water microemulsions. *Langmuir* 21: 5738–5746.

34 Thota, B.N.S., Berlepsch, H.v., Böttcher, C. et al. (2015). Towards engineering of self-assembled nanostructures using non-ionic dendritic amphiphiles. *Chem. Commun.* 51: 8648–8651.

35 Thompson, M.P., Chien, M.-P., Ku, T.-H. et al. (2010). Smart lipids for programmable nanomaterials. *Nano Lett.* 10: 2690–2693.

36 Peterson, A.M. and Heemstra, J.M. (2015). Controlling self-assembly of DNA-polymer conjugates for applications in imaging and drug delivery. *WIREs Nanomed. Nanobiotechnol.* 7: 282–297.

37 Kedracki, D., Safir, I., Gour, N. et al. (2013). DNA–polymer conjugates: from synthesis, through complex formation and self-assembly to applications. *Adv. Polym. Sci.* 253: 115–150.

38 Shimada, T., Sakamoto, N., Motokawa, R. et al. (2012). Self-assembly process of peptide amphiphile worm-like micelles. *J. Phys. Chem. B* 116: 240–243.

39 Xu, H., Wang, J., Han, S. et al. (2009). Hydrophobic-region-induced transitions in self-assembled peptide nanostructures. *Langmuir* 25: 4115–4123.

40 Ku, T.-H., Chien, M.-P., Thompson, M.P. et al. (2011). Controlling and switching the morphology of micellar nanoparticles with enzymes. *J. Am. Chem. Soc.* 133: 8392–8395.

41 Boyer, C., Huang, X., Whittaker, M.R. et al. (2011). An overview of protein–polymer particles. *Soft Matter* 7: 1599–1614.

42 Landsmann, S., Lizandara-Pueyo, C., and Polarz, S. (2010). A new class of surfactants with multinuclear, inorganic headgroups. *J. Am. Chem. Soc.* 132: 5315–5321.

43 Li, W. and Gunton, J.D. (2013). Self-assembly of Janus ellipsoids II: Janus prolate spheroids. *Langmuir* 29: 8517–8523.

44 Wen, J., Yuan, L., Yang, Y. et al. (2013). Self-assembly of monotethered single-chain nanoparticle shape amphiphiles. *ACS Macro Lett.* 2: 100–106.

45 Yu, X., Zhang, W.-B., Yue, K. et al. (2012). Giant molecular shape amphiphiles based on polystyrene–hydrophilic [60] fullerene conjugates: click synthesis, solution self-assembly, and phase behavior. *J. Am. Chem. Soc.* 134: 7780–7787.

46 Yu, X., Li, Y., Dong, X.-H. et al. (2014). Giant surfactants based on molecular nanoparticles: precise synthesis and solution self-assembly. *J. Polym. Sci., Part B: Polym. Phys.* 52: 1309–1325.

47 Zhang, W.-B., Yu, X., Wang, C.-L. et al. (2014). Molecular nanoparticles are unique elements for macromolecular science: from "nanoatoms" to giant molecules. *Macromolecules* 47: 1221–1239.

48 Lin, Z., Lu, P., Hsu, C.-H. et al. (2014). Self-assembly of fullerene-based Janus particles in solution: effects of molecular architecture and solvent. *Chem. Eur. J.* 20: 11630–11635.

2

Self-Assembly into Branches and Networks

Alexey I. Victorov

Institute of Chemistry, St. Petersburg State University, Universitetsky prospect 26, St. Petersburg 198504, Russia

2.1 Introduction

The micelles in solution may grow into long and flexible wormlike aggregates. Enhanced growth and entanglement of such aggregates leads to the formation of a spatial network. Viscoelastic behavior of these solutions [1] resembles the behavior exhibited by solutions of overlapping polymer chains. In contrast to polymers having frozen distribution of chain lengths, the wormlike micelles, called living polymers [2], perpetually break and recombine. The hypothesis that wormlike micelles may branch was suggested more than 20 years ago [3, 4] based on interpretation of the rheology experiment. Surprising fluidity observed for micellar networks was known to be typical of bicontinuous microemulsions and sponge phases that are Newtonian, low-viscosity fluids. Today the existence of branched micellar networks is well established. Branched micelles are clearly seen in numerous transmission electron microscopy (TEM)-pictures [5–7], Figure 2.1.

Networks of branched micelles have been found in a variety of systems that include aqueous solutions of nonionic surfactants, solutions of ionic (both cationic and anionic) surfactants in presence of a simple salt or of a hydrotrope, aqueous mixtures of cationic + anionic surfactants, mixtures of nonionic + ionic surfactants, and reverse micellar systems in nonaqueous media. Comprehensive lists and classification of wormlike micelles are given in recent reviews [6, 8–10]. Specific examples of systems with branched micellar networks include:

- Nonionic surfactant hexaethylene glycol monododecyl ether ($C_{12}E_6$) in water [11, 12], where branching is observed upon increasing temperature;
- Aqueous solutions of classical ionic surfactants in presence of simple or penetrating salts, such as cetylpyridinium chlorate ($CPyClO_3$) with $NaClO_3$ [13],

Self-Assembly: From Surfactants to Nanoparticles, First Edition. Edited by Ramanathan Nagarajan.

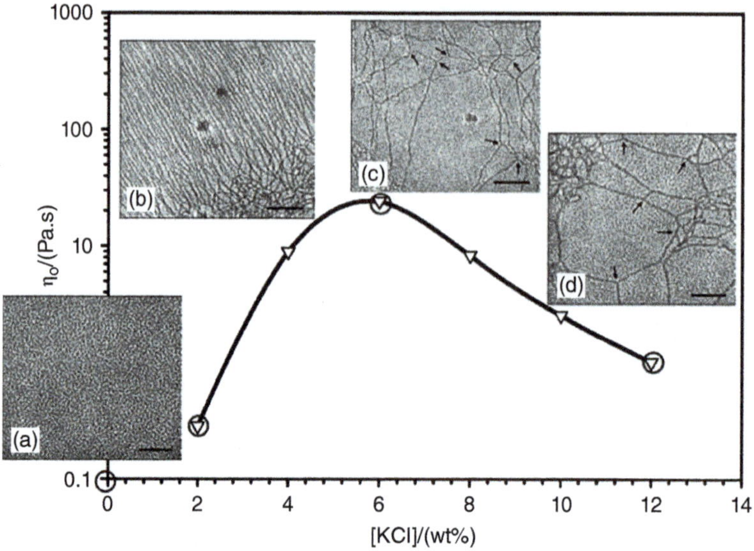

Figure 2.1 Low-shear viscosity versus salinity (KCl) for a solution of 1.5 wt% erucyl bis (hydroxyethyl) methylammonium chloride (EHAC). Cryo-TEM images (given for compositions indicated by circles: 0, 2, 6, and 12 wt% KCl) show the transition from spherical micelles to wormlike micelles and to a branched network. *Source*: Reproduced from Ref. [6] with permission of The Royal Society of Chemistry.

cetyltrimethylammonium bromide (CTAB) with KBr [14], or cetylpyridiniumchloride (CPyCl) with sodium salicylate (NaSal) [15];

- Aqueous mixture of the double-tailed anionic surfactant sodium bis(2-ethylhexyl)sulfosuccinate (NaAOT) with the ionic liquid 1-butyl-3-methylimidazolium bromide (bmimBr) [16];
- Aqueous solutions of CTAB and aqueous solutions of sodium dodecylsulfate (SDS) in presence of the inorganic salt (hydrazine nitrate), where addition of the same salt induces micellar growth and branching for both of these cationic and anionic surfactants [17];
- Aqueous mixture of cationic and anionic surfactants CTAB + sodium oleate (NaOL) [18];
- Strong micellar growth and branching are typical of dimeric and trimeric surfactants, where individual amphiphilic chains are separated by the spacer groups attached to the heads [19]. For example, the so called Gemini 12-2-12 dimeric surfactant (dimethylene-1,2-bis(dodecyl dimethylammonium bromide) forms micelles in water that rapidly grow and branch at relatively low concentration [20];

- Solutions of nonionic triblock copolymer Pluronic P123 ($EO_{20}PO_{70}EO_{20}$) in mixed solvent; aggregates of different morphologies, including bicontinuous structures have been found depending on the composition of solvent [21];
- Aqueous mixture of nonionic surfactant (polyoxyethylene phytosterol $PhyEO_{30}$) with nonionic monoglyceride (monolaurin or monopalmitin) cosurfactant [12], where branching is observed at elevated temperatures upon increasing concentration of the cosurfactant;
- Wormlike reverse micelles form in mixtures of lecithin in oil in presence of water [10]. These micelles may be branched or unbranched, depending on the oil composition: cyclohexane favors disconnected reverse micelles while isooctane promotes the formation of branches [22]. Networks of branched reverse micelles and gels have also been found in mixtures of lecithin with a multivalent carboxylic (e.g. citric) acid in oil (decane) [23].

Systems containing networks of wormlike micelles have been extensively studied experimentally and theoretically in the last two decades. Branching of wormlike micelles has also been observed in molecular dynamic simulations [24, 25]. Particularly, it has been shown [25] that Y-shaped junctions between the micelles are formed as the result of an applied external tension.

The experimental techniques include rheology, dynamic light scattering (DLS) [26], small-angle X-ray scattering (SAXS) [12, 23] and small-angle neutron scattering (SANS) [18, 27], nuclear magnetic resonance (NMR) [28, 29], fluorescence recovery after bleaching [13, 30, 31], and TEM [5]. Most of the authors perform rheological measurements that are interpreted in terms of the existing network theories and often complemented by structural data from scattering experiments or TEM, see [6, 32] and references therein. For wormlike micelles, an excellent review on extracting structural information from scattering experiment has been given by Dreiss [6]. The problem is that the scattering techniques may not distinguish between a multiconnected network of branched micelles and an entangled micellar network. For aqueous systems, the suitable method is cryo-TEM but studies of micelles in organic solvents are problematic. The technique called Fourier Transform Pulsed field Gradient Spin-Echo NMR (FT PGSE NMR) measures directly the self-diffusion of surfactant over long distances and helps to distinguish between the curvilinear surfactant motion along the body of a disconnected wormlike micelle (with the mean square displacement proportional to $t^{1/2}$) from its three-dimensional Gaussian motion (with the mean square displacement proportional to t) at high concentration of micellar connections in the saturated network [22].

Combined rheological and structural data show that appearance of intermicellar connections is accompanied by a decrease of the zero-shear viscosity (Figure 2.1) and by a decrease of the terminal relaxation time. These experimental findings support an earlier theoretical conclusion that appearance of sliding intermicellar connections in the entangled micellar network should increase the reptation diffusion and reduce the viscosity.

Theoretical studies of branched micellar networks include a description of network kinetics [33, 34], phase behavior at thermodynamic equilibrium, and percolation [35–38]. General basic features in the behavior of the spatial network have been established [37, 39] that are common for reversible networks in physical polymer gels, dipolar hard sphere fluids, bicontinuous sponge phases in block copolymers and microemulsions, microemulsion droplets crosslinked by polymer chains, and for the networks of aggregating colloid patchy (Janus) particles. The nature of specific system is reflected by the energetic cost of forming a branching part versus that of the terminal part of the aggregate. These costs depend on details of the microscopic picture.

Reversible response of a micellar network to various external stimuli is very attractive for numerous practical applications that require smart materials with tunable (viscoelastic) properties [40]. These topics are discussed in several reviews [6, 8, 9].

We start this chapter by discussing the rheological and kinetic properties of a branched living network and then review the theories of an equilibrium spatial network. The models of the free energy of the micellar branching portion are discussed next. We conclude by summarizing problems that await solutions and outline major trends in current research.

2.2 Rheology and Structure of Solutions Containing Wormlike Micelles

2.2.1 Viscoelasticity of Entangled Wormlike Micelles

For dilute solutions of wormlike micelles, the rheologic behavior is quite complex and includes a number of perplexing phenomena. Shear-induced time-dependent structures may form [41, 42], and shear thickening is sometimes observed owing to these temporal gel-like structures.

For semidilute solutions of entangled micelles, the Newtonian behavior at small shear rates is typically followed by shear thinning above a certain shear rate (Figure 2.2) in the steady shear flow experiments. The inverse of this critical shear rate gives an estimate of the longest micellar structural relaxation time, the terminal time τ_R. The Newtonian plateau is characterized by a zero-shear viscosity, defined for a Maxwell fluid as $\eta_0 = G_0\tau_R$, where G_0 is the shear (plateau) modulus.

The viscosity at zero shear rate dramatically increases when a change of conditions (e.g. an increase of salinity, surfactant, or additive concentration or temperature) induces micellar growth and leads to formation of an entangled

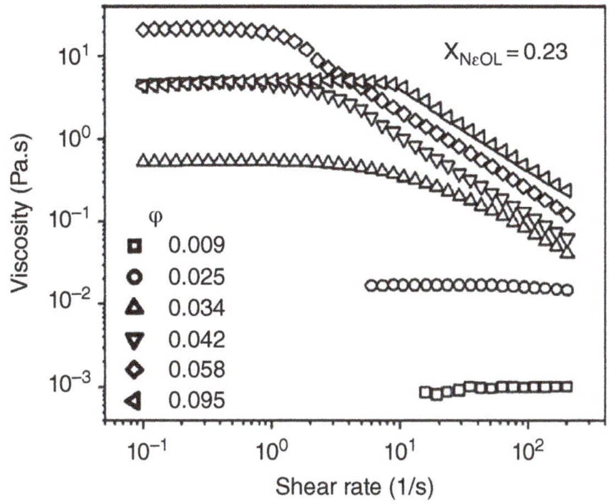

Figure 2.2 Viscosity vs shear rate for mixed micelles of cetyltrimethylammonium bromide (CTAB) + sodium oleate (NaOL) at a fixed mole fraction of NaOL, $x_{NaOL} = 0.23$, and varying total surfactant volume fraction φ. *Source*: Reprinted with permission from Ref. [15]. Copyright (2011) American Chemical Society.

network of wormlike micelles. Viscoelastic fluid formed by long, flexible micelles is reminiscent of a polymer solution. The dynamics of an entangled polymer are described in terms of the reptation model where polymer chains slither by a snake-like motion (reptation) through imaginary curvilinear tubes formed by entanglements of the neighboring chains [43, 44]. The reptation model of polymer relaxation has been extended to chains that are continuously breaking and recombining [45, 46]. This model introduces two time scales: the reptation time τ_{rep} and the breaking time τ_{break}, where τ_{rep} is the time for the wormlike micelle of the average length $\overline{L}_c$ to pass through the entanglement tube and τ_{break} is the time for the micelle of length $\overline{L}_c$ to break in two pieces. It is assumed that the reversible scission of the micelle is characterized by a temperature-dependent rate constant that does not depend on surfactant concentration: $\tau_{break} = (k\overline{L}_c)^{-1}$. The contour length of wormlike micelles may vary from nanometers to micrometers [6, 10]. A typical breaking time of wormlike micelles is on the order of milliseconds [6], although there are exceptions [10].

When $\tau_{break} \gg \tau_{rep}$ the reptation motion is the dominant mechanism of stress relaxation; the dynamic properties obey the laws of the classical polymer solutions with an exponential polydispersity of chain lengths [43]. In this regime we have $\tau_R = \tau_{rep}$ for the terminal relaxation time.

In the limit of frequent micellar breakage (the so-called fast-breaking limit), $\tau_{break} \ll \tau_{rep}$, chain breakage and recombination occur before the chain reptates out of its tube segment. When a chain breaks, two daughter chains become uncorrelated and recombine with a micellar end in a random way. Before a given tube segment relaxes, the chain in this segment undergoes so many breaking and fusing reactions that there is no memory of the initial length nor of the position of the chain in the tube. The stress relaxation function reduces to a single exponential. Fast kinetics explains a surprisingly simple viscoelastic behavior of highly polydisperse solution of micellar chains in contrast to polydisperse polymers that show more complex relaxation of stress. For $\tau_{break} \ll \tau_{rep}$, the viscoelastic behavior of a micellar network (at low-frequency, $\omega\tau_{break}^{-1} \ll 1$) is described by the Maxwell model with a single stress relaxation time

$$\tau_R = (\tau_{rep}\tau_{break})^{1/2} \tag{2.1}$$

For the Maxwell model, the elasticity modulus $G'(\omega)$ and the viscous modulus $G''(\omega)$ at the oscillatory shear frequency ω are given by

$$G'(\omega) = \frac{\omega^2\tau_R^2}{1+\omega^2\tau_R^2}G_0,\ G''(\omega) = \frac{\omega\tau_R}{1+\omega^2\tau_R^2}G_0 \tag{2.2}$$

From this model, $G'(\omega)$ approaches the plateau modulus, G_0, in the high-ω limit. The relaxation time may be estimated from the crossover frequency, $\tau_R = \omega^{-1}$, where $G'(\omega) = G''(\omega)$. Once τ_R and G_0 are available η_0 may be calculated from $\eta_0 = G_0\tau_R$.

The frequency dependence of the storage and loss moduli is measured in the oscillatory shear experiment, as shown in Figure 2.3. Direct experimental determination of G_0 is often problematic because the storage modulus does not level off at high frequencies, where fast relaxation mechanisms become dominant and lead to non-Maxwellian behavior as explained next.

The Cole–Cole plot in Figure 2.4 shows the imaginary part $G''(\omega)$ of the frequency-dependent shear modulus as function of the real part $G'(\omega)$. This plot may also be used to estimate the relaxation time. The Maxwell model gives Cole–Cole plot of a semicircular shape. This shape is observed in experiment for the left, low-frequency part of this plot. Deviations from this semicircular shape become significant at frequencies ω of the order of the breaking time of the micelles. When reptation is the dominant relaxation mechanism on the time scale less than τ_{break}, the departure from the semicircular shape at higher frequencies is manifested as a linear dependence of G'' on G' with a slope of −1 [49]. For higher-frequency regimes that involve even shorter time scales, the dominant motion of chains is not reptation but either breathing (here $G'' - G'$ slope of −2.4 arises from fluctuations of tube's length owing to Rouse-like stretches of the chain's contour length) or faster Rouse-like motion that arises from stretches of the chain shorter than the entanglement length, l_e [49], i.e. the average distance between entanglement points along a wormlike micelle. This and other characteristic lengths of a micellar entangled network are shown in Figure 2.5.

The high-frequency regime is well understood for polymer systems. The chain does not see any entanglements and follows simple Rouse's three-dimensional motions. The effect of these short time-scale motions is visible in the experimental results: the real and imaginary parts of the

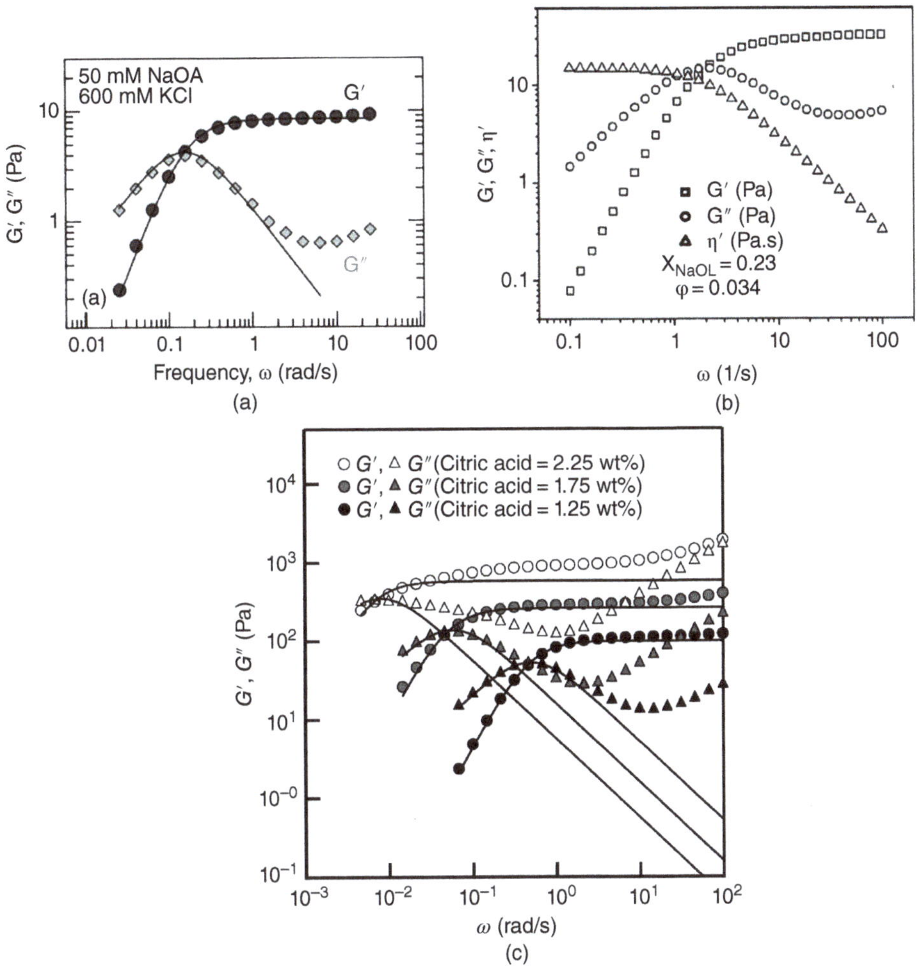

Figure 2.3 Frequency dependence of the storage and loss moduli. Symbols: experiment; curves: fit by the Maxwell model. (a) Solution of 50 mM sodium oleate in presence of 600 mM KCl at 298 K. *Source*: Reprinted with permission from Ref. [47]. Copyright (2005) American Chemical Society. (b) Solution of mixed micelles cetyltrimethylammonium bromide (CTAB) + sodium oleate (NaOL) at 298 K. Total surfactant volume fraction is 0.034; the mole fraction of NaOL $x_{NaOL} = 0.23$. The real part of the complex viscosity η' is also shown. *Source*: Reprinted with permission from Ref. [15]. Copyright (2011) American Chemical Society. (c) Lecithin + citric acid in n-decane at 298 K. Lecithin concentration is 10 wt.%. *Source*: Reprinted from Ref. [23], Copyright (2013), with permission from Elsevier.

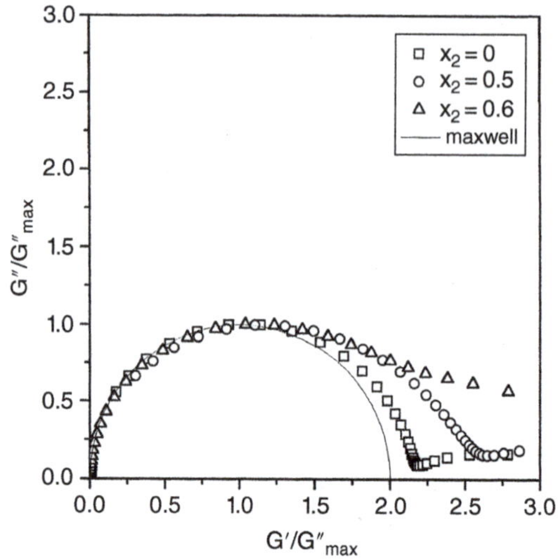

Figure 2.4 Normalized Cole-Cole plots for the mixture of cetyl trimethylammonium 3-hydroxy naphthalene 2-carboxylate (CTAHNC) with cetyl trimethylammonium tosylate (CTAT). G''_{max} is the maximum value of the loss modulus (measured at the origin of the semicircle). Points: experiment; curve: calculated from the Maxwell model. The total surfactant concentration is 100 mM, x_2 is the mole fraction of CTAHNC, $T = 303$ K. *Source*: Reprinted with permission from Ref. [48]. Copyright (1998) American Chemical Society.

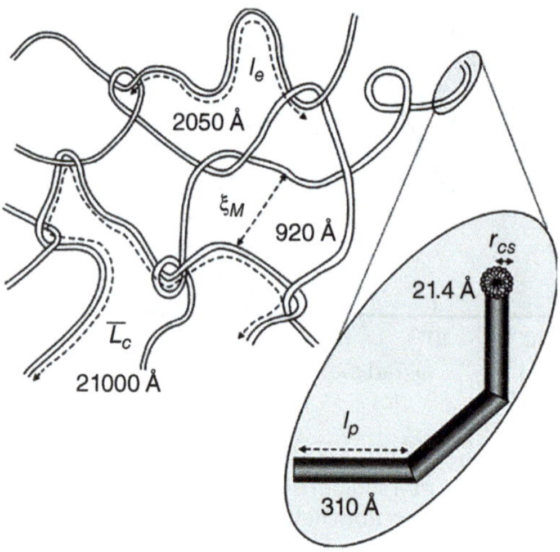

Figure 2.5 Different length scales in an entangled network of wormlike (unbranched) micelles: the average contour length, $\overline{L}_c$; the entanglement length, l_e; the mesh size, ξ_M; the persistence length, l_p; the cross-sectional radius, r_{cs}. Shown are measured values in solution of mixed surfactants cetyltrimethylammonium tosylate (CTAT) and sodium dodecyl benzylsulfonate (SDBS) with added hydrotropic salt (Na tosylate) in presence of a screening electrolyte (NaCl); 97:3 ratio of CTAT:SDBS at 1.5% total surfactant and 0.10% NaTosylate. *Source*: Reprinted with permission from Ref. [50]. Copyright (2003) American Chemical Society.

complex modulus start to turn up at high frequencies, whereas Maxwell's model predicts a constant asymptotic value for the real part and a decreasing imaginary part, see Figures 2.3 and 2.4.

The upturn of both G' and G'' at high frequencies results in a minimum in the Cole–Cole plot, Figure 2.4. The depth of this minimum may be used to estimate the number of entanglements per chain, $\overline{L}_c/l_e$. It is then possible to estimate the average micellar length in the entangled system as explained below.

The frequency ω_{min} at the dip of the Cole–Cole curve is of the order of $1/\tau_e$, where $\tau_e \sim l_e^2$ is the Rouse time of the chain's fragment having an entanglement length [43]. The depth of the dip depends strongly on the relative values of the breaking time, the reptation time, and τ_e. In the limit $\tau_{rep} \gg \tau_{break} \gg \tau_e$, the value of $G''(\omega)$ at the dip obeys the relation [46, 49].

$$G''_{\min}/G'_{\infty} \cong l_e/\overline{L}_c \tag{2.3}$$

where G'_{∞} is an estimated value of the plateau modulus (obtained by some authors from a linear extrapolation of the Cole–Cole graph to the abscissa [18]). The entanglement length l_e must be much larger than the persistence length l_p (Figure 2.5); otherwise Eq. (2.3) is not applicable.

The dip becomes shallower and Eq. (2.3) is no longer valid when any of the conditions $\tau_{rep} \gg \tau_{break}$, $\tau_{break} \gg \tau_e$ is violated. The second condition may be violated in systems where micelles grow at very low concentrations (large l_e) and hence τ_e may become large ($\tau_e \sim \tau_{break}$). When τ_{break} is very small, $\tau_R = (\tau_{rep}\tau_{break})^{1/2}$ is small and coupled reptation and Rouse motions may give overlapping contributions in contrast to ordinary polymers where we always have $\tau_R \gg \tau_e$.

The shallower the dip in the Cole–Cole plot, the larger is the uncertainty in the extrapolated G'_{∞}. When the dip is shallow, the terminal relaxation time may be estimated from the crossover frequency $\tau_R = \omega_R^{-1}$, where $G'(\omega_R) = G''(\omega_R)$ [18]. The plateau modulus is estimated as $G'_{\infty} = 2G'(\omega_R) \cong 2G''_{max}$.

The average micellar contour length $\overline{L}_c$ may be calculated from Eq. (2.3) using the known values for G''_{min}, G'_{∞} and l_e. From the theory of entangled polymer chains, the entanglement length l_e is related to the persistence length, l_p, and the mesh size, ξ_M [44].

$$l_e = \xi_M^{5/3}/l_p^{2/3} \tag{2.4}$$

The mesh size is related to the elasticity modulus by the theory of rubber elasticity [44]:

$$G_0 = kT/\xi_M^3 \tag{2.5}$$

Using G'_{∞} as an estimate of G_0 in this equation we find from Eq. (2.4).

$$G'_{\infty} = kT/\xi_M^3 = kT/(l_e^{9/5} l_p^{6/5}) \tag{2.6}$$

Eqs. (2.6) and (2.3) together give l_e and $\overline{L}_c$ from the measured values of G'_∞ and G''_{min} and known (measured) l_p. Thus, the structural characteristics of an entangled micellar network may be found from the rheology experiment.

Concentration dependence of the viscosity and of the terminal relaxation time is estimated from the classical mean-field or scaling theories of polymer solutions. For the reptation time of a flexible polymer in semidilute solution, we have [43] $\tau_{rep} \sim \overline{L}_c^3 \varphi^{3/2}$. Because τ_{break} is inversely proportional to $\overline{L}_c$, we have for the terminal relaxation time

$$\tau_R = (\tau_{rep}\tau_{break})^{1/2} \sim \overline{L}_c \varphi^{3/4} \tag{2.7}$$

in the fast-breaking regime, and

$$\tau_R \approx \tau_{rep} \sim \overline{L}_c^3 \varphi^{3/2} \tag{2.8}$$

in the slow-breaking regime.

The mean-field theory gives $G_0 = kT\varphi^{9/4}$, independent of the micelle length $\overline{L}_c$. Hence, for the viscosity, $\eta_0 = G_0\tau_R$, we have

$$\eta_0 \sim \overline{L}_c \varphi^3 \text{ (fast-breaking)} \tag{2.9}$$

$$\eta_0 \sim \overline{L}_c^3 \varphi^{15/4} \text{ (slow-breaking)}$$

Even for slowly breaking micelles, the scaling of the viscosity (and of the terminal relaxation time) with concentration is different from that for polymers because the average length of micellar chains depends on the concentration of surfactant.

2.2.2 Growth of Nonionic Micelles

To determine the concentration dependence of the viscosity, we need the dependence of the average micellar length on the surfactant concentration.

For nonionic micelles or for ionic micelles in the excess of salt where the electrostatic interactions are screened, a simple mean-field model of the free-energy may be applied to find equilibrium distribution of chain lengths [45, 51]

$$f/kT = \sum_L C(L)[\ln\{C(L)v^*\} - 1 + E] \tag{2.10}$$

where $C(L)$ is the number density of chains of length L, E is the scission energy needed to cut a wormlike micelle in two parts thus creating two spherical endcaps, v^* is the volume of a single monomer. Equilibrium distribution of chain lengths, and their average, $\overline{L}_c = \frac{\sum_L LC(L)}{\sum_L C(L)}$, is obtained by minimizing the free energy with respect to $C(L)$ at fixed surfactant volume fraction, $\sum_L LC(L) = \varphi$. Because minimization is at constant volume fraction, adding any arbitrary function of φ (e.g. the excluded volume interaction between the chains) may not alter the chain length distribution. The equilibrium distribution is exponential [45, 51], $C(L) \sim \exp(-L/\overline{L}_c)$, yielding

$$\overline{L}_c \sim \varphi^{1/2} \tag{2.11}$$

This gives

$$\eta_0 \sim \varphi^{3.5} \text{ and } \tau_R \sim \varphi^{1.25} \tag{2.12}$$

for frequently breaking micelles, $\tau_{break} \ll \tau_{rep}$. For the regime of infrequently breaking micelles, Eqs. (2.9) and (2.11) give: $\eta_0 \sim \varphi^{5.25}$ and $\tau_R \sim \varphi^{3.0}$.

When the mean-field relations are replaced by those from the scaling polymer theory the exponents in the concentration dependence $\sim \varphi^{\alpha}$ change [31, 45]. We have $\alpha = -0.77$ for the mesh size ξ_M, $\alpha = 2.31$ for the elasticity modulus, $\alpha = 0.6$ for the average micelle length and $\alpha = 3.42$ for the reptation time of the micelle of length $\overline{L}_c$. In the fast-breaking regime, we have $\tau_R \sim \varphi^{1.41}$ and $\eta_0 \sim \varphi^{3.72}$.

2.2.3 Growth of Ionic Micelles

The behavior of the ionic micelles is quite different. The scission energy includes repulsion between the charged surfactant heads that helps to break the micelles. The micelle growth depends on the entropy of counterions that prefer to condense near the cylindrical body of the micelle rather than near the spherical endcap. For the ionic micelles there are three different regimes of growth [51]:

(1) In dilute solution, the range of electrostatic interactions is larger than the micelle size. In this regime the micelle length increases very slowly with concentration

$$\overline{L}_c \sim \frac{1}{l_B r_{cs} \nu^2}\left[E + \ln\left(\frac{\varphi}{\overline{L}_c}\right)\right] \tag{2.13}$$

where l_B is the Bjerrum length, ν is the effective charge per unit length of the micelle and r_{cs} is the cross-sectional radius of the wormlike micelle. The micelles are nearly monodisperse in this regime.

(2) In semidilute solution the range of electrostatic interactions is shorter than the average micelle size. The micelle growth with surfactant concentration is very rapid:

$$\overline{L}_c \sim 2\varphi^{1/2} \exp\left\{\frac{1}{2}\left[E - \frac{l_B r_{cs} \nu^2}{\varphi^{1/2}}\right]\right\} \tag{2.14}$$

The scission energy is effectively reduced by the electrostatic repulsion of charges on the micellar surface, as reflected by the second term in the exponential. Because of this decrease of the scission energy the polydispersity

is broad in this regime, similar to neutral micelles but the average size is smaller. The growth is more rapid and may not be described by a power law.
(3) In the high concentration regime, the electrostatic contribution is dominated by the entropy of counterions near the endcaps. The micellar growth is described by an effective power law $\overline{L}_c \sim \varphi^{(1+\Delta)/2}$. The exponent $\Delta/2$ is related to the net charge of the endcap and depends only weakly (logarithmically) on φ. For large excess of salt or high surfactant concentration, all ionic contributions to the scission energy vanish and the size distribution reduces to that for nonionic system, where $\overline{L}_c \sim \varphi^{1/2}$.

Thus, in contrast to nonionic systems, where the growth follows a power law in both the dilute and semidilute solutions, the electrostatic effects for ionic micelles are manifested most clearly in an abrupt crossover from the dilute regime where the micelle growth is very weak to the semidilute regime of a very rapid growth. When the electrostatic effects are screened (regime 3) the results for nonionic systems, Eqs. (2.11) and (2.12), are reproduced.

For high salinity, the exponents that describe the concentration dependence of the terminal relaxation time and of the viscosity are in good agreement with experiment for CTAB + KBr and CPyCl + NaSal + NaCl systems [41].

In the semidilute regime, the viscosity of an entangled network of linear ionic micelles increases faster with surfactant concentration than that in the nonionic/screened system [52]:

$$\eta_0 \sim \varphi^{7/2} \exp\left\{\frac{E}{2}\left[1-\left(\frac{\varphi^*}{\varphi}\right)^{1/2}\right]\right\} \tag{2.15}$$

where $\varphi^* = (l_B r_{cs} \nu^2/E)^{1/2}$ is the crossover concentration between the dilute and the semidilute regimes. This crossover between regimes and the subsequent rapid growth reflected by Eqs. (2.14) and (2.15) have been confirmed in experiment for many ionic surfactants, see e.g. [53], and references therein. Nevertheless, the discrepancies have been found between the theory and experiment on the growth of ionic micelles in quite a number of systems. The reason for such discrepancies is not yet clearly understood [6].

2.2.4 Persistence Length of an Ionic Micelle

Viscoelasticity of a micellar network depends on the micelle's persistence length (l_p, Figure 5.5) that determines the flexibility of a wormlike particle. Much theoretical and experimental effort has been devoted to this key structural parameter; see reviews [6, 54], which also contain summaries of experimental data. For ionic micelles, an important contribution to the flexibility comes from the electrostatic component of the persistence length, l_p^{el}. Most of the authors estimate l_p^{el} from the classical Odijk-Skolnick-Fixman (OSF) theory [55, 56] that was originally developed for polyelectrolyte chains.

This theory employs the linearized Poisson-Boltzmann (LPB) approximation for charged chains of a vanishing thickness. However, in contrast to polymers the diameter of wormlike micelles is quite large [32]. Extension of the classical OSF theory to wormlike particles of a finite thickness has been obtained based on the asymptotic expansion of the LPB problem for toroidal geometry [57]. The electrostatic persistence length is given by a remarkably simple formula

$$l_p^{el} = l_B \nu^2 r_{cs} \kappa_D \left[1 + \frac{\kappa_D}{4 r_{cs}}\right] \tag{2.16}$$

where κ_D is the Debye length. For a threadlike micelle, $r_{cs} \to 0$, this formula reduces to the classical OSF expression while it gives a linear dependence on the Debye length, $l_p^{el} \sim \kappa_D$, for thick micelles at high salinity where $\kappa_D/(4r_{cs}) \ll 1$.

Figure 2.6 shows excellent agreement between Eq. (2.16) and the numerical solution of the LPB equation for wormlike particles of varying thickness in a large interval of salinity. Figure 2.7 illustrates good performance of Eq. (2.16) for a real solution of weakly charged wormlike micelles.

2.2.5 Networks of Branched Micelles

The discussion so far has been limited to networks of entangled linear micelles that do not branch. Branching of micelles is often observed upon the further change of the parameter that triggers the micellar growth, e.g. composition or temperature. The same factors that favor the lower curvature of the cylindrical micellar body over that of the endcaps and promote micellar growth are likely to promote formation of micellar junctions that have even smaller curvature. A branch may easily slip along the fluid body of a wormlike micelle and help to release the stress more quickly. Therefore, the branched wormlike micelles are likely to have lower viscosity than that of entangled linear wormlike micelles. A maximum of viscosity is then observed, as shown in Figure 2.1.

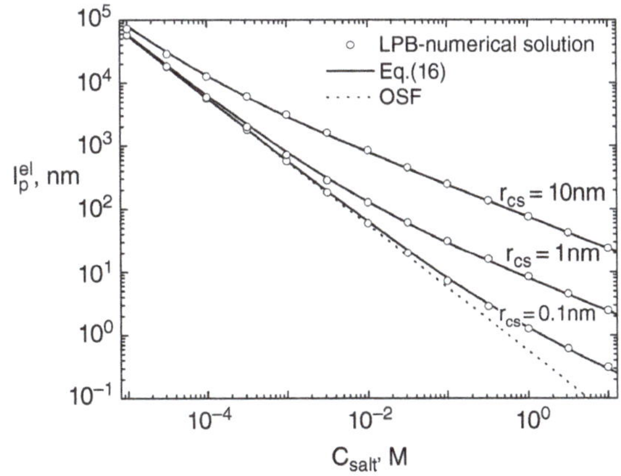

Figure 2.6 Electrostatic persistence length of a wormlike particle vs concentration of 1:1 electrolyte. Circles: calculated from numerical solution of LPB equation; solid curves: Eq. (2.16); dotted: OSF theory. The results are shown for different particle radii, r_{cs}, and the micelle's effective linear charge density of 5.86 nm^{-1}. *Source*: Reprinted with permission from Ref. [57, 58]. Copyright (2010) AIP Publishing LLC.

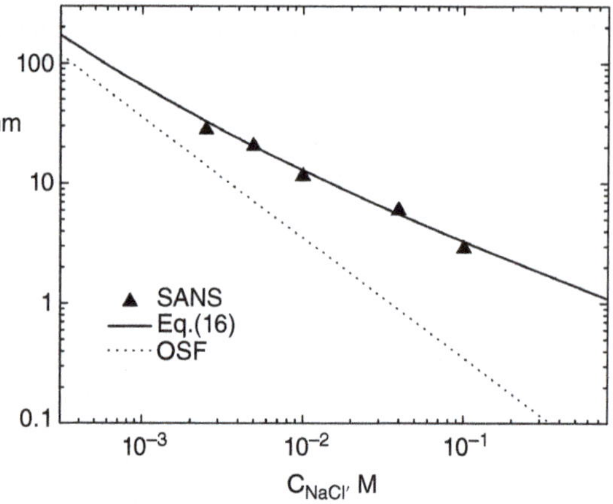

Figure 2.7 Electrostatic persistence length versus salinity for mixed wormlike micelles of hexaethyleneglycol n-hexadecyl ether ($C_{16}E_6$) + 1-hexadecane sulfonic acid $C_{16}SO_3Na$ in aqueous solution of NaCl. $C_{16}E_6$ to $C_{16}SO_3Na$ weight ratio is 0.06. Symbols: SANS data of Sommer et al. [27] Curves: calculated; $\nu = 1.46\ nm^{-1}$ and $r_{cs} = 2$ nm.

Formation of branches leads to a distinct dynamic behavior. Nevertheless, the equations of the Cates reptation model [45] retain their form [14, 34] (particularly, the monoexponential relaxation law with $\tau_{break} \sim \overline{L}_c^{-1}$ and Eq. (2.3)) if the average micelle length is redefined by the equation

$$\overline{L}_c = l_p\rho/(\rho_e + 2\rho_{jun}) \tag{2.17}$$

where ρ_e is the concentration of endcaps, ρ_{jun} is the concentration of threefold micellar junctions and ρ is the concentration (proportional to the volume fraction of surfactant φ) of micelle's linear segments (of persistence length) that form the wormlike bodies.

For the composition-independent scission energy (screened electrostatic interactions), the scission-recombination equilibrium requires that $\rho_e \sim \varphi^{1/2}$ and $\rho_{jun} \sim \varphi^{3/2}$ [14, 35]. Hence, for the solution that contains only nonbranched micelles, we have $\overline{L}_c \sim \varphi^{1/2}$. In the opposite limit of a saturated network $\rho_e = 0$ and $\overline{L}_c$ is equal to the average between the neighboring branches. Hence, from Eq. (2.17) we have $\overline{L}_c \sim \varphi^{-1/2}$. Thus, for a nonionic saturated network, the terminal relaxation time and the zero shear viscosity in the fast-breaking regime are given by

$$\tau_R \sim \varphi^{0.25} \text{ and } \eta_0 \sim \varphi^{2.5} \tag{2.18}$$

respectively. These exponents are significantly less than those for a fully entangled network, Eq. (2.12). Hence, an increasing proportion of micellar junctions results in a change of the scaling exponents that may not be related to the virtual decrease of the micellar length. An intermediate behavior owing to the presence of both the endcaps and the junctions is very much expected. Thus, the exponent for $G''_{\min}/G'_{\infty} \sim l_e/\overline{L}_c$ is expected to be less than −7/4, the value predicted for entangled linear micelles in good solvent (where $l_e \sim \xi_M^{5/3}$ and $\xi_M \sim \varphi^{-3/4}$). This exponent is predicted to approach −3/4 in the limit of predominant intermicellar connections [14]. An exponent of −0.77 found from experiment for

CTAB + KBr at high salinity is very similar indeed [14]. From the same experiment, the dependence of τ_R and η_0 on surfactant's concentration is also consistent with the picture of a branching network. For the network of reverse micelles of lecithin in water + isooctane [29], the scaling exponent for $\overline{L}_c$ is $\alpha = -0.55 \pm 0.07$, in good agreement with theoretical prediction for a saturated network.

Branching explains a decrease of the terminal relaxation time in salt-free solutions of cetyltrimethylammonium hydroxynaphthalene carboxylate at high surfactant concentration [53]. Addition of nonbinding salts to dilute solution of this surfactant also leads to branched micelles. In the presence of salt, the increase in viscosity is much smoother than in the absence of salt and may be fitted by a power law with an exponent 2.3, close to that of Eq. (2.18).

Although the overall success of the theory has been demonstrated quite widely, there is also experimental evidence of its limitations. In contrast to the theory that gives the same scaling for the plateau modulus in the entangled and in the branching regimes, different scaling exponents have been found in experiment for an aqueous mixture of cationic and anionic surfactants (CTAB + NaOL) [18]. For branched micelles, the scaling exponents of the viscosity and of the plateau modulus agree with the results from the theory. For the entangled micelles, the scaling exponent of the plateau modulus is significantly larger than predicted from theory. Similar results have been obtained in experiment for the reverse micelles of lecithin + multivalent carboxylic acids in decane [23].

There are other examples of systems that may not be described by the simple Maxwell model with a single relaxation time. Some surfactant solutions reversibly turn into strong gels [9] whose elastic response is almost independent of the shear rate; some micellar networks exhibit a spectrum of relaxation times [10]. We also note that the concepts explained in this chapter are no longer viable for networks of stiff micelles ($l_p \sim l_e$) and for strongly interacting aggregates that assemble into rigid orientationally ordered structures [9].

2.2.6 Ion-Specific Effect on Micellar Growth and Branching

The effect of salt on micellar growth and on the shape of the viscosity-salinity curve is highly ion-specific [17, 48, 59]. The counterions that contain hydrophobic groups, e.g. salicylates or benzoates, strongly bind to a micellar surface, inducing micellar growth and branching. For some systems, dispersion of micelles at high content of a penetrating salt has been observed [60]. In such systems, the maximum of the aggregation number causes a viscosity maximum. Thus, a decrease and an increase of the aggregate's preferential curvature may both be responsible for the observed maxima of the viscosity-salinity curves. For CTAB and CPyCl with added NaSal, a second viscosity maximum is observed at high salinity [15, 59]. This result is explained by a nonmonotonic change of the branching density.

A systematic study of various nonpenetrating salts + CTAB [59] shows that as the counterion binds weaker to the micelle the viscosity maximum shifts to higher salinity, the variation of the viscosity is weaker, and the micellar persistence length is larger.

A remarkable effect of specific ion on the micelle has been observed for aqueous mixtures of NaAOT + ionic liquid: for bmimBF4, the micelles are only spherical but branched and bicontinuous structures are found for bmimBr [16].

Understanding and description of the impact of a specific molecular "portrait" on aggregation is a very hard problem [10, 58]. Nevertheless, certain success has recently been achieved. Experimental observations of the specific effect of counterions on the formation and growth of ionic micelles may be explained [61] by a phenomenological concept of Collins [62]. The ion pair formed between a counterion and a surfactant head is more compact when the counterion and the head are both chaotropes or both cosmotropes, whereas the ion pairs between the cosmotropes and chaotropes are less compact. Packing of ionic pairs in micelle's corona control the preferential curvature of the aggregates and hence their growth and branching. This concept has been quantified in recent molecular thermodynamic models that describe the micelle formation [63, 64] and a subsequent growth of large wormlike micelles [65, 66] in solutions of cationic and anionic surfactants in presence of simple 1:1 salts. In these models, the key factors in manifesting the ion's specificity are the ionic polarizability, size, and the hydration of ion pairs in the micellar coronae. In line with the concept of Collins, these hydration effects differ for specific combinations "mobile ion-surfactant head" affecting the compactness of the ion pairs and the preferential curvature of the aggregate. The models reflect the subtle balance of opposing factors in the aggregation equilibrium and predict that some salts may only promote micellar growth whereas for other salts, the initial micellar growth may be followed by dispersion of the aggregates at high salinity [65, 66]. With the aid of one of these models the effect of coions on micellar growth has been described and a qualitative explanation has been found [66] to an observed strong effect of a coion on the location of the viscosity peak.

2.3 Branching and Equilibrium Behavior of a Spatial Network

2.3.1 The Entropic Network of Chains

Systems that contain branched aggregated structures often exhibit phase separation [67, 68]. Phase separation may be induced by adding salt, changing temperature, surfactant concentration, or composition of the mixed solvent. The equilibrium phases may consist of a dense and a dilute network. Thus, networks of branched aggregates are found by cryo-TEM close to phase separation loop, e.g. in the water-rich mixtures $C_{12}E_5$ + water + *n*-octane [68]. Similar phase behavior and similar topology of phases are exhibited by very different systems that include wormlike micelles, microemulsions and dispersions of dipolar particles [39]. The concept of entropic networks [37, 39] explains this similarity in the behavior of self-assembling linear chains. The building blocks of these linear chains may be quite different (e.g. a dipolar hard sphere, a patchy colloid particle, or amphiphilic molecules that form a wormlike micelle) but the energetically favorable state of these blocks is a linear aggregated structure. These linear aggregates may terminate only because the presence of the end-points (called defects) increases the entropy of the system. The second type of defect is the junction of two or more chains, i.e. the branching point of a linear chain.

Both topological defects have higher energy than the energy state in an infinite chain. Nevertheless both of the defects increase the entropy and therefore both defects exist at any finite temperature. When the energy of the chain end is larger than that of a junction, junctions dominate at low enough temperature and the system assembles in a large-scale connected network (physical gel).

A simple mean-field theory of an entropic network proposed by Tlusty and Safran [37] is outlined below. Let $\psi(\vec{r})$ denote the probability that a chain starts or terminates at a point $\vec{r}$ [44]. The probability for a chain to cross through $\vec{r}$ is proportional to $\psi^2(\vec{r})$, because the ends of two subchains must meet at that point to form a longer continuous chain. For the volume fraction of chains (surfactant), we have $\varphi \sim \psi^2$ within the mean-field approach. Hence the concentration of ends is given by

$$\varrho_e \sim \psi e^{-\varepsilon_e/T} \sim \varphi^{1/2} e^{-\varepsilon_e/T} \tag{2.19}$$

where the Boltzmann factor accounts for the energetic penalty, ε_e, of this defect. Similarly, for concentration of the threefold junctions, where the ends of three linear chains meet together we have

$$\varrho_{jun} \sim \psi^3 e^{-\varepsilon_{jun}/T} \sim \varphi^{3/2} e^{-\varepsilon_{jun}/T}. \tag{2.20}$$

Minimizing the ideal-gas free energy of noninteracting defects

$$f = kT\varrho_d[\ln \varrho_d - 1] + \varepsilon_d \varrho_d \tag{2.21}$$

where ϱ_d is the density of the defects and ε_d their energy, we obtain the optimal density $\varrho_d^* = e^{-\varepsilon_d/kT}$ and the minimal free energy $f^* = -kT\varrho_d^*$. Hence, the defect of any type (chain end or junction) contributes $-kT$ to the free energy. Therefore, the equilibrium free energy of a system that contains aggregating linear chains is given by

$$\frac{f}{kT} = -\varrho_e - \varrho_{jun} + \frac{1}{2}\varphi^2 \tag{2.22}$$

where the third term accounts for the excluded volume of chains.

The osmotic pressure is found from Eqs. (2.19), (2.20), and (2.22).

$$P = \varphi^2 \frac{\partial (f/\varphi)}{\partial \varphi} = \frac{1}{2}(\varrho_e - \varrho_j) + \frac{1}{2}\varphi^2 \tag{2.23}$$

This equation contains topologically induced contributions: the repulsion owing to the ends and the attraction owing to the junctions. We note that these interactions are not the microscopic forces acting between the junctions (ends) but the effective thermodynamic forces between all chain monomers in the system. Phase separation occurs when the attraction is strong enough, below the critical temperature T_{crit} where $\frac{\partial^2 f}{\partial \varphi^2} = 0$ and $\frac{\partial^3 f}{\partial \varphi^3} = 0$. This gives

$$T_{crit} \sim \varepsilon_e - 3\varepsilon_j \tag{2.24}$$

The phase separation is analogous to demixing in a binary fluid consisting of junctions and ends. In the low-density phase there are more ends than junctions, whereas in the high-density phase there is more junctions than ends. The translational entropy of chains is lower in the dense phase; the total entropy of the system nevertheless increases owing to an increase of the density of junctions in the high-density phase.

Junctions and ends are thermal defects. Their densities decrease exponentially on cooling, Eqs. (2.19) and (2.20). This gives longer branches. Owing to their lower energy, junctions disappear at a slower rate than the ends. At temperatures slightly below the critical point, the two-phase region becomes wider. Further decrease of temperature leads to a shift of the two-phase coexistence curves to the zone of low density [39].

This behavior – an entropically driven phase separation in dilute solution of crosslinked chains – has been predicted by Drye and Cates [33] and later by other authors for different systems. Such phase behavior has been found in computer simulation of spheres with specific sites (that may mimic Janus micelles or interacting colloids) that may stick together forming linear and branched chains [69]. In good agreement with computer simulation are predictions from the thermodynamic perturbation theory of Wertheim; such predictions have been performed [69] with the version of this theory widely known as statistical associating fluid theory (SAFT) [70]. Figure 2.8 shows typical reentrant phase diagram stretching to low densities. At low temperatures, sticky particles of low functionality may form liquid network of a very low density (the "empty" liquid) [69, 71], see Figure 2.9.

Reversible spatial networks and similar type of phase behavior have been predicted for solutions of sticky polymer chains from the mean-field theory of Semenov and Rubinstein [72–74]. The relation between this theory and the SAFT has been established [74]; relative locations of the percolation line and phase loops has been studied [72, 74]. The classical Flory-Stockmayer equation for the percolation threshold [75] has recently been extended to networks of chains with multiple types of sticking specific sites [71].

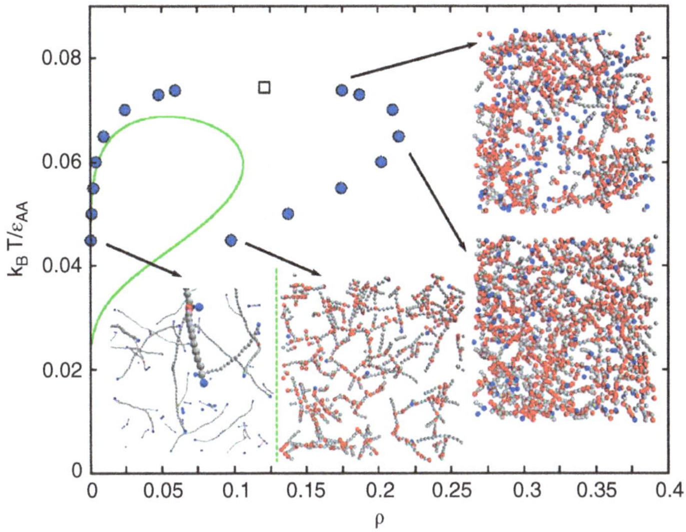

Figure 2.8 Phase coexistence for a fluid of associating spheres that have patches (association sites) of type A and type B. Each sphere carries two A-patches on its poles and nine B-patches evenly distributed along the equator. Bonds A-A account for the association in linear chains of spheres. Chain junctions are formed via the A-B bonds; no bonding between B-sites is introduced. The chain scission energy is $\in_{AA}$. For each of the two produced endcaps, the energy cost is $\varepsilon_e = \in_{AA}/2$. For the junction, we have $\varepsilon_j = \in_{AA}/2 - \in_{AB}$. $\in_{AB} = 0.37\ \in_{AA}$. Symbols: MC-simulation data; the open square is the estimated critical point. Curve: the binodal calculated from the Wertheim TPT-1 theory (SAFT). Snapshots from simulation are shown for selected points along the numerical coexistence line as indicated by arrows. Chain free ends and junctions are colored blue and red, respectively. *Source*: Reprinted (Figure 2.3) with permission from Ref. [69]. Copyright (2011) by the American Physical Society.

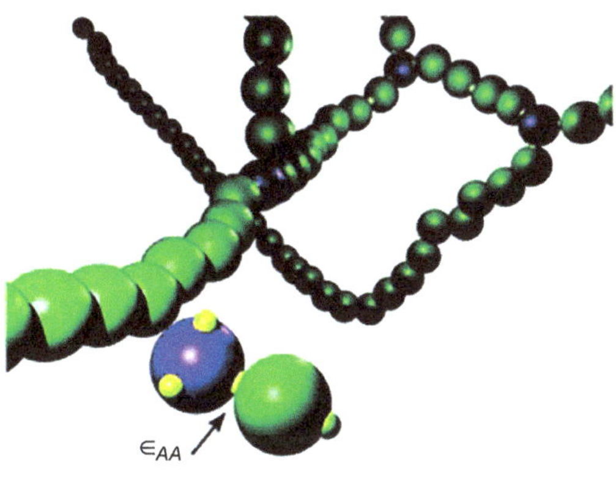

Figure 2.9 Network fluid of a low density in the binary mixture of spheres with two and three identical bonding sites (A). A bond between two sites lowers the energy by $\in_{AA}$. *Source*: Reproduced from Ref. [71] with permission of The Royal Society of Chemistry.

Rigorous statistical mechanical theory of branching entropic networks has also been developed [35, 36, 38, 76]. This theory employs the magnetic-polymer analogy [44] and maps the partition function of a solution of branched, self-assembling, mutually avoiding clusters onto that of a Heisenberg magnet in the mathematical limit of the zero number of spin components. This mapping accounts for all possible cluster configurations, except for closed rings.

The most complete version of this theory has been given by Zilman and Safran [38]. Junctions have an arbitrary functionality. Although the theory is mathematically involved, it gives simple results in the mean-field limit [38]. For systems with threefold junctions, Eqs. (2.22) and (2.23) are reproduced by this theory while different numerical prefactors appear in Eqs. (2.19), (2.20), and (2.24). The critical volume fraction is given by [38]

$$\varphi_{crit} \sim e^{(\varepsilon_j - \varepsilon_e)/T_{crit}} \tag{2.25}$$

For low critical temperatures, φ_{crit} is small. The phase transition is reentrant.

Presence of junctions leads to an attraction between the chains that reduces the effective excluded volume of chain segments [38, 72, 74]. Phase separation is observed for systems with threefold junctions. The attraction induced by four-fold junctions (and junctions of a higher functionality) is not strong enough to give phase transition. These high functionality junctions may only lead to percolation. Therefore, when phase separation is observed in such a system it is not topologically induced but comes from interactions between the chains.

For systems with threefold junctions, the average chain length between consecutive defects (end or junction) is given by [38]

$$\bar{l}_d = l_p 2\varphi/(\varphi_e + 3\varphi_j) \tag{2.26}$$

For a disconnected system, the ends dominate. Eqs. (2.19) and (2.26) give the mean chain length between endpoints: $\bar{l}_d \sim \varphi^{1/2}$. For a saturated network where junctions dominate, $\bar{l}_d \sim \varphi^{-1/2}$ is the mean chain length between the junctions. These results agree with those from an earlier mean-field theory, Eq. (2.17) [34]. $\bar{l}_d$ first increases and then decreases with the volume fraction. The maximum is located at the percolation threshold where $\varphi_e = 3\varphi_j$. For small volume fractions, the clusters become larger but more dilute as φ is increased. A further increase of φ above the percolation threshold makes the network denser. The average size of the largest cluster of chains [38]

$$\bar{R} \cong [2\varphi/(\varphi_e - 3\varphi_j)]^{1/2} l_p \tag{2.27}$$

diverges at this threshold.

The Zilman-Safran theory accounts for the spatial density correlations. It gives the scattering structure factor and the correlation length, the quantities accessible from scattering experiments.

2.3.2 The Shape of Micellar Branch and the Free Energy

The endcap and the junction energies depend on the microscopic characteristics of the system. On the mean-field level of description these energies are substituted by the averages such as the potentials of mean force and the free energies. These free energies may be estimated from molecular thermodynamic models of aggregation [77–81] or from a phenomenological model [11, 82]. The aggregation free energies are controlled by the preferential packing of surfactant molecules in the aggregates of different size and curvature and depend on temperature and composition. The curvature dependence of the free energy may be estimated from the Helfrich phenomenological expression [82]

$$f_{curv} = \frac{k}{2}\left(\frac{1}{R_1} + \frac{1}{R_2} - C_0\right)^2 + \bar{k}\frac{1}{R_1}\frac{1}{R_2} \tag{2.28}$$

where R_1 and R_2 are local radii of curvature, k and $\bar{k}$ are the bending and the saddle-splay moduli, respectively, and C_0 is the spontaneous curvature, i.e. the curvature that the surfactant film would adopt locally when free from global constraints. Eq. (2.28) is a curvature expansion and may only be applied to aggregates of small curvature, e.g. the microemulsion droplets, or the coronae of the micelles with large cores. The endcap (junction) free energy is the difference between the free energies per surfactant molecule for the endcap (junction) and for the cylindrical part of the micelle. Eq. (2.28) shows that the energies of the endcaps (having positive curvature $R_1 = R_2$) and junctions (having a saddle-like curvature, $\frac{1}{R_1}\frac{1}{R_2} < 0$) may depend in a subtle manner on the microscopic parameters of the system that affect the spontaneous curvature and the elasticity moduli [10, 83].

Eq. (2.28) has been used to predict percolation and phase diagram in the aqueous mixture of nonionic surfactant $C_{12}E_6$ having bulky heads [11]. It has been assumed that the temperature dependence of the curvature free energy is dominated by spontaneous curvature; the variation of the elasticity moduli with temperature may be neglected.

Figure 2.10 shows that the endcap energy increases with temperature, while the junction energy decreases. This result reflects the dehydration of the surfactant heads with increasing temperature. As a consequence, an increase of temperature makes junctions more favorable than the endcaps that have larger curvature [11]. This explains the inverted phase diagram with a lower critical temperature observed in experiment and calculated with the aid of the Zilman-Safran theory with the temperature-dependent energies of the defects [11]. Figure 2.11 shows this diagram. Percolated networks are predicted in the high temperature region of this diagram.

The phenomenological description of the curvature free energy has been used to determine the equilibrium shape of junctions and endcaps with the aid of a numerical minimization technique [83, 84]. The optimal shapes are shown

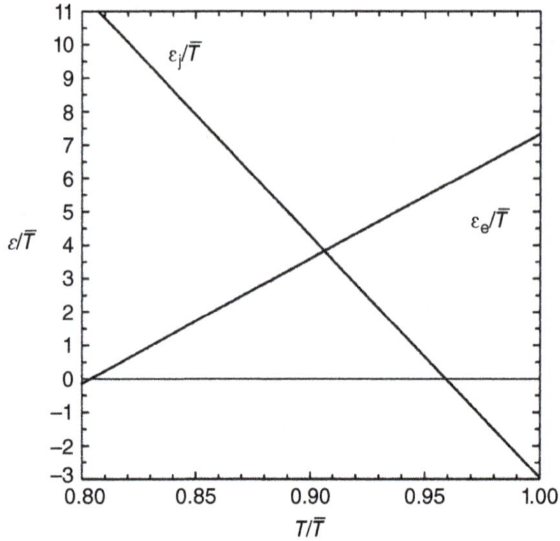

Figure 2.10 Temperature dependence of the energies (reduced units, $\overline{T} = 341$ K) of the endcaps and junctions for the aggregates of nonionic surfactant hexaethylene glycol monododecyl ether ($C_{12}E_6$) in water. This temperature dependence is dictated by the temperature dependence of the spontaneous curvature that has been determined from experiment. *Source*: Reprinted with permission from Ref. [11]. Copyright (2004) American Chemical Society.

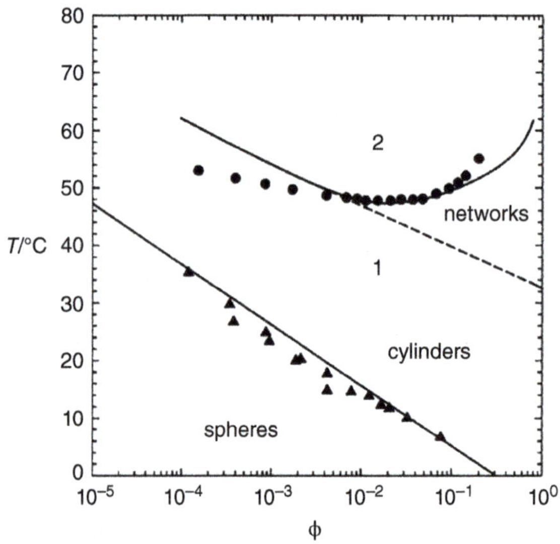

Figure 2.11 The diagram for the $C_{12}E_6$ + water mixture in the temperature – surfactant volume fraction plane. Symbols: experiment, curves: theoretical prediction. The lower line is the sphere-to-cylinder transition; the upper curve is the predicted spinodal of the phase separation between a dense network and a dilute phase (points show the experimental binodal); the dashed line is the predicted percolation threshold. Above this line, the system contains a network of branched micelles, spanning the whole volume. *Source*: Reprinted with permission from Ref. [11]. Copyright (2004) American Chemical Society.

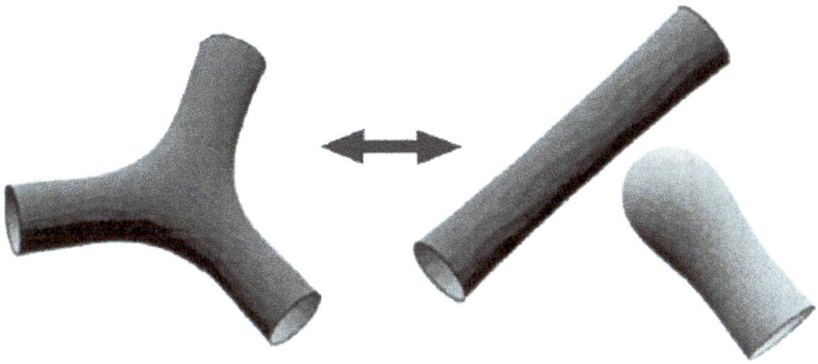

Figure 2.12 Formation of a threefold Y-shape junction from the cylindrical micellar body and the endcap. The equilibrium shapes are found by numerical minimization of the curvature free energy using the triangulation technique (the Evolver software). The junction has nearly lamellar bilayer patch in the middle. The cylinder terminates with an enlarged cap. *Source*: Reprinted with permission from Ref. [11]. Copyright (2004) American Chemical Society.

in Figure 2.12. An optimal micellar junction resembles its TEM images and is very similar to the Y-shaped model first suggested by Porte et al. [3]. The junction contains an almost planar bilayer patch in the middle surrounded by nearly semitoroidal rims. Junctions of this shape have been considered in molecular thermodynamic models of branched micelles [80, 81, 85].

May et al. [86] developed a molecular-thermodynamic model of the aggregation free energy for micellar junctions. They considered several contributions to the total aggregation free energy: a contribution from the bare interfacial tension, a contribution from the repulsion between surfactant's heads, and a contribution from the packing of chains in the micellar core. The latter was calculated numerically from the single-chain self-consistent field theory [87]. The model of May et al. gives an equilibrium shape of a threefold junction between wormlike micelles. This model predicts that junctions are always energetically unfavorable to cylinders and endcaps. This model is of limited use, particularly for ionic surfactants, because it does not provide an explicit dependence of the electrostatic contribution to the aggregate free energy on solution salinity and on the surfactant's molecular parameters.

A simple expression has been obtained for the electrostatic contribution to the free energy of a micellar junction based on approximate analytical solution of the LPB equation for toroid geometry [80]. This electrostatic contribution has been combined with other terms of a classical molecular thermodynamic micellization model [77–79]: bare interfacial, steric repulsion of heads, stretching of surfactant tails, and a hydrophobic (shape-independent) term. The model shows that at high salinity the branches may have lower aggregation energy than the endcaps. Moreover, there is a narrow salinity interval where the branch has the lowest free energy among the aggregates of all other shapes. This interval corresponds to the onset of bicontinuous structures; the lamellae become stable at a higher salinity [80].

The endcap's and junction's free energies from this model [80] serve as input parameters in the Zilman-Safran theory for predicting phase equilibrium, percolation, and other structural details as function of concentrations of surfactant and salt [80, 85]. For solution of erucyl bis(hydroxyethyl)methylammonium chloride (EHAC) + NaCl, such prediction shows that phase separation may be triggered by adding salt or surfactant [85]. Nevertheless, addition of surfactant at low salinity leads to gelation only; no demixing is predicted. Addition of salt to this gel results in phase separation. The equilibrium phases are two spatial networks (gels) differing in the density of junctions. Further addition of salt leads to sol–gel equilibrium. The rare network breaks up into separate clusters while the denser phase remains in the gel state.

The Zilman-Safran theory predicts strong branching and percolation for aggregates that have close endcap and junction energy: $\varepsilon_e \cong \varepsilon_j$ [83, 84]. This prediction has been used to estimate the conditions of expected branching for diblock copolymer micelles with a weak polyelectrolyte corona. Figure 2.13 shows the branching portion of such micelle. Swollen polyelectrolyte corona is sensitive to the acidity and salinity of the surrounding medium. The molecular thermodynamic description of this system has been derived from the analytical self-consistent field theory [88, 89] that has been extended to include branched micelles [81]. The aggregation free energy includes the chain-stretching contribution (for both the core and the corona), the bare interfacial term, the free energy of mixing for the swollen corona, and an ionic contribution that accounts for the presence of the fixed and the mobile charge in the aggregate's

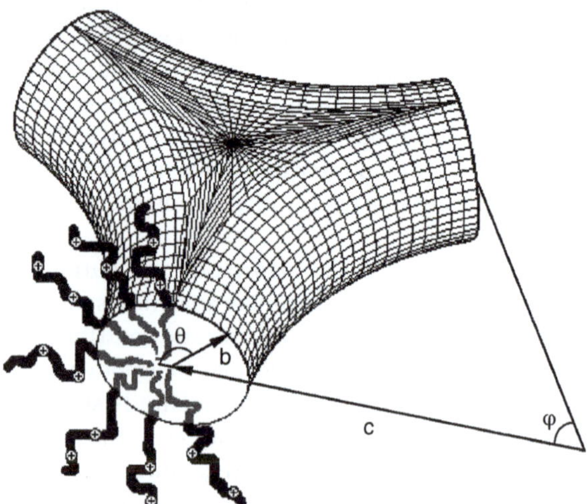

Figure 2.13 Geometry of a branch that connects three cylindrical micelles formed by an amphiphilic diblock copolymer. Weak polyacid subchains of the diblock (dark) form swollen corona. Hydrophobic subchains (light) form the core of the aggregate. *Source*: Reprinted with permission from Ref. [81]. Copyright (2010) American Chemical Society.

corona. The electrostatic problem is solved within the local electroneutrality approximation that makes possible direct calculation of the local electrostatic potential. This leads to the local form of the Donnan rule, thus avoiding a hard task of solving the Poisson-Boltzmann equation in complex geometry. The local equilibrium degree of dissociation is determined by minimizing the free energy. This leads to a local version of the mass action law. The resulting ionic contribution to the equilibrium-free energy accounts for the osmotic pressure of the mobile ions in the swollen brush and for the entropy of distribution of charges along the polyelectrolyte chain at dissociation equilibrium [81]. Within this model the shape and size of an optimal aggregate is controlled by several factors. The bare interfacial tension tends to contract the surface area and favors structures of a smaller curvature. Chain elasticity opposes their stretching away from the surface and hence favors larger surface area per chain. Repulsion caused by the electrostatic forces and by the excluded volume promotes chain-stretching in the charged corona but also expands the surface. The resulting balance of opposite trends defines the optimal aggregate shape.

Although this model is simple, a numerical calculation is required to obtain the equilibrium aggregate free energy. This calculation includes solution of the Euler-Lagrange equations that determine the equilibrium concentration profiles and equilibrium profile of the degree of ionization in the swollen corona [81]. This calculation also includes minimization of the free energy with respect to the curvature radii. A simpler way to determine optimal aggregate shape is obtained from the curvature expansion of the free energy of the swollen corona. Such curvature expansion is viable for crew cut aggregates, i.e. the aggregates where the corona thickness is much smaller than the radius of the core. This curvature expansion leads to analytical expressions for the elasticity moduli and the spontaneous curvature [90]. Combined with an explicit model of the micelle dry core this curvature expansion has been used for predicting morphology transitions. Figure 2.14 shows the results obtained for crewcut aggregates with the aid of two methods: (i) numerical calculation using complete (unexpanded) version of the model, and (ii) calculation using the curvature expansion for the corona's free energy. Both methods give similar results. As expected for micelles that contain weak polyelectrolyte [89], a peculiar sequence of structures has been predicted for varying salinity and pH around pK_a. The tendency toward branching and formation of bicontinuous structures is strong: the Y-shaped branching portion of a cylindrical aggregate is the most stable structure in the white area of the diagram. The location of the transition $S_1 = 1/2 \rightarrow S_2 = 2$ within the domain of stable cylindrical morphology shows the transition from branched to nonbranched wormlike micelles. Branched micelles appear as intermediate structures between the linear micelles and (perforated) bilayers. Transformation of wormlike micelles in a multiconnected perforated structure has indeed been observed in experiment. Figure 2.15 illustrates such transformation for anionic/nonionic surfactant

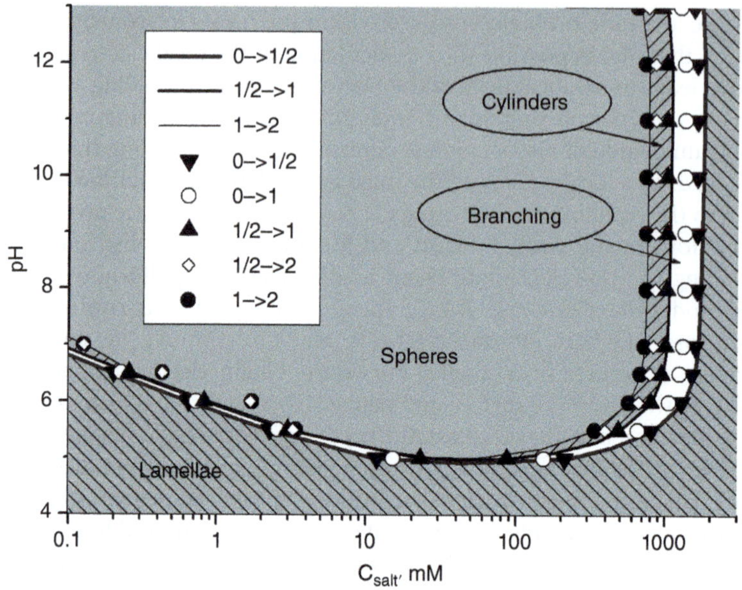

Figure 2.14 Predicted regions of stable aggregate shapes as a function of pH and salinity for solution of diblock copolymer that contains a hydrophobic subchain of 380 monomers and a weak polyacid subchain of 20 monomers. Shaded areas and curves: stability zones and transition curves obtained numerically from the self-consistent field model [81]. Symbols: morphology transitions calculated [90] from the curvature expansion of this model for the swollen aggregate coronae. $S_1 \rightarrow S_2$ denotes the transition between morphology S_1 and morphology S_2, where $S_1, S_2 = 0, ½, 1, 2$ for plane, branching portion of a cylindrical aggregate, cylinder, and sphere, respectively. $pK_a = 5$ for the dissociation constant of the ionic groups. *Source*: Reproduced from [90] with permission of The Royal Society of Chemistry.

mixture sodium dodecyl sulfate (SDS) + N-dodecyl β-D-glucopyranoside (DDGP). Adding DDGP to spherical micelles of SDS induces formation of long branched threadlike micelles. At higher content of DDGP, the density of branches increases, and a saturated network is formed.

2.4 Conclusions

Branched micelles and micellar networks have been found in a variety of surfactant solutions. Extensive rheological studies of the early 1990s [1, 4, 13–15] and the development of cryo-TEM techniques within the next decade [5] played a crucial role in characterizing the dynamics and structure of micellar networks, particularly when combined with various scattering experiments [6, 32]. For wormlike reverse micelles in nonaqueous media, the Fourier-transform

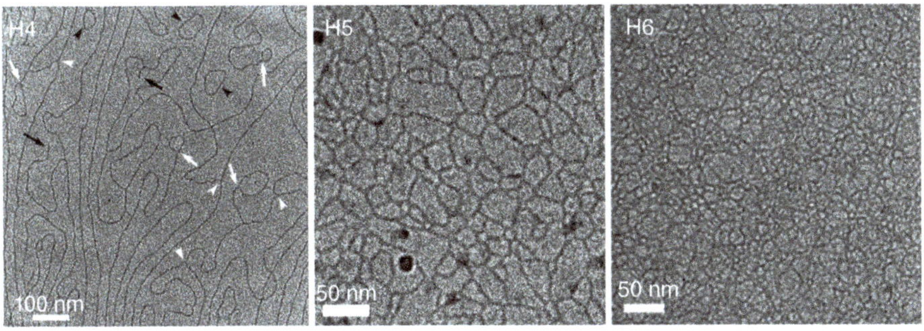

Figure 2.15 Cryo-TEM images of branched aggregates and saturated networks in anionic/nonionic mixtures of sodium dodecyl sulfate (SDS) and N-dodecyl β-D-glucopyranoside (DDGP) at fixed total surfactant content and an increasing proportion of the nonionic surfactant from H4 to H6. (H4) A variety of "solutions" to avoid endcaps (black arrows), including threefold junction points (white arrowheads), short segments closing into rings and connecting to the main body (white arrows) and small closed rings (black arrowheads). (H5) The saturated network (20:80 SDS:DDGP) with threefold junctions. (H6) High branching density at 8 wt.% SDS; the network transformed to a perforated structure where multiple connections emanate from small rings. *Source*: Reprinted from Ref. [5], Copyright (2012), with permission from Elsevier.

pulse-gradient NMR technique has been successfully applied [22] to distinguish the diffusion of surfactants in the entangled network from those in the branched network.

Theories describing kinetics of micellar networks – from their first versions based on the wormlike micelle – polymer analogy introduced by Cates et al. [2, 45, 46], – to their extensions for branched micellar networks [11, 14, 34] – provided greater understanding of observed viscoelastic behavior in nonionic and ionic systems, as well as the effects of concentration of salt and surfactant. The scaling exponents predicted from these theories are in qualitative agreement with experiment. However, the experimental exponents for the plateau modulus and for the entanglement length may substantially differ from theoretical predictions for both nonionic [12] and ionic surfactant systems [18]. Thus, the slope of the plateau modulus vs. volume fraction is substantially different for the regimes of linear and branched micellar networks [18]. For micelles with electrostatic interactions in semidilute regime, the plateau modulus is unexpectedly low [32]. Clearly, more systematic rheological and structural data are needed for a range of surfactant systems to identify the scaling exponents. Better understanding of the dynamical shear-induced structures in dilute solutions is another challenge for future work [42].

Statistical-mechanical theories of the entropic networks [35, 36, 38, 39, 69] helped to explain general features of phase behavior observed for solutions of branching micelles, microemulsions, physical polymer gels, and for some association colloids. The universal features of entropic networks are also manifested in the kinetic properties of branching chains and in their rheologic behavior. The specific chemistry of systems affects the regimes of the aggregate's growth, the characteristic relaxation times, and the system's response to different stimuli such as changing temperature, composition, and applied external field [9, 58].

Classical molecular thermodynamic theories [77–79] developed for dilute micelle solution may not be applied directly to overlapping micelles in the semidilute and entangled regimes. Nevertheless, the models of this type provide the estimates of the junction and the endcap free energies from a microscopic picture of the micellar aggregate [80, 81, 86, 90]. These estimates may be used as input in the theory that describes an ensemble of branching micelles. Molecular thermodynamic models of the aggregate free energy have also been used directly to estimate the range of molecular parameters and the interval of external conditions where a given aggregate's morphology (including branched wormlike micelles and bicontinuous structures) becomes preferable [81, 90]. Rich morphologic response of the aggregates to the change of salinity and pH has been predicted [81, 88–90] for the crew-cut diblock copolymer ionic micelles having weak polyelectrolyte coronae.

Further development of molecular thermodynamic models requires a more detailed description of the local composition of the micelle interior and of the specific and non-specific interactions between different molecular fragments. Such work is particularly important for mixed micelles and is in progress [91–93].

The experiment shows strong effect of ion specificity on the micellar growth and branching [16, 59, 60]. Existing molecular thermodynamic models that take into account this effect are limited to spherical and globular [63, 64] or wormlike [65, 66] aggregates in solution of simple salts. Description of branched aggregates in solutions of salt with penetrating ions remains yet an unsolved problem.

Dynamics of micellar branching is an emerging important field of research. Recent field theoretical study [94] shows that dynamical branching at the surface of a growing micelle is an important kinetic pathway for the formation of branched wormlike micelles. Thus, the micelle's shape may be controlled not only by thermodynamics but also by a dynamic instability of the aggregated structure. More modeling and experimental studies of this phenomenon are expected in the near future.

Self-assembled active networks, self-propulsion and spontaneous motion is a fascinating new area belonging to a rapidly growing field of Active Soft Matter [95]. For instance, very unusual percolating active network has been recently produced from microtubules [96]. This network exhibits internally generated chaotic flows, spontaneous fractures and self-healing. In contrast to classical networks that respond only passively to external stress, the internal generation of flows continued in these active networks, limited only by the available chemical fuel adenosine triphosphate (ATP).

Acknowledgments

Financial support from Saint-Petersburg State University (project #12.38.199.2014) and from the Russian Foundation for Basic Research (project #18-03-00698-a) is acknowledged.

References

1 Hoffmann, H. (1994). Viscoelastic surfactant solutions. In: *ACS Symposium Series*, vol. 578, 2–31.
2 Cates, M.E. (1987). Reptation of living polymers: dynamics of entangled polymers in the presence of reversible chain-scission reactions. *Macromolecules* 20: 2289–2296.
3 Porte, G., Gomati, R., El Haitamy, O. et al. (1986). Morphological transformations of the primary surfactant structures in brine-rich mixtures of ternary systems general discussion. *J. Phys. Chem.* 90: 5746–5751.
4 Appell, J., Porte, G., Khatory, A. et al. (1992). Static and dynamic properties of a network of wormlike surfactant micelles (cetylpyridinium chlorate in sodium chlorate brine). *J. Phys. II* 2: 1045–1052.
5 Danino, D. (2012). Cryo-TEM of soft molecular assemblies. *Curr. Opin. Colloid Interface Sci.* 17: 316–329.
6 Dreiss, S.A. (2007). Wormlike micelles: where do we stand? *Soft Matter* 3: 956–970.
7 Croce, V., Cosgrove, T., Maitland, G. et al. (2003). Rheology, cryogenic transmission electron microscopy, and small-angle neutron scattering of highly viscoelastic wormlike micellar solutions. *Langmuir* 19: 8536–8541.
8 Ezrahi, S., Tuval, E., and Aserin, A. (2006). Properties, main applications and perspectives of worm micelles. *Adv. Colloid Interf. Sci.* 128–130: 77–102.
9 Chu, Z., Dreiss, S.A., and Feng, Y. (2013). Smart wormlike micelles. *Chem. Soc. Rev.* 42: 7174–7203.
10 Palazzo, G. (2013). Wormlike reverse micelles. *Soft Matter* 9: 10668–10677.
11 Zilman, A., Safran, S.A., Sottmann, T. et al. (2004). Temperature dependence of the thermodynamics and kinetics of micellar solutions. *Langmuir* 20: 2199–2207.
12 Sharma, S.C., Shrestha, L.K., Sakai, K. et al. (2010). Viscoelastic solution of long polyoxyethylene chain phytosterol/monoglyceride/water systems. *Colloid Polym. Sci.* 288: 405–414.

13 Khatory, A., Kern, F., Lequeux, F. et al. (1993). Entangled versus multiconnected network of wormlike micelles. *Langmuir* 9: 933–939.

14 Khatory, A., Lequeux, F., Kern, F. et al. (1993). Linear and nonlinear viscoelasticity of semidilute solutions of wormlike micelles at high salt content. 9: 1456–1464.

15 Rehage, H. and Hoffmann, H. (1988). Rheological properties of viscoelastic surfactant systems. *J. Phys. Chem.* 92: 4712–4719.

16 Murgia, S., Palazzo, G., Mamusa, M. et al. (2011). Aerosol-OT in water forms fully-branched cylindrical direct micelles in the presence of the ionic liquid 1-butyl-3-methylimidazolium bromide. *Phys. Chem. Chem. Phys.* 13: 9238–9245.

17 Lu, T., Xia, L., Wang, X. et al. (2011). A dually effective inorganic salt at inducing obvious viscoelastic behavior of both cationic and anionic surfactant solutions. *Langmuir* 27: 9815–9822.

18 Koshy, P., Aswal, V.K., Venkatesh, M. et al. (2011). Unusual scaling in the rheology of branched wormlike micelles formed by cetyltrimethylammonium bromide and sodium oleate. *J. Phys. Chem. B* 115: 10817–10825.

19 Zana, R.J. (2002). Dimeric (gemini) surfactants: effect of the spacer group on the association behavior in aqueous solution. *Colloid Interface Sci.* 248: 203–220.

20 Bernheim-Groswasser, A., Zana, R., and Talmon, Y. (2000). Sphere-to-cylinder transition in aqueous micellar solution of a dimeric (gemini) surfactant. *J. Phys. Chem. B* 104: 4005–4009.

21 Denkova, A.G., Bomans, P.H.H., Coppens, M.-O. et al. (2011). Complex morphologies of self-assembled block copolymer micelles in binary solvent mixtures: the role of solvent–solvent correlations. *Soft Matter* 7: 6622–6628.

22 Angelico, R., Amin, S., Monduzzi, M. et al. (2012). Impact of branching on the viscoelasticity of wormlike reverse micelles. *Soft Matter* 8: 10941–10949.

23 Imai, M., Hashizaki, K., Taguchi, H. et al. (2013). A new reverse worm-like micellar system from a lecithin, multivalent carboxylic acid and oil mixture. *J. Colloid Interface Sci.* 403: 77–83.

24 Karaborni, S., Esselink, K., Hilbers, P.A.J. et al. (1994). Simulating the self-assembly of gemini (dimeric) surfactants. *Science* 266: 254–256.

25 Briels, W.J., Mulder, P., and den Otter, W.K. (2004). Simulations of elementary processes in entangled wormlike micelles under tension: a kinetic pathway to Y-junctions and shear induced structures. *J. Phys.ics Condens. Matter* 16: S3965–S3974.

26 Buhler, E., Munch, J.P., and Candau, S.J. (1995). Dynamical properties of wormlike micelles: a light scattering study. *J. Phys. II* 5: 765–787.

27 Sommer, C., Pedersen, J.S., Egelhaaf, S.U. et al. (2002). Wormlike micelles as "equilibrium polyelectrolytes": light and neutron scattering experiments. *Langmuir* 18: 2495–2505.
28 Angelico, R., Olsson, U., Palazzo, G. et al. (1998). Surfactant curvilinear diffusion in giant wormlike micelles. *Phys. Rev. Lett.* 81: 2823–2826.
29 Ambrosone, L., Angelico, R., Ceglie, A. et al. (2001). Molecular diffusion in a living network. *Langmuir* 17: 6822–6830.
30 Narayanan, J., Urbach, W., Langevin, D. et al. (1998). Self-diffusion in wormlike micelles networks with electrostatic interactions: a universal behavior? *Phys. Rev. Lett.* 81: 228–232.
31 Schmitt, S. and Lequeux, F. (1998). Surfactant self-diffusion in wormlike micelles. *Langmuir* 14: 283–287.
32 Lequeux, F. (1996). Structure and rheology of wormlike micelles. *Curr. Opin. Colloid Interface Sci.* 1: 341–344.
33 Drye, T.J. and Cates, M.E. (1992). Living networks: the role of cross-links in entangled surfactant solutions. *J. Chem. Phys.* 96: 1367–1375.
34 Lequeux, F. (1992). Reptation of connected worms. *Europhys. Lett.* 19: 675–681.
35 Elleuch, K., Lequeux, F., and Pfeuty, P. (1995). Crosslink effects on equilibrium polymers. *J. Phys. I France* 5: 465–474.
36 Cristobal, G., Rouch, J., Curely, J. et al. (1999). Phase separation in living micellar networks. *Phys. A* 268: 50–64.
37 Tlusty, T. and Safran, S.A. (2000). Defect-induced phase separation in dipolar fluids. *Science* 290: 1328–1331.
38 Zilman, A.G. and Safran, S.A. (2002). Thermodynamics and structure of self-assembled networks. *Phys. Rev. E* 66: 051107–051128.
39 Tlusty, T. and Safran, S.A. (2001). Entropic networks in colloidal self-assembly. *Philos. Trans. R. Soc. A* 359: 879–881.
40 Shi, H., Ge, W., Oh, H. et al. (2013). Photoreversible micellar solution as a smart drag-reducing fluid for use in district heating/cooling systems. *Langmuir* 29: 102–109.
41 Larson, R.J. (1999). *The Structure and Rheology of Complex Fluids (Chapter 12 "Surfactant Solutions")*, 663. New York, NY: Oxford University Press.
42 Cates, M.E. and Fielding, S.M. (2006). Rheology of giant micelles. *Adv. Phys.* 55: 799–879.
43 Doi, M. and Edwards, S.F. (1986). *The Theory of Polymer Dynamics*, 391. Oxford: Clarendon Press.
44 de Gennes, P.G. (1979). *Scaling Concepts in Polymer Physics*, 324. Ithaca, NY: Cornell University Press.
45 Cates, M.E. and Candau, S.J. (1990). Statics and dynamics of worm-like surfactant micelles. *J. Phys. Condens. Matter* 2: 6869–6892.
46 Granek, R. and Cates, M.E. (1992). Stress relaxation in living polymers: results from a poisson renewal model. *J. Chem. Phys.* 96: 4758–4767.

47 Kalur, G.C. and Raghavan, S.R. (2005). Anionic wormlike micellar fluids that display cloud points: rheology and phase behavior. *J. Phys. Chem. B* 109 (18): 8599–8604.

48 Hassan, P.A., Candau, S.J., Kern, F. et al. (1998). Rheology of wormlike micelles with varying hydrophobicity of the counterion. *Langmuir* 14: 6025–6029.

49 Turner, M.S. and Cates, M.E. (1991). Linear viscoelasticity of living polymers: a quantitative probe of chemical relaxation times. *Langmuir* 7: 1590–1594.

50 Schubert, B.A., Kaler, E.W., and Wagner, N.J. (2003). The microstructure and rheology of mixed cationic/anionic wormlike micelles. *Langmuir* 19 (10): 4079–4089.

51 MacKintosh, F.C., Safran, S.A., and Pincus, P.A. (1990). Self-assembly of linear aggregates: the effect of electrostatics on growth. *Europhys. Lett.* 12: 697–702.

52 Kern, F., Lequeux, F., Zana, R. et al. (1994). Entangled versus multiconnected network of wormlike micelles. *Langmuir* 10: 1714–1723.

53 Narayanan, J., Kern, F., Lequeux, F. et al. (1997). Linear viscoelasticity of wormlike micellar solutions found in the vicinity of a vesicle-micelle transition. *Langmuir* 13: 5235–5243.

54 Magid, L.J. (1998). The surfactant – polyelectrolyte analogy. *J. Phys. Chem. B* 102: 4064–4074.

55 Odijk, T. (1977). Polyelectrolytes near the rod limit. *J. Polym. Sci., Part B* 15: 477–483.

56 Skolnick, J. and Fixman, M. (1977). Electrostatic persistence length of a wormlike polyelectrolyte. *Macromolecules* 10: 944–948.

57 Andreev, V.A. and Victorov, A.I. (2010). Electric potential and bending rigidity of a wormlike particle in electrolyte solution. *J. Chem. Phys.* 132: 054902/9.

58 Victorov, A.I., Voznesenskiy, M.A., and Safonova, E.A. (2015). Spatial networks in solutions of wormlike aggregates: universal behavior and molecular portraits. *Russ. Chem. Rev.* 84: 693–711.

59 Oelschlaeger, C., Suwita, P., and Willenbacher, N. (2010). Effect of counterion binding efficiency on structure and dynamics of wormlike micelles. *Langmuir* 26: 7045–7053.

60 Magid, L.J., Han, Z., Li, Z. et al. (2000). Tuning microstructure of cationic micelles on multiple length scales: the role of electrostatics and specific ion binding. *Langmuir* 16: 149–156.

61 Vlachy, N., Jagoda-Cwiklik, B., Vácha, R. et al. (2009). Hofmeister series and specific interactions of charged headgroups with aqueous ions. *Adv. Colloid Interface Sci.* 146: 42–47.

62 Collins, K.D., Neilson, G.W., and Enderby, J.E. (2007). Ions in water: characterizing the forces that control chemical processes and biological structure. *Biophys. Chem.* 128: 95–104.

63 Moreira, L. and Firoozabadi, A. (2010). Molecular thermodynamic modeling of specific ion effects on micellization of ionic surfactants. *Langmuir* 26: 15177–15191.

64 Lukanov, B. and Firoozabadi, A. (2014). Specific ion effects on the self-assembly of ionic surfactants: a molecular-thermodynamic theory of micellization with dispersion forces. *Langmuir* 30: 6373–6383.

65 Koroleva, S.V. and Victorov, A.I. (2014). Modeling of the effects of ion specificity on the onset and growth of ionic micelles in a solution of simple salts. *Langmuir* 30: 3387–3396.

66 Koroleva, S.V. and Victorov, A.I. (2014). The strong specific effect of coions on micellar growth from molecular-thermodynamic theory. *Phys. Chem. Chem. Phys.* 16: 17422–17425.

67 Raghavan, S.R., Edlund, H., and Kaler, E.W. (2002). Cloud-point phenomena in wormlike micellar systems containing cationic surfactant and salt. *Langmuir* 18: 1056–1064.

68 Bernheim-Groswasser, A., Tlusty, T., Safran, S.A. et al. (1999). Direct observation of phase separation in microemulsion networks. *Langmuir* 15: 5448–5453.

69 Russo, J., Tavares, J.M., Teixeira, P.I.C. et al. (2011). Reentrant phase diagram of network fluids. *Phys. Rev. Lett.* 106: 085703.

70 Vega, L.F. and Jackson, G. (2011). SAFT. *Fluid Phase Equilib.* 306: 1–3.

71 de las Heras, D., Tavares, J.M., and Telo da Gama, M.M. (2011). Phase diagrams of binary mixtures of patchy colloids with distinct numbers of patches: the network fluid regime. *Soft Matter* 7: 5615–5626.

72 Semenov, A.N. and Rubinstein, M. (1998). Thermoreversible gelation in solutions of associative polymers. 1. Statics. *Macromolecules* 31: 1373–1385.

73 Rubinstein, M. and Semenov, A.N. (1998). Thermoreversible gelation in solutions of associating polymers. 2. Linear dynamics. *Macromolecules* 31: 1386–1397.

74 Tcyrulnikov, S. and Victorov, A.I. (2013). Molecular thermodynamic modeling of gelation and demixing in solution of cross-associating chains. *Macromolecules* 46: 4706–4715.

75 Rubinstein, M. and Colby, R.H. (2003). *Polymer Physics*. Oxford: Oxford University Press.

76 Zilman, A., Kieffer, J., Molino, F. et al. (2003). Entropic phase separation in polymer-microemulsion networks. *Phys. Rev. Lett.* 91: 015901–015904.

77 Nagarajan, R. and Ruckenstein, E. (1991). Theory of surfactant self-assembly: a predictive molecular thermodynamic approach. *Langmuir* 7: 2934–2969.

78 Nagarajan, R. and Ruckenstein, E. (2000). Self-assembled systems. In: *Equations of State for Fluids and Fluid Mixtures* (ed. J.V. Sengers, R.F. Kayser, C.J. Peters and H.J. White), 589–749. Amsterdam: Elsevier Science.

79 Goldsipe, A. and Blankschtein, D. (2006). Molecular-thermodynamic theory of micellization of ph-sensitive surfactants. *Langmuir* 22: 3547–3559.

80 Andreev, V.A. and Victorov, A.I. (2006). Molecular thermodynamics for micellar branching in solutions of ionic surfactants. *Langmuir* 22: 8298–8310.

81 Victorov, A.I., Plotnikov, N.V., and Hong, P.-D. (2010). Molecular thermodynamic modeling of the morphology transitions in a solution of a diblock copolymer containing a weak polyelectrolyte chain. *J. Phys. Chem. B* 114: 8846–8860.

82 Safran, S.A. (1999). Curvature elasticity of thin films. *Adv. Phys.* 48: 395–448.

83 Tlusty, T., Safran, S.A., and Strey, R. (2000). Topology, phase instabilities, and wetting of microemulsion networks. *Phys. Rev. Lett.* 84: 1244–1247.

84 Tlusty, T. and Safran, S.A.J. (2000). Microemulsion networks: the onset of bicontinuity. *Phys. Condens. Matter* 12: A253–A262.

85 Andreev, V.A. and Victorov, A.I. (2007). A simple analytical approach to electrostatics for nanoparticles with torus-like elements in 1:1 electrolyte solution. *Mol. Phys.* 105: 239–247.

86 May, S., Bohbot, Y., and Ben-Shaul, A. (1997). Molecular theory of bending elasticity and branching of cylindrical micelles. *J. Phys. Chem. B* 101: 8648–8657.

87 Szleifer, I. and Carignano, M.A. (1996). Tethered polymer layers. In: *Advances in Chemical Physics: Polymeric Systems*, vol. 94 (ed. I. Prigogine and S. Rice), 165–260. New York, NY: Wiley.

88 Zhulina, E.B. and Borisov, O.V. (2005). Theory of morphological transitions in weakly dissociating diblock polyelectrolyte micelles. *Macromolecules* 38: 6726–6741.

89 Zhulina, E.B., Birshtein, T.M., and Borisov, O.V. (2006). Curved polymer and polyelectrolyte brushes beyond the daoud-cotton model. *Eur. Phys. J. E* 20: 243–256.

90 Victorov, A. (2012). Curvature elasticity of a weak polyelectrolyte brush and shape transitions in assemblies of amphiphilic diblock copolymers. *Soft Matter* 8: 5513–5524.

91 Victorov, A.I. (2014). Modeling of micelle-solution equilibria for mixed nonionic micelles with strong specific interactions in coronae: group-contribution approach. *J. Chem. Eng. Data* 59: 2995–3002.

92 Mehling, T., Ingram, T., Storm, S. et al. (2013). Estimation of LPC/water partition coefficients using molecular modeling and micellar liquid chromatography. *Colloids and Surf. A* 431: 105–113.

93 Mehling, T., Ingram, T., and Smirnova, I. (2012). Experimental methods and prediction with cosmo-rs to determine partition coefficients in complex surfactant systems. *Langmuir* 28: 118–124.

94 Tang, M. and Carter, W.C. (2013). Branching mechanisms in surfactant micellar growth. *J. Phys. Chem. B* 117: 2898–2905.

95 Cates, M.E. and MacKintosh, F.C. (2011). Active soft matter. *Soft Matter* 7: 3050–3051.

96 Sanchez, T., Chen, D.T.N., DeCamp, S.J. et al. (2012). Spontaneous motion in hierarchically assembled active matter. *Nature* 491: 433.

3

Self-Assembly of Responsive Surfactants

Timothy J. Smith and Nicholas L. Abbott

University of Wisconsin, Department of Chemical and Biological Engineering, Madison, WI, 53706 USA

3.1 Introduction

Surfactants are used in a wide array of applications ranging from coatings to pharmaceuticals due to the influence of surfactants on both bulk solution and interfacial properties. The ability to achieve active and reversible control over surfactant-based properties of aqueous systems through the application of a broad range of external triggers has attracted significant attention, and several review articles have been published on the topic [1–4]. Responsive surfactant systems have been designed to undergo changes in self-assembly triggered by redox potential, pH, light, CO_2, enzymes, and magnetic fields [3]. In this chapter, we focus on surfactant systems that respond to redox state and light because these two triggers are particularly versatile and applicable to a broad range of contexts. We describe, for example, how redox-active and light-responsive surfactants can be used to bring about reversible changes in interfacial and bulk properties of aqueous solutions, to associate in a stimuli-dependent manner with DNA and thus assist in DNA delivery to cells, and to achieve spatial control over surfactant-based phenomena in microfluidic systems.

3.2 Redox-Active Surfactants

While the earliest redox-active surfactant systems used viologen derivatives as a redox group within the surfactant molecules [5–8], most of the more recent work performed with redox-active surfactants has been based on surfactants containing the ferrocene moiety. In contrast to viologens, ferrocenyl surfactants are largely insensitive to oxygen. Ferrocenyl surfactants were first synthesized by Saji et al. [9, 10] and have been the subject of extensive research

Self-Assembly: From Surfactants to Nanoparticles, First Edition. Edited by Ramanathan Nagarajan.

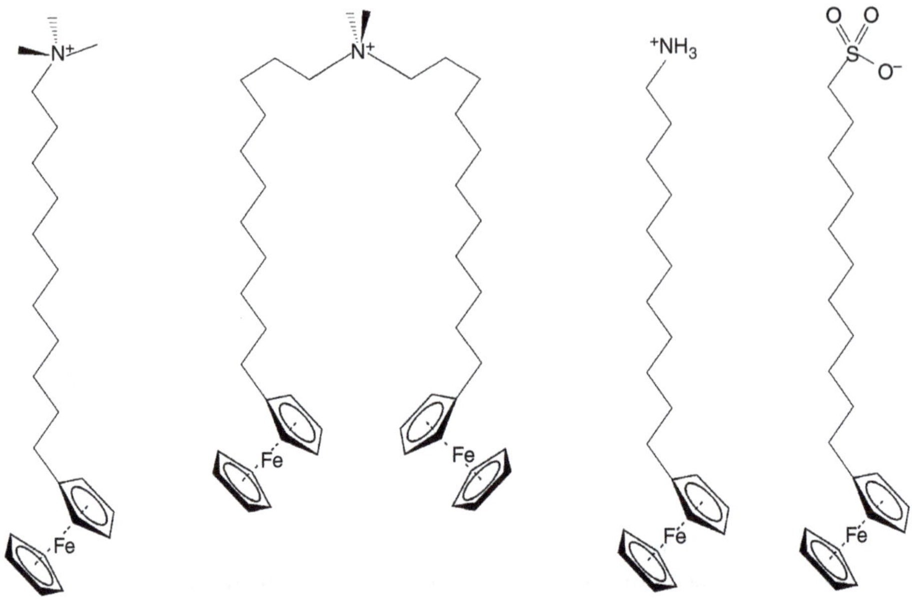

Figure 3.1 Molecular structures of ferrocenyl surfactants addressed in this chapter: 11-ferrocenylundecyl trimethylammonium bromide (FTMA), bis(11-ferrocenylundecyl) dimethylammonium bromide (BFDMA), 11-ferrocenylundecylammonium bromide (11-FAB), and sodium (11-ferrocenylundecyl) sulfonate (SFS) (counter ions not shown).

since that time [2]. Although ferrocenyl surfactants with varied molecular architectures have been synthesized, this chapter focuses on amphiphiles with the structures shown in Figure 3.1: 11-ferrocenylundecyltrimethylammonium bromide (FTMA), bis(11-ferrocenylundecyl) dimethylammonium bromide (BFDMA), 11-ferrocenylundecylammonium bromide (11-FAB), and sodium (11-ferrocenylundecyl) sulfonate (SFS).

3.2.1 Reversible Changes in Interfacial Properties

Redox-active surfactants have been used extensively to drive reversible changes in interfacial adsorption and associated properties such as surface tension. The effect of a change in oxidation state of ferrocene on interfacial properties depends strongly on the architecture of the surfactant that hosts the ferrocene. To illustrate this point, here we contrast studies of the changes in surface tension resulting from the oxidation and reduction of two redox-active surfactants that have been studied extensively: FTMA [9, 11–13] and SFS [14]. FTMA (see Figure 3.1) has the architecture of a conventional cationic surfactant with a +1 charge in its reduced state and, upon oxidation to a +2 state, it transforms to a cationic bolaform surfactant-like structure. Gallardo et al. [11, 12] measured the surface tensions of aqueous solutions of FTMA in 100 mM Li_2SO_4 as a function of FTMA concentration. The plots of surface tension vs. concentration measured for reduced FTMA (see Figure 3.2a) are seemingly typical of a conventional cationic surfactant: the surface tension decreases until a critical micelle concentration (CMC) (~0.1 mM for reduced FTMA) is reached, above which the surface tension maintains a value of 49 mN m^{-1}. Oxidized FTMA also causes a decrease in the surface tension of water but a higher concentration is necessary to see a significant effect. Oxidized FTMA, however, does not exhibit a break in the surface tension vs. concentration plot up to concentrations of at least 30 mM. At concentrations below 1 mM FTMA, oxidation of FTMA thus leads to an increase in surface tension of up to 23 mN m^{-1} (see Figure 3.1). At higher concentrations (above 10 mM), in contrast, oxidation of FTMA leads to a decrease in surface tension of the aqueous solution.

To provide insight into the above observations, the excess surface concentrations of FTMA, Γ_s, (and area per molecule) were calculated from measured changes in surface tension, γ, with concentration, C_s, using the Gibbs adsorption isotherm:

$$\Gamma_s = -\frac{1}{kT}\frac{d\gamma}{d[\ln Cs]} \tag{3.1}$$

where k is the Boltzmann constant and T is the temperature. Evaluation of the surface excess led to the proposal that both oxidized and reduced forms of FTMA exist in a looped configuration at the air–water interface. Specifically, for reduced FTMA, the limiting area per molecule at the interface was 85 ± 5 Å^2 per molecule, a value that is substantially larger than for conventional cationic surfactants (dodecyltrimethylammonium bromide, DTAB, for example, has a limiting area per molecule of 63 ± 2 Å^2 per molecule) [12]. The large value of the limiting area per molecule for reduced FTMA supports the interpretation of the looped configuration and also gives rise to the interesting conclusion that the dominant mechanism by which reduced FTMA lowers surfactant tension is not via electrostatics (as is the case with conventional ionic surfactants) but instead via an entropic term associated with constraints on the configuration of the molecules at the interface. The existence of the looped configuration also underlies the higher surface tension plateau (49 mN m^{-1}) measured with reduced FTMA than with DTAB (40 mN m^{-1}) [15, 16], as elucidated by a molecular-thermodynamic model of FTMA [13]. Interestingly, because oxidized FTMA does not form micelles in bulk solution (see below for additional discussion), the limiting area per molecule of oxidized FTMA at the surface of the aqueous solution was 75 ± 5 Å^2 per molecule. This small value of the area per molecule (as compared to reduced FTMA), when combined with the +2 charge, explains the relatively lower surface tensions generated by oxidized FTMA as compared to reduced FTMA at high bulk concentrations.

Whereas the results above apply to the cationic ferrocenyl surfactant FTMA, an anionic ferrocenyl surfactant, SFS, has also been shown to permit redox-induced, reversible changes in surface tension [14]. SFS is similar to

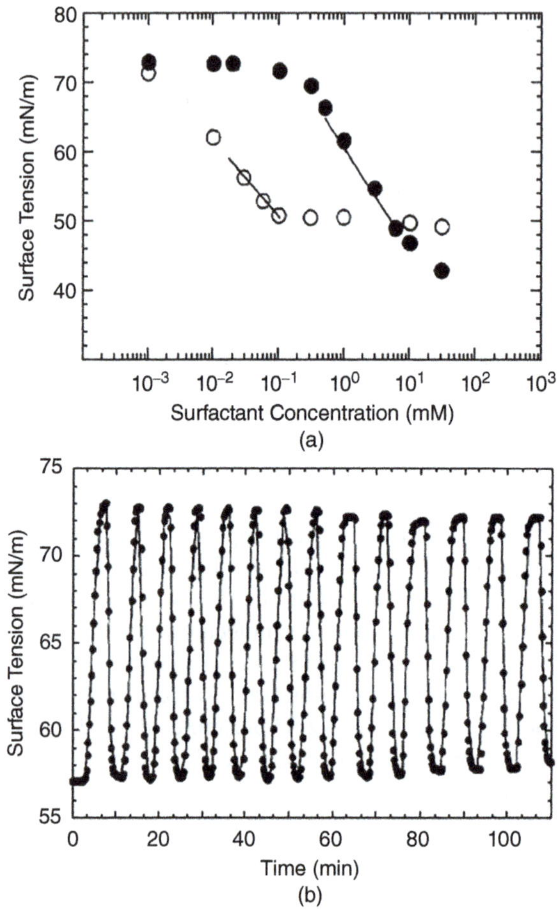

Figure 3.2 (a) Equilibrium surface tensions of aqueous solutions (0.1 M Li_2SO_4, pH 2, 25 °C) of reduced FTMA (○) and oxidized FTMA (•). (b) Dynamic surface tension of an aqueous solution of 0.3 mM reduced/oxidized FTMA (0.1 M Li_2SO_4, pH 2, 20 °C) measured during the repeated cycling of the surfactant between oxidation states. (c) Equilibrium surface tensions of aqueous solutions of reduced SFS (○) and oxidized SFS (□) (10 mM Li_2SO_4, pH 5, 37 °C). (d) Dynamic surface tensions of an aqueous solution of 0.6 mM oxidized and reduced SFS (10 mM Li_2SO_4, pH 5, 37 °C) measured during the repeated cycling of the surfactant between oxidation states. In (b) and (d), the high values of surface tension correspond to a solution of oxidized surfactant, and the low values correspond to a solution of reduced surfactant. *Source*: (a) and (b) reproduced from [11] and (c) and (d) reproduced from [14].

Figure 3.2 (*Continued*)

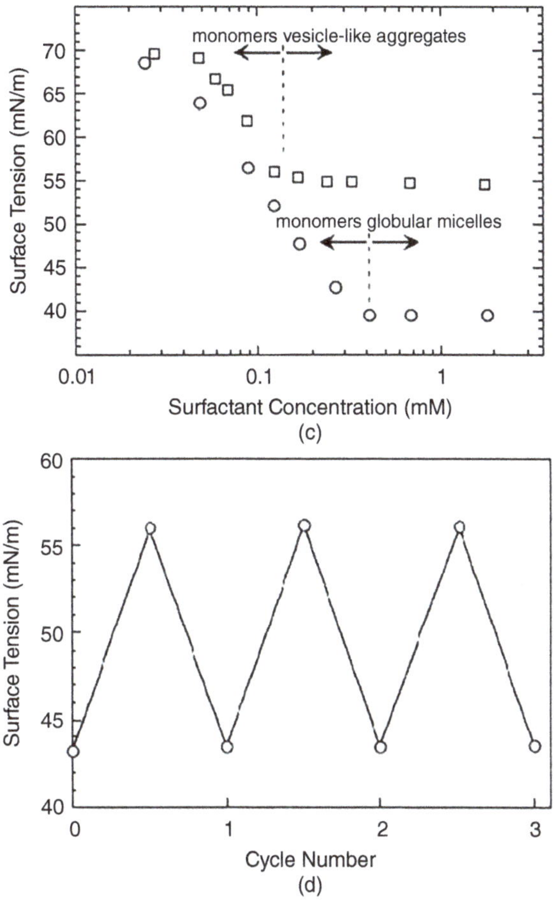

FTMA except that the trimethylammonium group is replaced by a sulfonate group. This change in the head group charge, however, leads to significantly different response of SFS to oxidation of the ferrocene as compared to FTMA. First, inspection of Figure 3.2c reveals that oxidation of SFS results in a decrease in the critical aggregation concentration (CAC) of the SFS in solution (recall that the CMC of FTMA increased with its oxidation). As discussed below, the morphologies of the assemblies formed by SFS are also strikingly different from FTMA. Second, at concentrations below which either oxidized or reduced SFS self-assembles in bulk solution, we observe that both states of SFS lower the surface tension. Oxidized FTMA, at these concentrations, did not measurably lower surface tension. In contrast to FTMA, however, oxidation of reduced SFS (within a monolayer formed at the surface of the solution) leads to a decrease in both the electrostatic free energy of the monolayer (as the resulting zwitterionic species does not have a net charge) and the hydrophobicity of SFS that drives formation of the monolayer. These two effects appear to largely counterbalance, with the net result being only a small change in surface tension upon oxidation of SFS at low concentrations. Below we discuss differences in the bulk solution behaviors of FTMA and SFS that underlie observations reported above.

3.2.2 Reversible Changes in Bulk Solution Properties

Underlying the changes in interfacial properties described in the previous section are, of course, changes in bulk solution self-assembly that are triggered by the redox reactions. As inferred from the surface tension vs. concentration plots, for concentrations from 0.1 to 10 mM, FTMA forms aggregates in its reduced state but not in its oxidized state. By using dynamic light scattering measurements [14] (see Table 3.1) and small angle neutron scattering (SANS), it has been shown that reduced FTMA forms globular micelles with a hydrodynamic diameter, D_h, of 6 ± 2 nm, similar to what is typically observed for a conventional ionic surfactant. Reduced SFS behaves similarly to reduced FTMA, forming globular micelles with $D_h = 6 \pm 2$ nm. Interestingly, however, oxidized SFS has the architecture of a zwitterionic bolaform and forms vesicle-like aggregates with $D_h = 70 \pm 8$ nm. This behavior is consistent with that seen for many conventional zwitterionic surfactants [17]. We note also that the CAC of oxidized SFS is lower than that of reduced SFS (in contrast to the effects of oxidation of ferrocene on the CAC of FTMA).

The bulk solution properties of 11-FAB, which differs from FTMA by having an ammonium moiety rather than trimethylammonium as its headgroup, have also been characterized [18]. The smaller headgroup of reduced 11-FAB leads to the formation of vesicles in aqueous solutions (at low pH) rather than micelles. Upon oxidation, similar to FTMA, the reduced 11-FAB aggregates disassemble into oxidized 11-FAB monomers (see Figure 3.3).

Table 3.1 Hydrodynamic diameters, Dh, of assemblies formed by ferrocenyl surfactants (0.6 mM) within aqueous solutions of 10 mM Li_2SO_4 for the SO_3^--based surfactants and 100 mM Li_2SO_4 for the $N^+(CH_3)_3$-based surfactants, all at 37 °C.

Surfactant	D_h (nm)	Concentration (mM)
$Fc(CH_2)_{11}SO_3^-$	6 ± 2	>0.4
$Fc^+(CH_2)_{11}SO_3^-$	70 ± 8	>0.15
$Fc(CH_2)_{11}N^+(CH_3)_3$	6 ± 2	>0.1
$Fc^+ (CH_2)_{11}N^+(CH_3)_3$	Does not aggregate	<10

Source: Table reproduced from [15].

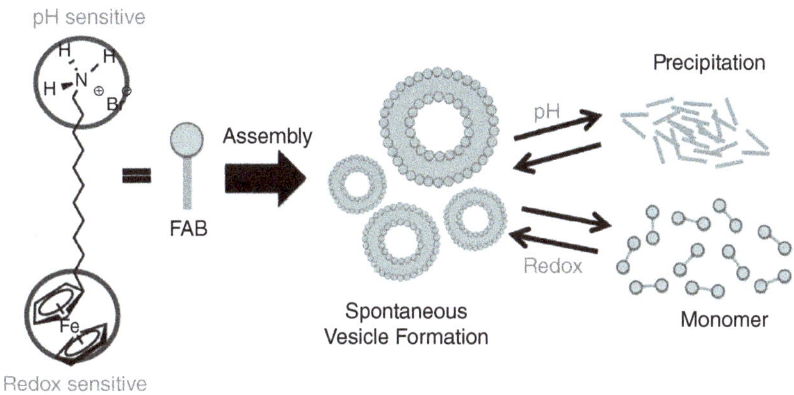

Figure 3.3 Possible mechanism leading to the formation and disruption of vesicles by pH variation and redox chemistry. Pure 11-FAB spontaneously forms vesicles in aqueous solutions. Disappearance of charge associated with the ammonium headgroup with increasing pH leads to disruption of vesicles and formation of precipitates in water. Vesicles reform by decreasing pH again. Reduced 11-FAB also forms vesicles in electrolyte solutions. Vesicles are disassembled by electrochemical oxidation, and oxidized 11-FAB behaves as a bola surfactant. Subsequently, chemical reduction of oxidized 11-FAB brings about reformation of vesicles. *Source*: Figure reproduced from [18].

However, 11-FAB is also responsive to pH changes. Under basic conditions, for example, reduced 11-FAB will precipitate out of solution, thus providing a convenient means to recover 11-FAB from solution.

In addition to the formation of globular micelles or vesicles, redox-active surfactants have also been used to form wormlike micelles and thus provide control over bulk solution properties such as viscosity. For example, Abe et al. [19] found that reduced FTMA forms giant worm-like micelles with sodium salicylate. Solutions containing these wormlike micelles exhibit viscoelastic behavior and high zero-shear viscosities (around 15 Pa s). Upon oxidation of the ferrocene, the wormlike micelles break into ellipsoidal micelles, thus generating a Newtonian fluid with a viscosity of 2.5×10^{-3} Pa s, which is around 1/6000th the zero-shear-rate viscosity of the solution containing reduced FTMA (and salicylate).

FTMA has also been used to tune the interactions between surfactants and polymers such as ethyl(hydroxyethyl) cellulose (EHEC) [20]. Reduced FTMA, for example, was found to promote the aggregation of EHEC at low FTMA concentrations and disperse aggregates at high FTMA concentrations. In contrast, oxidized FTMA did not strongly associate with EHEC and thus had very little effect on the state of aggregation of the EHEC. The use of redox-active surfactants to tune the interactions and state of dispersion of polymers in solution has also been extended to a range of biopolymers, including DNA [21]. DNA adopts a coiled configuration in aqueous solution in the absence of surfactant.

The addition of 2–5 μM of reduced FTMA, however, was measured to generate coexisting populations of coils and compact globules of DNA. Upon further addition of reduced FTMA to a concentration of 30 μM, all of the coils were transformed to compact globules. In contrast, oxidized FTMA had no measurable effect on the DNA conformation at concentrations up to 75 μM, allowing for reversible control of DNA conformations via manipulation of the redox state of FTMA.

Several past studies have also reported on complexes (lipoplexes) formed by DNA and the double-tailed ferrocenyl surfactant BFDMA [22]. These studies were motivated by the widespread use of double-tailed cationic surfactants for the packaging and delivery of DNA to cells. BFDMA was measured to form lipoplexes with DNA in both redox states. These lipoplexes vary in both size and zeta potential as a function of BFDMA concentration (see Figure 3.4). As the concentration of reduced BFDMA increased, the hydrodynamic sizes of the lipoplexes passed through a maximum. Over this same range of concentrations, the zeta potentials of the complexes increased from −40 to +40 mV. Interestingly, for oxidized BFDMA, however, the lipoplex size passed through a minimum as a function of increasing BFDMA concentration and the zeta potential was measured to be negative across all BFDMA concentrations. These measurements led to the conclusion that redox state-dependent changes in the transfection efficiency of BFDMA lipoplexes (see the next section) likely reflect the dependence of the zeta potential of BFDMA-DNA complexes on the oxidation state of BFDMA.

3.2.3 Control of Biomolecule-Surfactant Assemblies

As noted above, DNA forms complexes (called lipoplexes) with many double-tailed cationic surfactants, and the resulting complexes can be transported across cell membranes, leading to "transfection" of cells. The capability to control the transport of DNA across cell membranes has the potential to be useful in a range of contexts such as gene therapy and, more broadly, biotechnology [23, 24]. As discussed in the previous section, BFDMA forms lipoplexes with DNA in both redox states. Recently, however, it was reported that reduced BFDMA forms complexes with plasmids that encode for enhanced green fluorescent protein (EGFP) and are transported into cells (COS-7 cell line), resulting in expression of the EGFP. In contrast, the complexes formed between the plasmids and oxidized BFDMA do not result in the delivery of the plasmids to the cells [25]. This difference in behavior between the two redox states allows the DNA to be transfected upon delivery of a trigger that reduces oxidized BFDMA. Examples of chemical reducing agents that have been used as a trigger include glutathione [26] and ascorbic acid [27, 28].

One such demonstration is shown in Figure 3.5, in which lipoplexes of oxidized BFDMA and DNA (encoding for EGFP) are dispersed in culture media

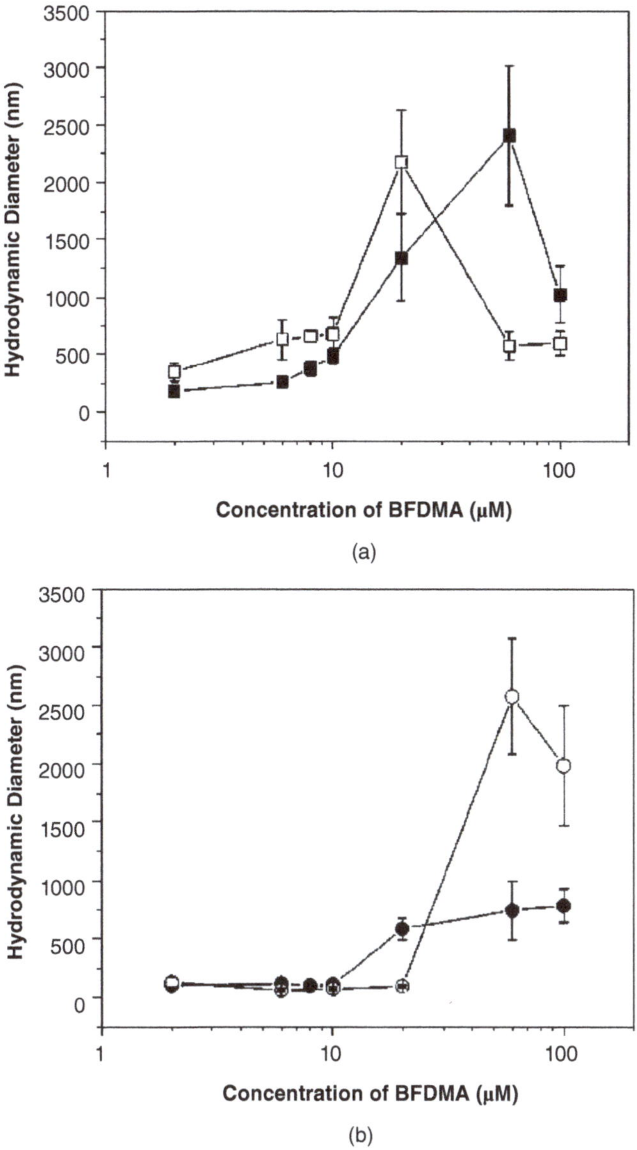

Figure 3.4 Hydrodynamic diameters of complexes formed from reduced and oxidized BFDMA as a function of concentration of BFDMA, as calculated from dynamic light scattering measurements. (a) Lipoplexes formed by reduced BFDMA and DNA in 1 mM Li_2SO_4 (□) and OptiMEM (■). (b) Complexes formed by oxidized BFDMA and DNA in 1 mM Li_2SO_4 (○) and OptiMEM (•). *Source*: Figure reproduced from [22].

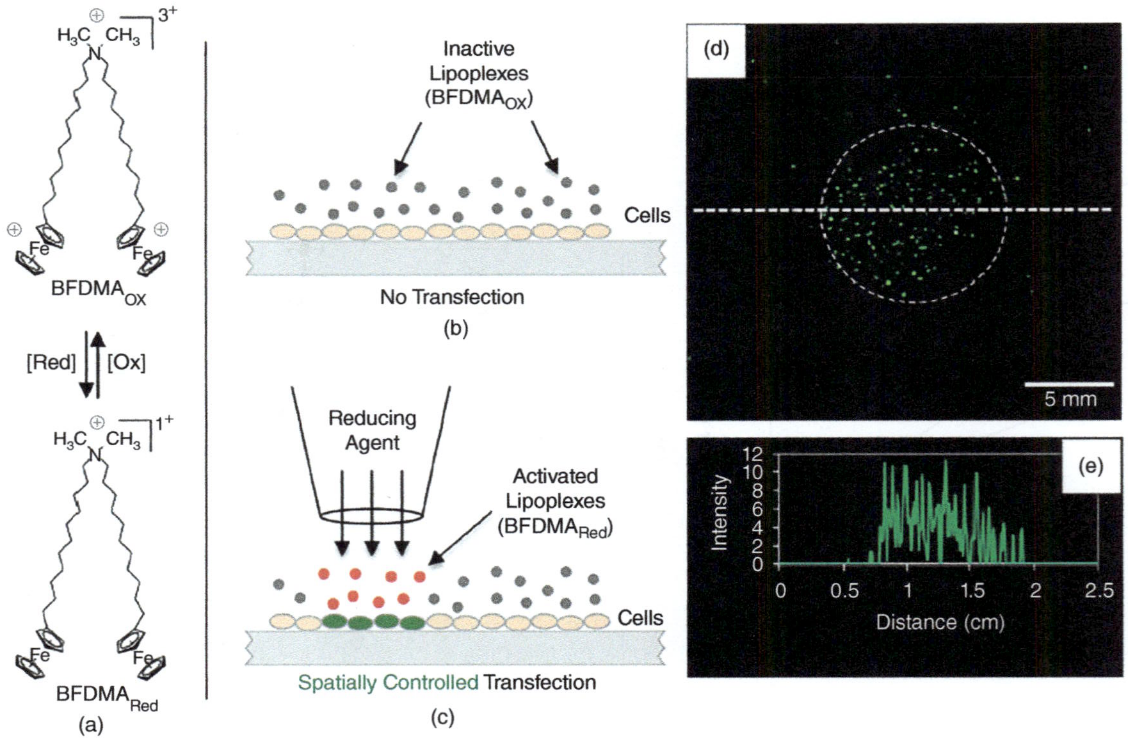

Figure 3.5 (a) Structure of BFDMA. (b, c) Schematic illustration demonstrating core principles: (b) Lipoplexes of oxidized BFDMA (gray) are inactive and can be distributed uniformly over cells without promoting transfection, (c) Controlled administration of soluble or solid-phase reducing agents results in localized activation of lipoplexes (red) and spatially controlled transfection (green); adjacent cells are unaffected. (d) Composite of fluorescence micrographs showing the area of a cell culture well containing inactive lipoplexes of oxidized BFDMA and a hollow cylinder (dotted circle) used to localize delivery of ascorbic acid. The lipoplexes within the cylinder are activated upon reduction and transfer plasmid DNA encoding enhanced green fluorescent protein (EGFP). (e) Line intensity profile of EGFP fluorescence measured along the dotted line

in which cells are growing. Ascorbic acid is added locally to the culture media (via a hollow cylinder). The BFDMA-containing lipoplexes exposed to ascorbic acid are reduced and the associated DNA is able to enter the cells. Fluorescence is detected only in the regions to which the ascorbic acid was delivered, demonstrating that transfection is limited to cells in the activated area.

3.2.4 Spatial Control of Surfactant-Based Properties

As illustrated by the previous example, localized delivery of reducing agents to solutions containing redox-active surfactants can be used to generate spatially localized changes in surfactant-based properties. This concept has also been extended to create controlled gradients in bulk and interfacial properties. Here we summarize, in particular, two types of spatial gradients that have been explored in redox-active surfactant systems: surface tension-driven (Marangoni) flows and spatial gradients in the concentrations of micelles in bulk solution.

Marangoni flows are generated by stresses associated with gradients in surface or interfacial tension. As discussed in Section 3.3.1, oxidation and reduction of redox-active surfactants can generate significant changes in the surface tension of aqueous solutions. These changes in surface tension can be exploited to generate surface tension gradients and thus drive interfacial flows. Gallardo et al. [29] demonstrated (see Figure 3.6) that Marangoni flows can be generated using FTMA in a fluidic network comprising two intersecting channels. Electrodes were placed at the ends of each of the channels. Liquid crystal (LC) droplets were used to visualize the interfacial flows. At each electrode, FTMA was either reduced or oxidized, thus leading to gradients in the concentrations of these species across the interface and thus to surface tension gradient-driven flows. In contrast to Marangoni flows driven by conventional surfactants, we emphasize that steady-state gradients in surface concentrations can be generated by redox-active surfactants (based on reversible changes in oxidation state) and that the magnitudes of the gradients can also be controlled by manipulation of the potentials applied to the electrodes. We comment also that the flows in Figure 3.6 are not electrokinetic in nature, but arise from gradients in surfactant concentration and thus surface tension.

Redox-active surfactants have also been used to generate gradients in the concentration of micelles in bulk solution. Here we illustrate the opportunity associated with such gradients in bulk solution [30, 31] by describing recent studies that have used microfluidic channels with opposing side walls that are each lined with an electrode.

Application of reducing and oxidizing potentials to the opposing electrodes leads to the rapid generation of steady-state gradients in reduced and oxidized FTMA concentrations across the channel. Since reduced FTMA forms micelles whereas oxidized FTMA does not (see above), a spatial gradient in

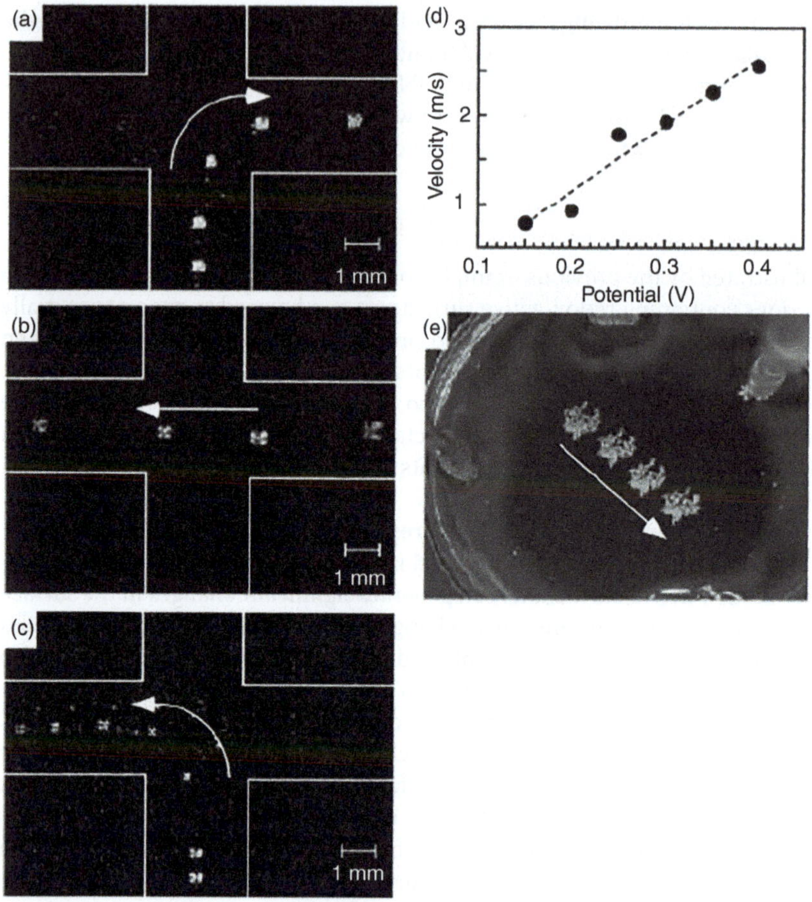

Figure 3.6 (a–c) Time-lapse images (4–6 frames obtained at ~0.5 Hz) of the pumping of liquid crystal (LC) droplets across the surface of an aqueous solution of 0.3 mM FTMA (0.01 M Li_2SO_4) contained within a simple fluidic network formed by four glass microscope slides affixed to a larger glass plate. Platinum (Pt) electrodes protrude through the surface of the solution at the ends of the left, right, and lower channels (widths 4 mm). The end of the top channel contains a saturated calomel reference electrode and Pt counterelectrode. (a) Droplet of LC dispensed into the bottom channel of the intersection is pumped into the right channel by application of a potential of −0.3 V to the bottom Pt electrode and +0.3 V to the right Pt electrode. (b) The droplet of LC in the right channel of the intersection is pumped into the left channel by application of a potential of −0.3 V to the right Pt electrode and +0.3 V to the left Pt electrode. (c) A droplet of LC dispensed into the bottom channel of the intersection is pumped into the left channel by application of a potential of −0.3 V to the bottom Pt electrode and + 0.3 V to the left Pt electrode. (d) The velocity of fluid motion measured in a straight channel (width 4 mm) as a function of applied potential. The *x*-axis indicates the magnitude of the anodic and cathodic potentials. (e) Time-lapse image (four frames obtained at 0.3 Hz) of sulfur microparticles supported on an unconfined surface of a solution of oxidized FTMA. Application of −0.3 V to two Pt electrodes placed at the top and left sides of the image caused the sulfur microparticles to be pumped in the direction indicated by the arrow. The horizontal dimension of the image is ~40 mm. *Source*: Figure reproduced from [29].

the concentration of micelles across the channel is also generated. Because sparingly water-soluble and amphiphilic solutes interact with the micelles within the microfluidic channel, the gradients in micelle concentration can be used to selectively partition solutes across the channel via their interaction with the micelles, thus offering the basis of a continuous separations process (the outlet stream is fractionated by splitting the flow into a series of small channels that sample the lateral concentration profile across the microfluidic system) (see Figure 3.7).

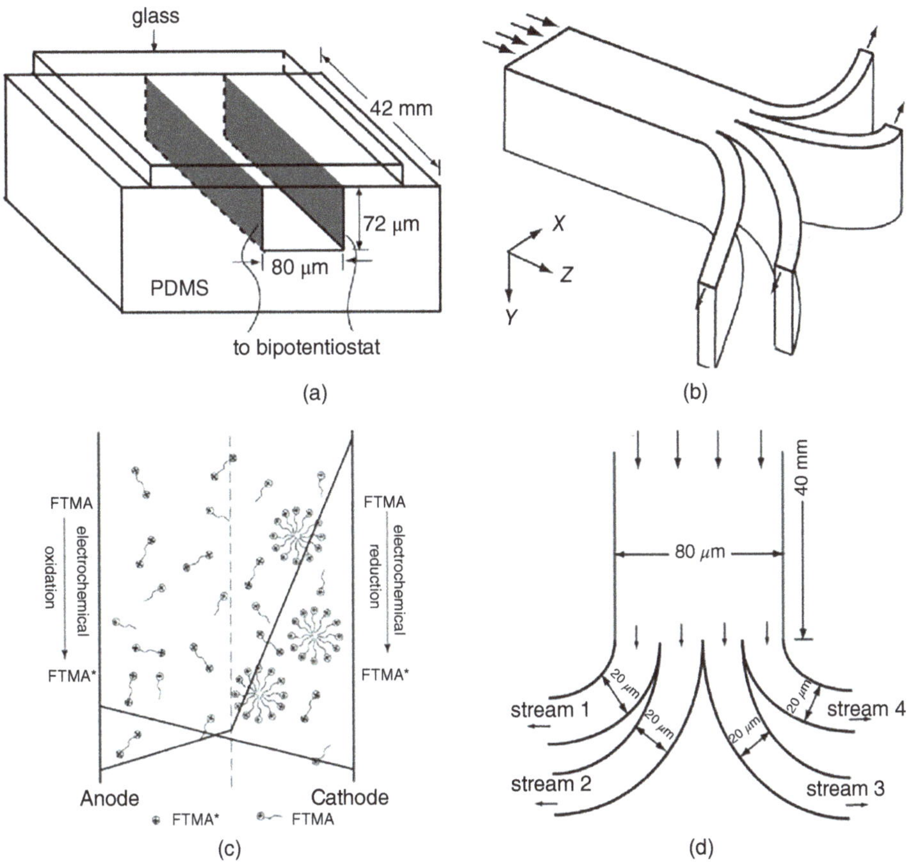

Figure 3.7 (a) Expanded view of a microfluidic channel with two independent electrodes lining the vertical walls of the channel. (b) Schematic illustration of the microfluidic channel used to electrochemically generate and measure lateral gradients in the concentration of redox-active surfactant. (c) Schematic illustration of electrochemically generated gradient in concentration of FTMA across the microfluidic channel. (d) Schematic illustration of top view of the outlet of the microfluidic channel. *Source*: (a) and (b) reproduced from [30] and (c) and (d) reproduced from [31].

3.3 Light-Responsive Surfactants

The second broad class of responsive surfactants addressed in this chapter are those that contain light-responsive functional groups. The most common functional group used in these studies is azobenzene [1, 3]. Azobenzene undergoes a *trans*-to-*cis*-photo-isomerization upon irradiation of light at different wavelengths. UV radiation (~360 nm wavelength) converts azobenzene to a *cis*-rich isomeric state while visible light (~460 nm wavelength) enriches the system in the *trans*-conformation [32]. Other photosensitive moieties that have been used include spiropyrans [33–36] and stilbenes [37–40].

3.3.1 Interfacial Properties

Similar to redox-active surfactants, light-responsive surfactants have been used to trigger changes in interfacial properties such as surface tension. One example of an azobenzene-containing surfactant that has been characterized in some detail is 4,4′-bis(trimethylammoniumhexyloxy)azobenzene bromide (BTHA) [41]. While changes in equilibrium surface tensions that result from photoisomerization of BTHA were found to be small, dynamic surface tensions of mixtures of BTHA and sodium dodecylsulfate (SDS) were observed to be strongly dependent on illumination. Specifically, the onset of illumination with UV light was observed to trigger a rapid change in dynamic surface tension of up to 25 $mN\,m^{-1}$ whereas in the absence of illumination, the surface tension evolved slowly over a matter of minutes or hours. We note also that in this system, the change in the equilibrium surface tension induced by illumination with UV light was modest (only 2 $mN\,m^{-1}$). However, because the surfactant solutions require minutes to hours to equilibrate, the capability to change dynamic surface tension on time-scales of seconds appears potentially useful for a range of applications.

3.3.2 Bulk Solution Properties

In addition to triggering a rapid decrease in the dynamic surface tension of an aqueous solution, illumination of a bulk aqueous solution of SDS and BTHA with UV light was also found to trigger reversible changes in the equilibrium morphologies of assemblies formed by BTHA and SDS. Specifically, at low total surfactant concentration (0.1 wt%), solutions rich in SDS (85 mol% of the total surfactant was SDS) were found to form vesicles (diameters of 135 ± 5 nm). However, illumination of these solutions of vesicles with UV light resulted in the transformation of the system to cylindrical aggregates (*cis*-BTHA) with radii of 1.9 ± 0.3 nm and a length of 24 ± 5 nm [42]. By performing small-angle neutron scattering measurements, the transformation was shown to be reversed by sequential illumination with UV, then visible wavelengths of light. Additional studies [43] of this surfactant system resulted in the mapping of the ternary phase diagram (see Figure 3.8). Inspection of Figure 3.8 reveals that the effect of UV light on the state of aggregation of the BTHA/SDS mixture is strongly dependent on the composition of the mixture.

Here we note that a second photo-responsive surfactant system [44], consisting of sodium oleate (NaOA) and an azobenzene-containing surfactant, 1-[2-(4-phenylazo-phenoxy)-ethyl]-3-methylimidazolium bromide (C_0AZOC_2IMB) has been reported to exhibit reversible formation of giant wormlike micelles (see Figure 3.9). In the *trans*-rich state, C_0AZOC_2IMB forms wormlike micelles with NaOA and the solution is highly viscoelastic with a zero-shear-rate viscosity of ~1300 Pa-s. However, upon UV irradiation, the *trans*-C_0AZOC_2IMB is partially converted to *cis*-C_0AZOC_2IMB and the wormlike micelles are transformed to ellipsoidal micelles, causing the solution to become Newtonian with a viscosity of 0.01 Pa-s.

3.3.3 Biomolecule-Surfactant Interactions

Photoresponsive self-assembly of amphiphiles has also been explored to control the interactions of amphiphiles and DNA in the context of delivery of DNA to cells. Specifically, the interactions of DNA and light-sensitive surfactants were investigated to determine if the photoresponsive interaction could be used to promote the disassembly of DNA-surfactant complexes after the complexes had crossed cell membranes. In the scheme described by Lee and coworkers [45], vesicles prepared from an azobenzene-containing cationic surfactant, azobenzene trimethylammonium bromide (azoTAB), and a nonphotoactive anionic surfactant were shown to complex with DNA and facilitate the transport of DNA across cell membranes. Following uptake by cells, UV illumination was used to photoisomerize the azoTAB and thereby release the DNA from the surfactant-DNA complex. When compared to a standard transfection agent, polyethylenimine (PEI), the azoTAB vesicles exhibited similar transfection efficiencies (35–40%) when irradiated with UV light and much lower efficiencies (15–20%) without UV illumination. Complementary studies have characterized the thermodynamics of azoTAB-DNA complexes [46] and the effects of surfactant tail-length and solution ionic strength on photoinduced changes in the properties of this system [47].

Finally, we note a variant of the above-described approach for photo-controlled delivery of DNA to cells that is based on an azoTAB-derived surfactant [48]. In this additional approach, the bromide counter ion of azoTAB was replaced by $[FeCl_3Br]^-$, a paramagnetic counterion (this combination is referred to as azoTAFe). As in previous studies, DNA was released from complexes containing azoTAFe upon UV irradiation. However, when using a paramagnetic counterion, the azoTAFe-DNA complexes could be

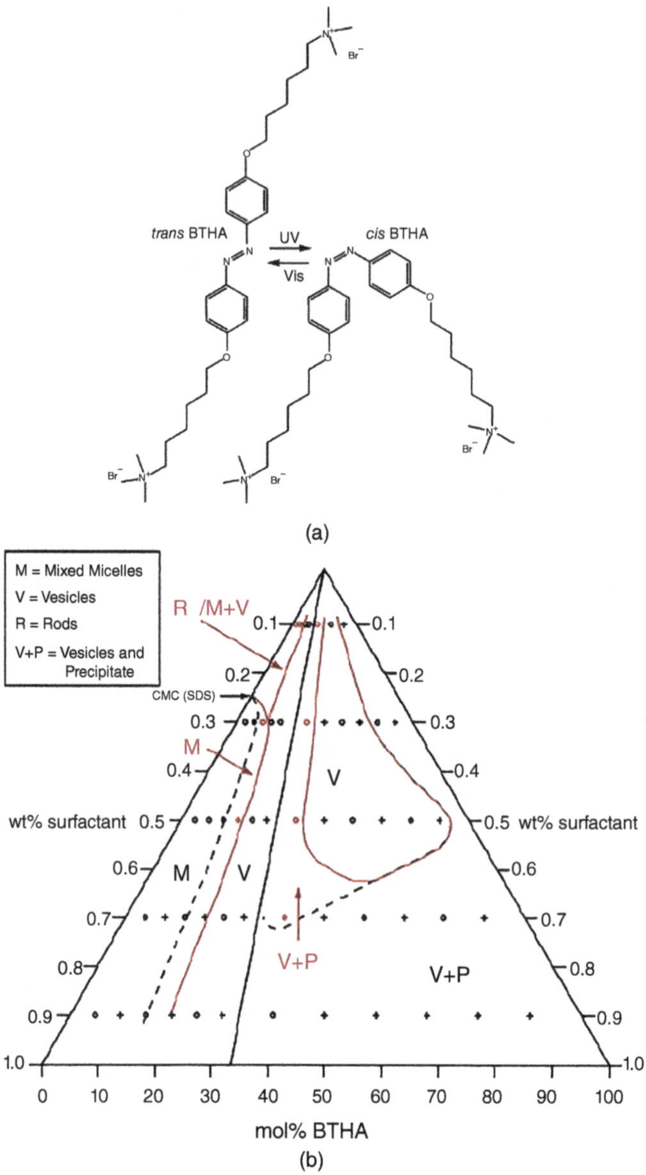

Figure 3.8 (a) Molecular structures of the *trans*- and *cis*- isomers of BTHA. (b) Microstructure diagram for SDS/BTHA/H_2O. Black lines represent the boundaries from the phase diagram with *trans*-BTHA, and the red lines indicate the new boundaries formed after isomerization to *cis*-BTHA. Samples that undergo a change in the type of microstructure upon isomerization are marked in red. SDS-rich (c) and BTHA-rich (d) samples of SANS data for a series of BTHA/SDS mixtures containing 0.5 wt% total surfactant. The markers represent the data, and the solid lines represent fits to the data using an appropriate form factor model. The data have been offset by the indicated factors for clarity. *Source*: Figure reproduced from [43].

Figure 3.8 (*Continued*)

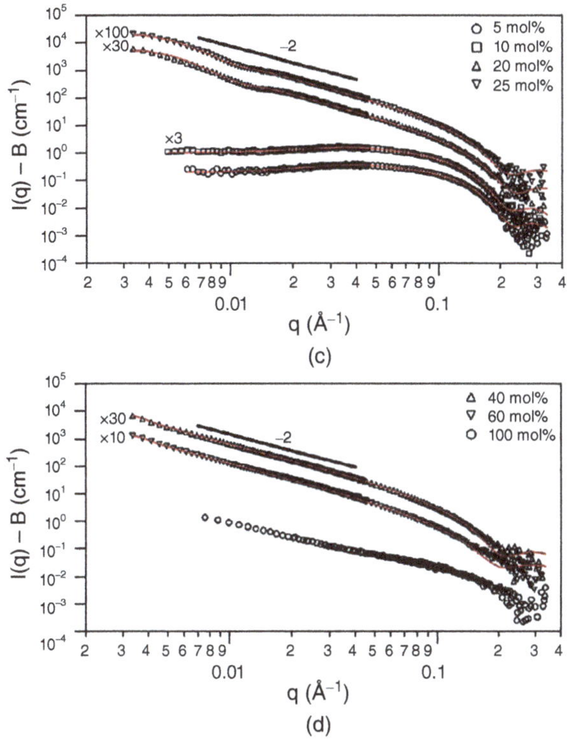

manipulated by an applied magnetic field. This offers the capability to position the azoTAFe-DNA complexes prior to DNA release by illumination.

3.3.4 Spatial Control of Surfactant-Based Properties Using Light

Photoresponsive surfactants have also been used to achieve spatial control of interfacial properties resulting in the generation of Marangoni flows [49]. This has been demonstrated by using azoTAB. *trans*-azoTAB preferentially adsorbs to the surface of water whereas, upon illumination with UV light, *cis*-azoTAB generated at the interface rapidly desorbs. Irradiation of a localized region of a surface using UV light converts the *trans*-azoTAB to *cis*-azoTAB, thus resulting in a region of high surface tension (where the *cis*-azoTAB has desorbed) and a Marangoni flow toward the illuminated spot (Figure 3.10).

3.4 Conclusion

This chapter presents a brief overview of select experimental studies that have incorporated redox-active and light-sensitive groups into surfactant

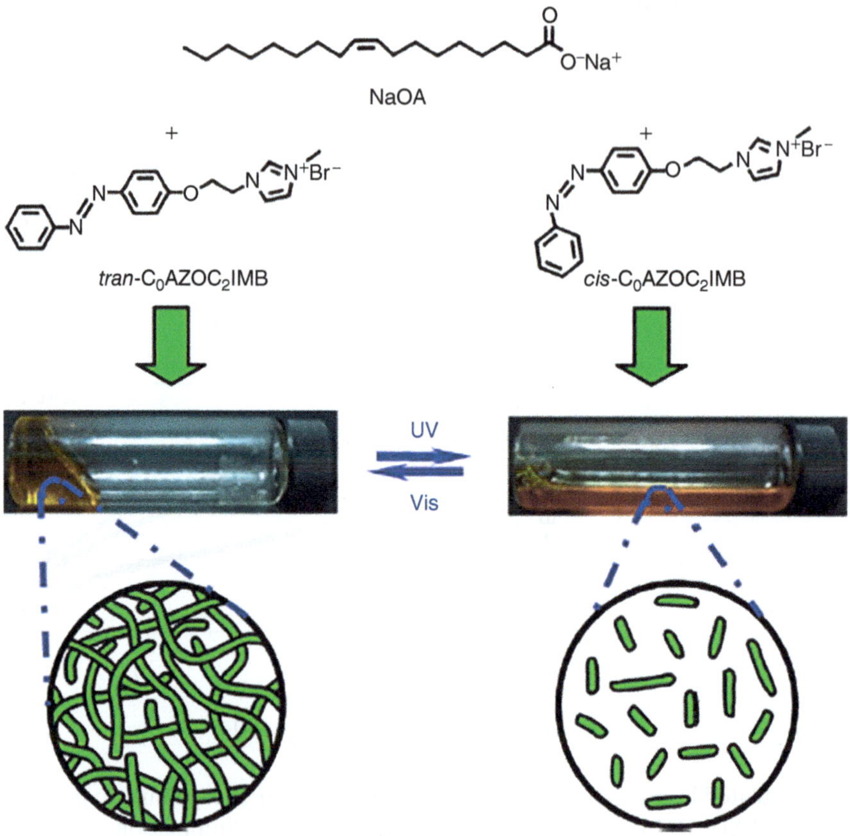

Figure 3.9 Representative light-reversible transition of NaOA and C_0AZOC_2IMB binary mixtures. *Source*: Figure reproduced from [44].

structures to achieve control over both interfacial and bulk solution properties. The "take-home" message of this chapter is that the use of redox- and light-triggerable surfactants provides exciting opportunities for both fundamental studies of surfactant systems (e.g. by providing a unique tool to investigate non-equilibrium properties) and for new surfactant-based technologies. For example, the transformation of reduced FTMA to oxidized FTMA can lead to large and reversible changes in surface tension and also to formation and disassembly of micelles in bulk solution. The difference between the surface tension of reduced and oxidized FTMA, which can be as large as $23\,mN\,m^{-1}$, can be exploited to generate flows that transport droplets (e.g. of an oil) along a channel. Alternatively, redox-based control of micellization can be used to generate concentration gradients of micelles laterally across

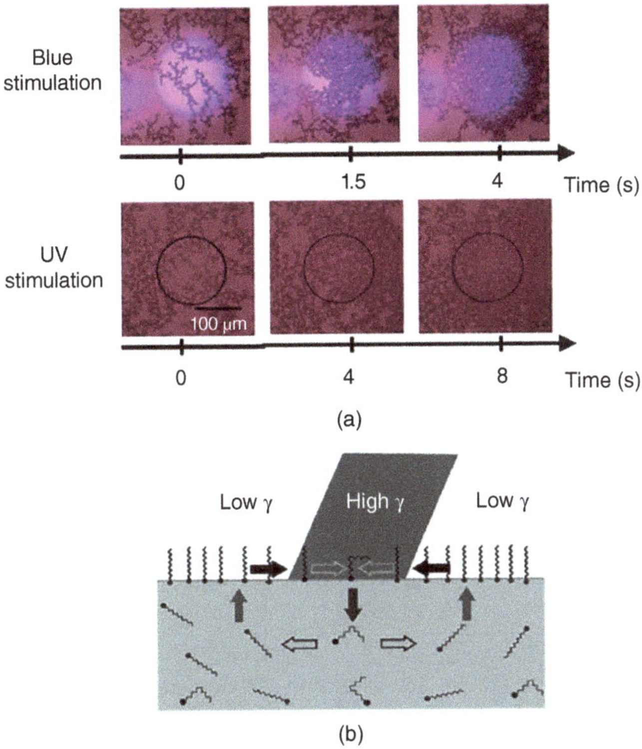

Figure 3.10 Marangoni effect – a solution containing photosurfactant is placed under a microscope and talc particles are spread at the surface. Blue or UV light is then passed through the objective of the microscope to create a light spot (ca. $I_{blue} \sim 100\,mW\,cm^{-2}$ $I_{UV} \sim 20\,mW\,cm^{-2}$). The total concentration of surfactants is 1.15 mM. Outside the spot a majority of *trans* isomers are adsorbed. Inside the spot, the surface tension rapidly increases because *trans* isomers are converted to *cis* isomers, which rapidly desorb. This surface tension gradient results in a Marangoni flow from the outside to the inside of the spot and gathers colloidal particles toward the spot. (a) Photograph of the fast aggregation of the talc particles under the UV or the blue spot. (b) Schematic diagram representing the mechanism causing the aggregation of the particles. The open arrows represent the photoconversion reaction and the filled arrows represent the transfer of molecules via Marangoni flows and adsorption/desorption fluxes. *Source*: Figure reproduced from [49].

a microfluidic channel, providing the basis for a new class of continuous, microscale separations processes. Another ferrocenyl surfactant, BFDMA, has been used to provide redox control over DNA delivery to cells. DNA forms lipoplexes with both reduced and oxidized BFDMA, but only lipoplexes with reduced BFDMA are able to cross the cell membrane. This permits the delivery of DNA to cells upon application of a redox trigger, such as ascorbic acid (vitamin C). Similarly, light-responsive surfactants have also been used to demonstrate photocontrol of both interfacial and bulk solution properties.

A particularly interesting demonstration is the use of photosurfactants to generate Marangoni flows via light-induced desorption of surfactant from the air-water interface. Photoresponsive surfactants have also been used to transport DNA into cells and to subsequently release DNA from lipoplexes within the intercellular environment by illumination. Overall, these studies illustrate the wide range of possibilities that result from the incorporation of redox-active and light-sensitive functional groups into surfactants.

Acknowledgments

This work was funded by the National Science Foundation through grant CBET- 1263970. Partial support is also acknowledged by the ARO (through W911NF-14-1-0140 and W911NF-11-1-0251) and the NSF (DMR-1121288). Use of shared experimental facilities funding by the Wisconsin MRSEC is also gratefully acknowledged.

References

1 Eastoe, J. and Vesperinas, A. (2005). Self-assembly of light-sensitive surfactants. *Soft Matter* 1: 338–347.
2 Liu, X. and Abbott, N.L. (2009). Spatial and temporal control of surfactant systems. *J. Colloid Interface Sci.* 339: 1–18.
3 Brown, P., Butts, C.P., and Eastoe, J. (2013). Stimuli-responsive surfactants. *Soft Matter* 9: 2365–2374.
4 Mura, S., Nicolas, J., and Couvreur, P. (2013). Stimuli-responsive nanocarriers for drug delivery. *Nat. Mater.* 12: 991–1003.
5 Baumgartner, E. and Fuhrhop, J. (1980). Vesicles with a monolayer, redox-active membrane. *Angew. Chem. lnl. Ed. Engl.* 19: 550–551.
6 Brugger, P. and Grätzel, M. (1980). Light-induced charge separation by functional micellar assemblies. *J. Am. Chem. Soc.* 102: 2461–2463.
7 Grätzel, M. (1980). Light-induced charge separation and water cleavage in microheterogeneous aqueous systems. *Farad. Discuss.* 70: 359–374.
8 Hoshino, K. and Saji, T. (1987). Electrochemical formation of thin film of viologen by disruption of micelles. *Chem. Lett.* 7: 1439–1442.
9 Saji, T., Hoshino, K., and Aoyagui, S. (1985). Reversible formation and disruption of micelles by control of the redox state of the surfactant tail group. *J. Chem. Soc. Chem. Commun.* 13: 865–866.
10 Saji, T., Hoshino, K., and Aoyagui, S. (1985). Reversible formation and disruption of micelles by control of the redox state of the surfactant tail group. *J. Am. Chem. Soc.* 107: 6865–6868.

11 Gallardo, B.S., Hwa, M.J., and Abbott, N.L. (1995). In situ and reversible control of the surface activity of ferrocenyl surfactants in aqueous solutions. *Langmuir* 11: 4209–4212.

12 Gallardo, B.S., Metcalfe, K.L., and Abbott, N.L. (1996). Ferrocenyl surfactants at the surface of water: principles for active control of interfacial properties. *Langmuir* 12: 4116–4124.

13 Aydogan, N., Gallardo, B.S., and Abbott, N.L. (1999). A molecular-thermodynamic model for gibbs monolayers formed from redox-active surfactants at the surfaces of aqueous solutions: redox-induced changes in surface tension. *Langmuir* 15: 722–730.

14 Aydogan, N. and Abbott, N.L. (2001). Comparison of the surface activity and bulk aggregation of ferrocenyl surfactants with cationic and anionic headgroups. *Langmuir* 17: 5703–5706.

15 Ansari, A.A., Kamil, M., and Kabir-ud-Din (2013). Polymer-surfactant interactions and the effect of tail size variation on micellization process of cationic ATAB surfactants in aqueous medium. *J. Dispersion Sci. Technol.* 34: 722–730.

16 Acosta, E.J., Mesbah, A., and Tsui, T. (2006). Surface activity of mixtures of dodecyl trimethyl ammonium bromide with sodium perfluorooctanoate and sodium octanoate. *J. Surfactants Deterg.* 9: 367–376.

17 Kaler, E.W., Herrington, K.L., Murthy, A.K. et al. (1992). Phase behavior and structures of mixtures of anionic and cationic surfactants. *J. Phys. Chem.* 96: 6698–6707.

18 Hata, S., Takahashi, H., Takahashi, Y. et al. (2014). Control of dual stimuli-responsive vesicle formation in aqueous solutions of single-tailed ferrocenyl surfactant by varying pH and redox conditions. *J. Oleo. Sci.* 63: 239–248.

19 Tsuchiya, K., Orihara, Y., Kondo, Y. et al. (2004). Control of viscoelasticity using redox reaction. *J. Am. Chem. Soc.* 126: 12282–12283.

20 Hays, M.E. and Abbott, N.L. (2005). Electrochemical control of the interactions of polymers and redox-active surfactants. *Langmuir* 21: 12007–12015.

21 Hays, M.E., Jewell, C.M., Lynn, D.M. et al. (2007). Reversible condensation of DNA using a redox-active surfactant. *Langmuir* 23: 5609–5614.

22 Hays, M.E., Jewell, C.M., Kondo, Y. et al. (2007). Lipoplexes formed by DNA and ferrocenyl lipids: effect of lipid oxidation state on size, internal dynamics, and ζ-potential. *Biophys. J.* 93: 4414–4424.

23 Naldini, L., Blömer, U., Gallay, P. et al. (1996). In vivo gene delivery and stable transduction of nondividing cells by a lentiviral vector. *Science* 272: 263–267.

24 Felgner, J.H., Kumar, R., Sridhar, C.N. et al. (1994). Enhanced gene delivery and mechanism studies with a novel series of cationic lipid formulations. *J. Biol. Chem.* 269: 2550–2561.

25 Abbott, N.L., Jewell, C.M., Hays, M.E. et al. (2005). Ferrocene-containing cationic lipids: influence of redox state on cell transfection. *J. Am. Chem. Soc.* 127: 11576–11577.

26 Jewell, C.M., Hays, M.E., Kondo, Y. et al. (2008). Chemical activation of lipoplexes formed from DNA and a redox-active, ferrocene-containing cationic lipid. *Bioconjugate Chem.* 19: 2120–2128.

27 Aytar, B.S., Muller, J.P.E., Golan, S. et al. (2012). Addition of ascorbic acid to the extracellular environment activates lipoplexes of a ferrocenyl lipid and promotes cell transfection. *J. Control. Release* 157: 249–259.

28 Aytar, B.S., Muller, J.P.E., Kondo, Y. et al. (2013). Spatial control of cell transfection using soluble or solid-phase redox agents and a redox-active ferrocenyl lipid. *ACS Appl. Mater. Interfaces* 5: 8283–8288.

29 Gallardo, B.S., Gupta, V.K., Eagerton, F.D. et al. (1999). Electrochemical principles for active control of liquids on submillimeter scales. *Science* 283: 57–60.

30 Liu, X. and Abbott, N.L. (2009). Electrochemical generation of gradients in surfactant concentration across microfluidic channels. *Anal. Chem.* 81: 772–781.

31 Liu, X. and Abbott, N.L. (2011). Lateral transport of solutes in microfluidic channels using electrochemically generated gradients in redox-active surfactants. *Anal. Chem.* 83: 3033–3041.

32 Fisher, E., Frankel, M., and Wolovsky, R.J. (1955). Wavelength dependence of photoisomerization equilibria in azocompounds. *Chem. Phys.* 23: 1367.

33 Mysliwiec, J., Sznitko, L., Bartkiewicz, S. et al. (2009). Amplified spontaneous emission in the spiropyran-biopolymer based system. *Appl. Phys. Lett.* 94: 241106.

34 Sakai, K., Imaizumi, Y., Oguchi, T. et al. (2010). Adsorption characteristics of spiropyran-modified cationic surfactants at the silica/aqueous solution interface. *Langmuir* 26: 9283–9292.

35 Sakai, H., Ebana, H., Sakai, K. et al. (2007). Photo-isomerization of spiropyran-modified cationic surfactants. *J. Colloid Interface Sci.* 316: 1027–1030.

36 Liu, S., Fujihira, M., and Saji, T. (1994). Formation of an organic thin film by photochemical isomerization of a surfactant with a spiropyran moiety. *J. Chem. Soc. Chem. Commun.* 16: 1855–1856.

37 Whitten, D.G. (1993). Photochemistry and photophysics of trans-stilbene and related alkenes in surfactant assemblies. *Acc. Chem. Res.* 26: 502–509.

38 Mooney, W.F. III, Brown, P.E., Russell, J.C. et al. (1984). Photochemical reactions in organized assemblies. 37. Photochemistry and photophysics of surfactant trans-stilbenes in supported multilayers and films at the air-water interface. *J. Am. Chem. Soc.* 106: 5659–5667.

39 Guo, X., Lin, L., and Guo, R. (2007). Effect of surfactant structure on catalysis of microemulsion for photoisomerization of trans-stilbene. *Chin. Chem. Lett.* 18: 1265–1268.

40 Eastoe, J., Dominguez, M.S., Wyatt, P. et al. (2004). UV causes dramatic changes in aggregation with mixtures of photoactive and inert surfactants. *Langmuir* 20: 6120–6126.

41 Shin, J.Y. and Abbott, N.L. (1999). Using light to control dynamic surface tensions of aqueous solutions of water soluble surfactants. *Langmuir* 15: 4404–4410.

42 Hubbard, F.P., Santonicola, G., Kaler, E.W. et al. (2005). Small-angle neutron scattering from mixtures of sodium dodecyl sulfate and a cationic, bolaform surfactant containing azobenzene. *Langmuir* 21: 6131–6136.

43 Hubbard, F.P. and Abbott, N.L. (2007). Effect of light on self-assembly of aqueous mixtures of sodium dodecyl sulfate and a cationic, bolaform surfactant containing azobenzene. *Langmuir* 23: 4819–4829.

44 Lu, Y., Zhou, T., Fan, Q. et al. (2013). Light-responsive viscoelastic fluids based on anionic wormlike micelles. *J. Colloid Interface Sci.* 412: 107–111.

45 Lu, Y.-C., Le Ny, A.-L.M., Schmidt, J. et al. (2009). Photo-assisted gene delivery using light-responsive catanionic vesicles. *Langmuir* 25: 5713–5724.

46 Zakrevskyy, Y., Roxlau, J.,.a., Brezesinski, G. et al. (2014). Photosensitive surfactants: micellization and interaction with DNA. *J. Chem. Phys.* 140: 044906.

47 Zakrevskyy, Y., Titov, E., Lomadze, N. et al. (2014). Phase diagrams of DNA–photosensitive surfactant complexes: effect of ionic strength and surfactant structure. *J. Chem. Phys.* 141: 164904.

48 Xu, L., Feng, L., Hao, J. et al. (2015). Controlling the capture and release of DNA with a dual-responsive cationic surfactant. *ACS Appl. Mater. Interfaces* 7: 8876–8885.

49 Chevallier, E., Mamane, A., Stone, H.A. et al. (2011). Pumping-out photo-surfactants from an air–water interface using light. *Soft Matter* 7: 7866–7874.

4

Self-Assembly and Primitive Membrane Formation: Between Stability and Dynamism

Martin M. Hanczyc[1,2] *and Pierre-Alain Monnard*[3]

[1] *Università degli Studi di Trento, Centre for Integrative Biology (CIBIO), Via Sommarive, 9, Trento, Italy*
[2] *Chemical and Biological Engineering, University of New Mexico, Albuquerque, NM, USA*
[3] *University of Southern Denmark, Department of Physics, Chemistry, and Pharmacy, Odense M, Denmark*

4.1 Introduction

Amphiphile molecules are molecules of a dual nature, named from the Greek "amphis" and "philia," which means "both" and "love," respectively. Their duality is due to the covalent bonding of a hydrophobic and a hydrophilic moiety in the same molecule. The hydrophobic part usually consists of hydrocarbon chemical structures (linear, branched, and cyclic, Figure 4.1-I), also called the tail, in which carbon atoms are linked to their neighbors via single (saturated) or double (unsaturated) bonds. The hydrophilic part is composed of polar or charged functional groups (Figure 4.1-II), also called the headgroup. This dual nature causes these molecules to spontaneously self-assemble in water into aggregates that expose hydrophilic (water-loving) headgroups to the water. At the same time the hydrocarbon moieties, which do not easily mix with water, are packed together away from the water as they are hydrophobic (water-fearing). Various classes of amphiphile are known that are characterized by the nature of their hydrophilic and hydrophobic moieties (e.g. size and charge of headgroups and number and length of hydrophilic hydrocarbons chains, respectively; Figure 4.1-II). We will, however, focus our chapter on the single hydrocarbon chain amphiphiles (hereafter SCAs).

Historically, the lipid membrane was considered an oily boundary that held together the living protoplasm of the living cell [1]. This view slowly evolved from a static and rather uninteresting wall-like structure into a fluid and dynamic membrane that included crucial functional molecules such as transmembrane proteins [2]. Once the true complexity of the cell membrane was appreciated, the coupling between the external complex environment and the internal complex cellular machinery could be understood and studied.

Self-Assembly: From Surfactants to Nanoparticles, First Edition. Edited by Ramanathan Nagarajan.

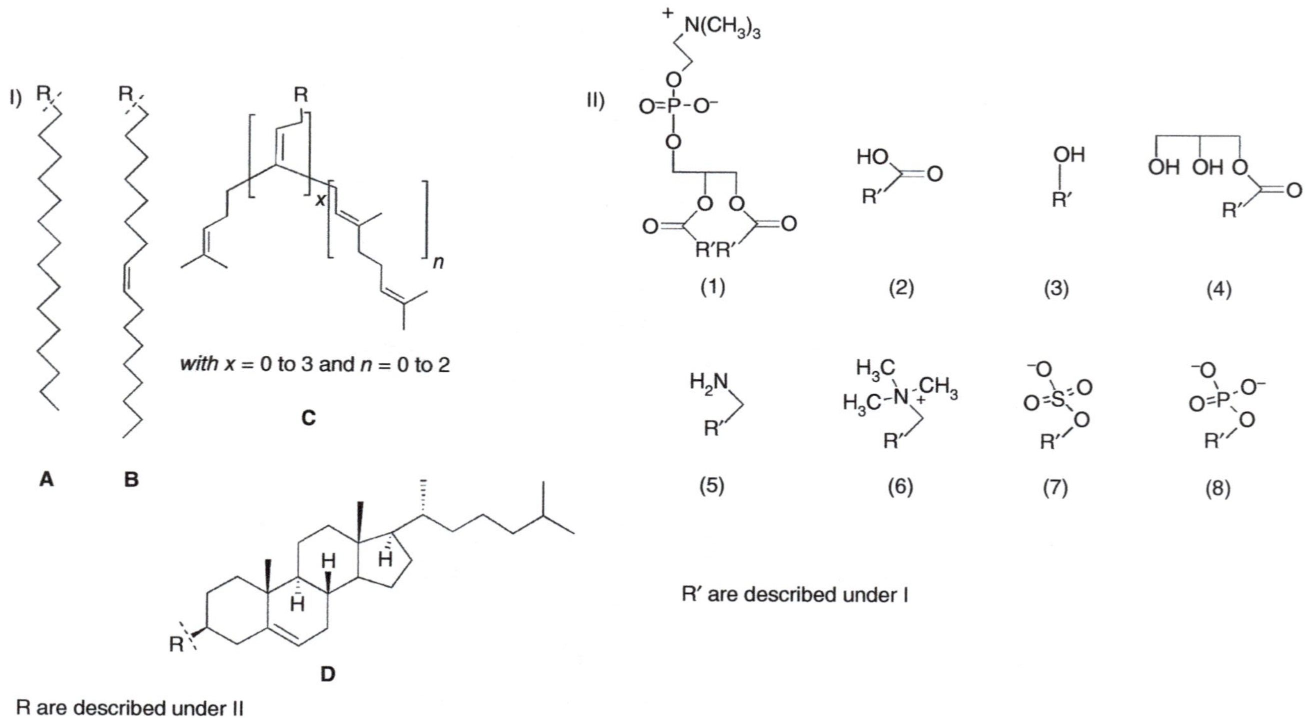

Figure 4.1 Chemical structures of amphiphile molecules. (I) Hydrophobic moieties (tails): **A** pentadecyl, **B** heptadec-8-enyl, **C** isoprenyl derivatives, **D** cholesteryl. (II) Hydrophilic functional groups (headgroups): **1** glycerol phosphocholine, **2** acid, **3** alcohol, **4** glycerol, **5** amine, **6** trimethyl ammonium, **7** sulfate, **8** phosphate. Note that the fatty acids are also classified according to their hydrocarbon chain length (i.e. the number of carbon atoms) as small-hydrocarbon-chain fatty acid (SCFA) when their aliphatic chain is less than 6 carbons, medium-hydrocarbon-chain fatty acid (MCFA) with 6–12 carbons, and long-hydrocarbon-chain fatty acid (LCFA) with more than 13 carbons.

Analytical techniques are now applied to the membranes of living cells in the new field of lipidomics. Among other insights, such investigations revealed both the high diversity of membrane components and how the membrane composition is of prime importance in many biological functions. Interestingly, it has also underlined the fact that lipid functions, such as the building of membranes, are in fact controlled by basic physical principles (e.g. hydrophobicity, electrostatic interactions), rather than directed processes encoded in the cellular genetic material. This nonencoded character of the amphiphile self-assembly is important for the understanding and modeling not only of extant biological membranes but also of the emergence of life on the early Earth along with the evolutionary steps that might have led to the emergence of contemporary cells.

In the *origins of life* research field, experimental models that capture some of the systemic and functional properties of a cell, albeit at a lower level of complexity, are called "protocells." Here we define the basic features of a protocell and by extension the functions protocell models should possess. This task is relatively difficult because this necessitates a definition of *life* itself, and a universal consensus definition of life [3, 4] is still elusive. Many groups of researchers have therefore adopted operational definitions of life to serve as criteria for their investigations toward creating a protocell model in the laboratory. In general, researchers agree that a protocell, by analogy to contemporary biological cells, should be composed of three main components: a compartment, a reaction network, and an information component. The integration of these components should then result in a chemical system that exhibits the emergent properties of self-maintenance, self-replication and evolvability – the hallmarks of contemporary living organisms. Such an operational description of a living cell may be followed to create a living protocell in the laboratory.

Many different molecules could act as amphiphiles to build the membrane of a protocell compartment [5, 6] although alternative compartments can also be considered [7–11]. However, in sharp contrast to living organisms that are capable of synthetizing their own lipids, e.g. phospholipids (Figure 4.1-II **1**), a protocell at least in the early stage of the emergence of life would have likely relied on amphiphiles available in the environment. Due to the lack of direct evidence, the exact composition of these primitive membranes is still an open question. Nonetheless, it seems likely that the early amphiphiles were simpler than modern phospholipids because of the complexity of their nonenzymatic synthesis [12].

The investigation of the carbon contents of carbonaceous chrondrite meteorites [13–20], have established that amphiphile molecules that are composed of a single hydrocarbon chain with a polar headgroup, mostly an acid function, were available, however, as minor components of rather complex mixtures. Of particular importance is the presence of monocarboxylic acids or fatty acids in these mixtures, as the modern phospholipids are composed of fatty acids. The presence of amphiphiles with different functional headgroups, like phosphate, phosphonate, sulfate, sulfonates, amines, and ammoniums, can be also inferred because traces of short alkyl chains with a functional group containing phosphorus or nitrogen have been found in carbonaceous meteorites even though these particular molecules cannot self-assemble [21–23]. To complement the results from meteorite anaylsis, experimental synthesis under plausible prebiotic conditions have yielded SCAs with acid functions [24, 25], as well as phosphate headgroups [26].

Therefore single-hydrocarbon chain amphiphiles are of interest in studying the formation of membranes in a prebiotic context because (i) they are constituents of modern lipids, (ii) they are present in organics samples extracted from extraterrestrial sources, (iii) they have been created in simulated early earth organic synthesis, and finally (iv) in some cases the single-hydrocarbon chain amphiphiles are sufficient for the formation of bilayer membranes. Single-hydrocarbon chain amphiphiles therefore are the molecules of choice for many research groups trying to create a protocell.

4.2 Basis of Self-Assembly of Single-Hydrocarbon-Chain Amphiphiles

As mentioned previously, many different single-hydrocarbon-chain amphiphiles suspended in an aqueous medium will spontaneously self-assemble into aggregates given the right conditions (pH, ionic strength, temperature, and pressure). The propensity for an amphiphile molecule to self-assemble is promoted by three main types of interactions in order of decreasing impact: (i) van der Waals forces between the hydrocarbon chains and hydrophobic effect; (ii) headgroup-to-headgroup interactions (either through H-bonds or electrostatic pairing); and (iii) interactions between the amphiphile headgroups and solutes, e.g. counterions, present in the aqueous environment or/and the water molecules themselves.

In addition, the assembly depends on geometrical constraints of amphiphilic molecules: head size, tail length, the degree of tail unsaturation, and conformational arrangements. All these factors, and most prominently temperature, determine an effective size and shape of the lipid molecules, which in turn determine the tendency of amphiphiles to form particular types of aggregates.

4.2.1 van der Waals Forces and Hydrophobic Effect

Water is the "matrix of life," and its involvement in cellular interactions goes beyond being a simple solvent. Water molecules form an extensive network of hydrogen bonds with their neighbors (on average, four at room temperature and 1 atm pressure) due to the sp^3 hybridization of oxygen [27]. The hydrogen bonds are constantly breaking and re-forming, which leads to a strong bulk stabilization effect by entropy. When oily hydrocarbons that only interact via the van der Waals (dipole–dipole) attractive forces are present in an aqueous solution, they disrupt the hydrogen-bond network; hence, the total energy of the system is raised. As a result, the system equilibrates such as to recover the network by diminishing the overall oil–water interfacial contact. This energy minimization of the system can eventually lead to phase separation, known as the hydrophobic effect.

The hydrophobic interactions for hydrocarbon chains are proportional to the length of the chains, as are the van der Waals attractive forces. Thus, the energy, which is required to take a hydrocarbon chain of an amphiphile out of a bulk hydrocarbon phase and insert it into water, can approximated as the sum of the contribution of *n* CH_2 groups and the CH_3 group [28].

4.2.2 Headgroup-to-Headgroup Interactions

The interactions between amphiphile headgroups will contribute to the self-assembly of amphiphile structures. Two main interaction types dominate the stabilizing interactions in the case of SCAs: hydrogen bonds and electrostatic pairs. The polarity of many simple amphiphile headgroups, illustrated in Figure 4.1-II, can be altered by changing the medium conditions. Indeed, acid, phosphate, and sulfate as well as amine headgroups can act as acids and base, respectively. These amphiphiles will co-exist in protonated/polar and deprotonated/charged states, with the relative abundance of each species at a given pH depending on the pK_a or pK_b of the amphiphile.

The state of protonation in an acidic amphiphile suspension is essential as it sets the conditions needed for structure formation: in a relatively narrow pH region corresponding to approximately the *apparent* pK_a of an acid (e.g. pH 7–9 for fatty acids with hydrocarbon chains of 8–18 carbons [29]), a tight packing of the amphiphiles can be achieved, resulting in bilayer structure formation [30, 31]. In these fatty acid structures, each $RCOO^-$ group is stabilized by strong hydrogen bonds to the neighboring neutral molecules, $RCOO^-$— $(RCOOH)_n$, an arrangement thermodynamically favored over the hydrated-ion/solvent assemblies, $RCOO^-$—$(H_2O)_n$ [18].

Strong hydrogen bonding in aggregates lowers the acidity of the fatty acid headgroups and therefore increases their pK_a as compared to the pK_a of acetic acid ($\approx$4.76). Hence, the shift of the pK_a is one of the characteristics of the formation of aggregates or protoaggregates in an aqueous medium.

The same process also applies to phosphate amphiphiles, however at lower pH ($pK_a \approx 2$), even though these amphiphiles are polyacidic [32–34]. The partial protonation of amine amphiphile does not lead to the formation of stabilizing hydrogen bonds between the headgroups. However, both phosphates and amine headgroups can stabilize bilayer structures by forming ion pairs with amphiphile headgroups of opposite charge [34–38]. In such a case, the stabilization can be ascribed to a reduction of the repulsive interactions between headgroups bearing the same charge through the intercalations of the oppositely charged amphiphiles in conjunction with the hydrocarbon-chain stabilization processes.

4.2.3 Interactions Between the Amphiphile Headgroups and Solute/Solvent Molecules

It is a common observation that divalent cations, such as those found in "hard water," can precipitate soaps, fatty acids, and other amphiphilic compounds that have the potential to form supramolecular structures. In general, any charged or H-bond forming solutes can potentially affect the self-assembly process [39, 40], depending on the nature of the solutes (charge sign, number of charges for polyelectrolytes, propensity to form H-bonds) and the investigation conditions (pH, pressure, and temperature) [41–44]. The observed effects range from structure formation support, enhanced structure stability, alterations of the aggregate morphology or type (local modification of the effective shapes of the lipids leading to changes in curvature, see below) to complete disruption of the aggregates and, in extreme cases, precipitation of the amphiphile-solute pairs out of the suspension.

4.3 Types of Structures

When dispersed in an aqueous medium under the right conditions, single-chain amphiphiles can assume various aggregation morphologies: micelles, stacked monolayers or bilayers in a crystalline state, vesicles, emulsions, and hexagonal phases (Figure 4.2). These aggregates span a large size range from a few nanometers (micelles) to tens of micrometers in diameter (vesicles). Alterations in the chemical and physical conditions can lead to the shape transformation or shifting between aggregate types, which are often reversible [39].

Two numerical parameters are used to characterize the two main types of aggregates. The first is the critical aggregate concentration (CAC). The CAC represents a phase transition and is a well-defined thermodynamic variable. This can be referred to as the CMC for *micelles* and CVC for *vesicles*. The second parameter is the packing parameter, which is a convenient, however putative, aggregate-dependent measure of the shape of amphiphile molecules [46]. These two parameters can define the phase space of supramolecular structures for both single amphiphiles and mixtures of amphiphiles.

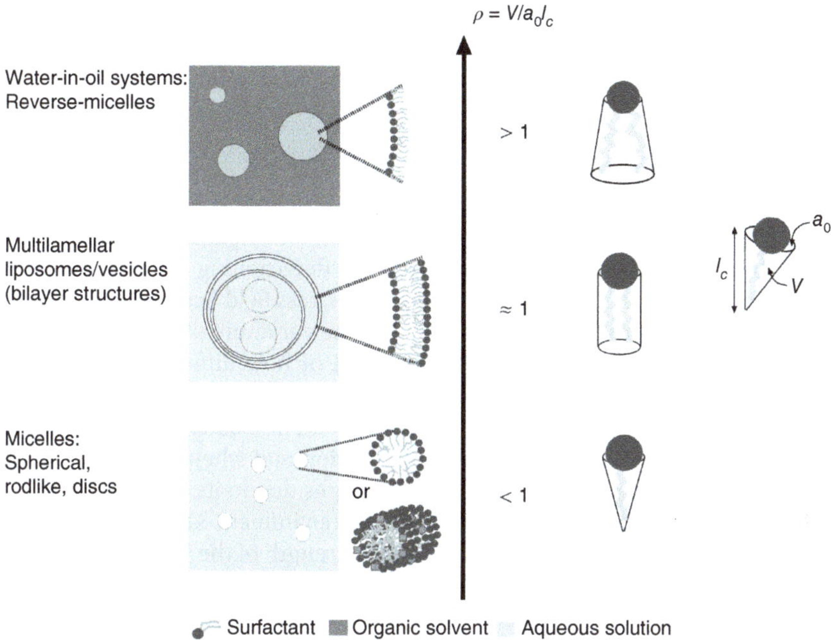

Figure 4.2 Schematic view of self-assembled structures of single-chain amphiphiles typically used for protocell design. *Source*: Adapted from [45]. Emulsion compartments, i.e. surfactant coated oil droplets can also be used as containers (not shown). The packing parameter is a putative, aggregate-dependent measure of the shape of lipid molecules. For a more detailed description of all possible amphiphile structures, see [45].

4.3.1 Critical Aggregate Concentration

The CAC is defined as the transition concentration of amphiphiles in an aqueous suspension needed to observe the formation of a type of aggregate. In the case of fatty acid vesicles, the critical concentrations for all types of fatty acids is relatively high (on the order of 100 μM to 100 mM) compared to phospholipids (10^{-12}–10^{-9} M). This high critical concentration will have a significant impact on the behavior/stability of the structures. For example, a vesicle suspension of 30 mM decanoic acid (CVC = 10.7 mM at pH 7, 100 mM phosphate buffer, room temperature [47]) will always have a constant concentration of free, single molecule or monomer amphiphiles around 10.7 mM, with the remaining 19.3 mM of decanoic acid bound in vesicle structures. However, the membrane bound fatty acids are not fixed permanently. Rather the fatty acid molecules will tend to rapidly exchange back and forth between the freely diffusing and the structure bound states.

4.3.2 Packing Parameter

A packing parameter that describes the effective shape of lipid amphiphiles is given by

$$\rho = \frac{V_o}{a_o l_o}$$

where V_0 is the surfactant tail volume, l_0 is the tail length, and a_0 is the equilibrium area per molecule at the aggregate surface (Figure 4.2). Although a subjective measure, the parameter is useful in determining the propensity of particular lipid molecules to affect a curvature stress field (e.g. a greater repulsion between headgroups generally increases the curvature of the surface). The packing parameter captures the general trend of how different amphiphiles might have a tendency to form certain types of supramolecular structure, but the term is not specific enough to be predictive. For example, the parameter does not adequately capture headgroup repulsion and when in the fluid state an amphiphile may adopt several different shapes due to its flexibility [48]. In addition the properties of the amphiphile are often influenced by the properties of the solvent, the temperature, and the ionic strength of the solvent. However, general trends can be captured using this parameter.

The packing parameter values $\rho < 1$, $\rho = 1$, and $\rho > 1$ correspond to conical, planar, and inverted conical shapes, respectively (Figure 4.2). Micelles are cylindrical or spherical structures with a pure hydrophobic core and a hydrophilic outer shell, typically of few nanometers in diameter. They are formed from ionic surfactants ($\rho < 1$) above the *CMC* and above the Krafft temperature. These amphiphiles are essentially fully deprotonated (high *pH*) and thereby experience a large electrostatic repulsion. Micelles rapidly form and dissociate (milliseconds to maximally minutes) [49].

Vesicles are closed bilayer structures, which are characterized by packing parameter values ρ close to 1. The more ρ deviates from 1 (towards conical shapes) the less stable the bilayers become as the curvature stress that builds up in each of the monolayers forces nonlamellar structures (Figure 4.2). Depending on the extent by which two monolayers differ in their curvatures, bilayers can attain a spontaneous curvature and an inherent tendency to bend. The resulting structure can be uni- or multilamellar, the latter having internal water core surrounded by several concentric amphiphile shells with a thin layer of water between each of them. The diameter of the vesicles ranges from 50 nm for small unilamellar vesicles (SUV) to tens of micrometers in size for giant vesicles (GV) [50, 51]. Various closed membrane shapes are theoretically predicted based on the minimization of total bending energy [52, 53].

In addition to the micelle and vesicle supramolecular structures, another common structure is the emulsion. Emulsion compartments can be made by mixing an aqueous solution and an apolar phase (e.g. organic solvent [54]), in the presence of amphiphilic molecules with $\rho > 1$. The interface between the two phases will be stabilized by the amphiphiles. Depending on the ratio of the two phases, water-in-oil or oil-in-water emulsion compartments will be produced when the main phase is the oil/apolar solvent or the aqueous solution, respectively. Their sizes range from several nanometers in microemulsions (also called reverse-micelles in the water-in-oil case) to several microns in macroemulsions. The resulting size depends on the ratio of water to surfactant, the specific properties of the amphiphile, the mechanical energy injected in the system (e.g. by stirring), and the temperature of the system. Fatty acids and other SCAs can be used to form stable emulsions between the two liquid phases.

4.4 Self-Assembly of a Single Type of Single-Hydrocarbon-Chain Amphiphile

4.4.1 Single Species of Single-Hydrocarbon-Chain Amphiphile

Among all possible amphiphiles, only few types of amphiphiles have been investigated as single amphiphile building blocks for protocell vesicles: fatty acids and acyl phosphates. Other SCAs, such as those containing amines, ammonium, and sulfate as headgroups, tend to preferentially form micelles structures.

Fatty acids and acyl phosphates have been shown to self-assemble into vesicles around their pK_a at pH 7–9 and 2, respectively [32, 33, 55]. At higher pH, the presence of micelles has been confirmed, while at lower pH in the case of fatty acid, the formation of oil droplets, which over time will coalesce into a separated oil phase, is reported due to the higher hydrophobicity of the protonated species (the predominant state of fatty acids).

The pH range for vesiculation clearly indicates the need for coexisting protonated and deprotonated species, i.e. the importance of H-bonding between headgroups. Interestingly, while significantly contributing to the stabilization of vesicles, these interactions are not sufficient to obviate the need for a strong stabilizing contribution of the hydrocarbon chain packing through van der Waals interactions. Indeed, the study of the dependence on the number of carbon in the hydrophobic moiety has revealed that vesicles can only be formed by fatty acids with 8 and higher number of carbons at prebiotically plausible concentrations. As the chain length decreases, the CVC required to the formation of vesicles increases extensively from 100 μM (oleic acid, 18 carbons with one unsaturated C—C bond) to 150 mM (octanoic acid, 8 saturated carbons) [47, 56, 57].

A same general behavior should be expected for phosphate amphiphiles, but only few studies have been reported. Phosphate headgroups have two pK_as, but vesiculation only occurs around the lower pK_a (pH − 2), a fact that underlines the importance of repulsive interactions between charged headgroups. Contrary to fatty acids, the second pK_a is associated with the presence of anionic and dianionic species, which seem to destabilize the vesicles [34, 58].

Fatty acid vesicle formation and stability was also investigated in terms of the influence of the chemical composition of the aqueous medium and its temperature, as well as pressure. It was established that high salt concentrations, even monovalent cations like sodium ions, can depending on the hydrocarbon chain length of the fatty acids, associate with the molecules, and induce morphological changes in the case of cesium [40] or outright precipitation of the amphiphiles in the case of potassium and to a lesser extent sodium [39]. The effect is stronger in the case of multivalent cations, such as magnesium and calcium. This detrimental influence of divalent cations can be observed in everyday life (limestone buildup on washing-machine coils). However, the effects of multivalent cations can be reversed by chelating the metal ions with multidentate ligands: the presence of citrate [59] or ethylenediaminetetraacetic acid (EDTA) [39] will both be able to preserve the integrity of medium chain fatty acid (MCFA, 8–10 carbons long) vesicles and allow fatty acids precipitated or flocculated by metal ions to reform bilayers. Similarly, metal ion complexes, such as ruthenium trisbipyridine [60, 61], will electrostatically interact with bilayers without disrupting them, at least at low concentrations that would be detrimental if the ruthenium atoms were present as uncoordinated cations (unpublished observations).

Temperature also affects the stability of the vesicles in a similar way: decanoic acid systems will precipitate at 45 °C and oleic acid systems at temperatures in excess of 55 °C. Even at 20 °C, decanoic acid vesicles will extensively exchange bilayer molecules. When mixing two vesicle systems, each containing fluorescent dyes, the resulting mixing of membrane components can be quantitated by Förster resonance energy transfer (FRET). For decanoic acid vesicles, mixing of up to 40% of the vesicle membranes components can occur within a day [47].

Other SCAs, such as those containing amines, ammonium [38] and sulfate as headgroups [30], tend to form micelles in aqueous medium. Their use as detergents has been long established, like that of sodium dodecylsulfate (SDS).

4.4.2 Mixtures of Single-Hydrocarbon-Chain Amphiphiles

As mentioned earlier, single-species structures were very unlikely on the early Earth as most plausible sources of amphiphilic materials would have delivered a collection of amphiphiles as well as other partially or completely hydrophobic molecules. This has prompted researchers to look at the self-assembly of mixtures of amphiphiles.

4.4.2.1 Mixtures of Amphiphiles with the Same Functional Headgroups

The effects of small amount of longer chain fatty acid (LCFA) on the self-assembly of MCFA can be assessed. In their investigations, Szostak et al. [57] established that the presence of oleic acid at low molecular fractions (less 0.2) will strongly stabilize the formation of vesicles from decanoic acid, as evidenced by the drop of CAC, even at an unfavorable pH, 1.2 units above the pKa of this fatty acid. This result is comparable to that obtained in oleic acid mixtures containing small amounts of phospholipids [62]. This implies that the stability of protocell containers could be gradually improved as systems themselves evolved to synthesize lipids of higher molecular weight.

Mixtures of small-hydrocarbon-chain fatty acids (SCFAs) and MCFAs are a better representation of plausible early fatty acid composition [63], and they also exhibit enhanced self-assembly patterns. Contrary to expectations, the presence of SCFAs, which are very poor amphiphiles themselves, seem to enhance the propensity of MCFAs (nonanoic and decanoic acids) to aggregate, again according to the observed decrease of CAC [64]. Furthermore, by increasing the number of SCFA types present, the participation of the ever-shorter MCFAs in vesicle formation was observed. Therefore, mixtures of prebiotically plausible monocarboxylic acids would benefit from the presence of short nonvesicle forming molecules that seem to promote vesicle formation from the medium-chain-length fraction.

4.4.2.2 Mixtures of Single-Hydrocarbon Chain Amphiphiles and Neutral Co-surfactants

The reliance of self-assembly process for protonable anionic amphiphiles on the formation of H bonds limits the pH range at which vesicle structures can be formed. This observation led to the exploration the impact of alkyl alcohols [30, 34, 39, 47, 56, 65–67] as a way to provide pH independent H bonds forming species. In the case of alkyl alcohols, e.g. 1-decanol and polyols, e.g. glycerol monodecanoate (GMD), it was established that the presence of 10 and 33 mol%, respectively, of alcohol co-surfactant led to a significant stabilization of the anionic surfactant (acid and phosphate) at pH well above their pK_a. This effect could also be observed in the decreasing CVC of the mixed systems. Above this threshold concentration, the formation of emulsions was often observed, a fact that points to a stabilization through potentially dynamic H bonds involving several anionic and one alcohol molecules. Unfortunately, the exact composition of the bilayers themselves remains to be determined. The existence of H bonds shared by multiple molecules is not without precedent, as Cistola et al. [31] already reported the formation of fatty acid vesicles in a range of deprotonated/protonated species of 3:1 to 1:1.

The mixed structures also were less sensitive to disruptions by metal ions, in the case of glycerol derivatives [39], and remained intact over a larger temperature range, both below and above the Krafft point [47]. Finally, the presence of alcohols in the bilayers reduced in most cases the passive diffusion of hydrophilic solutes [47].

4.4.2.3 Mixtures of Charged Single Hydrocarbon Chain Amphiphiles

The stabilization of SCA structures can also be achieved by reducing the electrostatic repulsion between headgroups bearing the same charge, e.g. by the insertion of co-surfactants with an opposite charge. The presence of ammonium/amine amphiphiles under prebiotic conditions has to be confirmed, but derivatives with shorter hydrocarbons have been identified in the meteorite carbon contents.

Amines and trimethyl ammonium surfactants have been investigated in mixtures containing either acid and phosphate amphiphiles, and several compositions have shown improved self-assembly and stability properties compared to single amphiphile systems. The nature of the charged species differs between both species: the amine group is protonable at pH above 9, whereas the trimethyl ammonium group charge is pH independent. Trimethyl ammonium is a rather bulky group, which might not insert in bilayers as efficiently as its smaller nonmethylated counterpart.

As expected, the protonation state of the amine, i.e. the formation of an ammonium, is determinant in the formation of stable structures. Below its pKa, but well above that of decanoic acid (pH 7.0–7.2), Namani and Deamer [36] could observe vesicles that are in a pH range where decanoic acid alone does not self-assemble into bilayers. The stability of these structures was significantly reduced around pH 7, as a clear phase separation could be observed, likely due to the formation of large oil inclusions in the bilayers or ammonium-coated oil-in-water emulsions. The surprising result was, however, the observation of vesicles at low pH, which cannot be explained by direct ion pair formation between the two headgroups, but might be due to more complex interactions mediated by the borate buffer molecules. In the case of decyl phosphate, the presence of decylamine led to the formation of vesicles over a large pH range of 2–12 [34].

In clear contrast to the amine case, alkyl trimethyl ammonium species do not always stabilize adequately protonable amphiphile bilayers, even though they, too, are positively charged. In the case of fatty acids, several reports point to the formation of stable mixed bilayers with large changes in CVC [38] and temperature stability [35, 37]. In the case of LCFAs, the insertion of long alkyl ammonium species into LCFA structures led to their stabilization at an interesting molar ratio of two acids for each trimethyl ammonium headgroup. The resulting vesicles were stable even during dialysis experiments [37], in clear contrast to the observations reported for decanoic acid and decyl trimethylammonium bromide (DTAB) [38], where DTAB seems to preferentially self-assemble into micelles. Even in cases where DTAB was found necessary for self-assembly, e.g. hexadecanedioic acid/DTAB, DTAB did not present a strong cosurfactant character.

In the case of decyl phosphate/DTAB mixture, investigations of the CVC even at pH 2, where the phosphate species self-assembles alone, revealed an increase in CVC compared to pure phosphate systems [34] Thus, this different behavior hints at a steric hindrance that would prevent a favorable packing of the amphiphiles in a bilayer configuration, although this repulsive interaction between the primary surfactant and the co-surfactant did not hinder the formation of vesicles at pH 2, whose mixed character has yet to be confirmed. Alternatively, the lack dissociation of DTAB into its ion-constituents (DTA^+ and Br), which may be as low as 26% [68], could significantly reduce the net positive charge available in the bilayer, thereby preventing the stabilization of the phosphate amphiphile in a bilayer structure.

4.4.2.4 Mixtures of Single-Chain Amphiphiles and Lipids

Even though prebiotic syntheses of phospholipids were demonstrated, but with low yields [12], the researcher community has largely abandoned the idea of early protocellular membranes composed of these lipids. However, the concept of evolutionary continuity (i.e. the principle that "random changes were selected as a result of their ability to give a selective advantage to the corresponding organisms at every stage of evolution from the very beginning" [69]) obliges us to consider how gradual alterations of pure SCA bilayers (refer to Figure 2 in Monnard and Deamer [70]) could have led to their replacement by phospholipids, and this besides the obvious chemical "filiation."

Szostak et al. [62] investigated mixed phospholipids/oleic acid bilayers, and established that the presence of a small amount of long-acyl-chain phospholipids such as 1,2-Dioleoyl-sn-glycero-3-phosphocholine (DOPC) and 3,4-dihydroxy-phenyl-alanine (DOPA) within a fatty acid membrane significantly reduced the efflux of the fatty acid molecules from the structure. In competition experiments, the presence of phospholipids in the mixed structures led to their more rapid growth by uptake of additional fatty acid molecules compared to the pure fatty acid structures. Moreover, the mixed membranes were less permeable. The authors therefore surmised that chemical systems adventitiously possessing mixed membranes would have had an evolutionary advantage, provided they could develop a more sophisticated "metabolism," but at a cost: the decrease of chemicals diffusing in from the environment. A contradiction to this phospholipid effect can also be found in the work of Chakrabarti et al. [71] that demonstrated that liposomes composed of dilauroylphosphatidylcholine (DLPC) and phosphatidylcholine (PC) with shorter hydrocarbon chains could not retain an enzyme.

Other interesting studies revolve around the question of cholesterol and its potential forebears. Indeed, cholesterol is an essential fluidity regulator in phospholipid membranes [72]. Even though cholesterol is not prebiotic, it was reported that cholesterol could help bilayer formation by LCFAs, such as palmitic acid, at high temperatures and preserve the structures once the samples were cooled down to room temperature [41]. Its effects are therefore comparable to those of alkyl alcohol. The prebiotic chemical mimic might be more difficult to surmise, but the study of mixed membranes with oxidized polycyclic aromatic hydrocarbons (PAHs, see below) derivatives has led to the assumption that these molecules could have been the functional substitute of cholesterol effect in prebiotic membranes [73]; however, a logical chemical filiation is not likely.

4.4.3 Mixtures of Single-Hydrocarbon-Chain Amphiphiles and Other Molecules

Small nonamphiphile molecules can influence SCA self-assembly. In the origins of life, two categories can be considered: molecules that are relevant for the development of precellular compartmentalization and reaction networks, and those that are simply "bystanders." In both cases, their effects depend on their chemical nature. Among the hydrophobic molecules, PAHs [36, 73] and nucleobases [74], both of which can be either found in meteorite and interstellar dust particle carbon content [75, 76] or synthesized in model synthesis schemes, have been shown to enhance the formation of bilayers and stabilize them significantly.

The interest in the PAHs, e.g. pyrene, chrysene, phenanthrenes, and benzopyrenes, is related to primitive light-harvesting systems [64, 77, 78] as these molecules are light active and could therefore function as pigments [79]. The formation of mixed systems from lipid aggregates with inserted PAHs occurs by self-assembly, as PAHs, which can be very hydrophobic, preferentially partition into the hydrophobic core of the lipid aggregates [36, 64, 73, 80]. The propensity to interact would have been strong as the incorporation of up to 10 mol% PAHs in decanoic acid bilayer models demonstrates [73]. Moreover, 1-hydroxypyrene and 9-anthracene carboxylic acid in fatty acid bilayers can significantly lower their permeability of small solutes (up to fourfold), thus exhibiting a cholesterol-like modulating effect. It should be noted here that fatty acid bilayers tend to better accommodate PAHs than the phospholipid counterparts (unpublished observation).

Nucleobases, in particular, the canonical ones, are obviously of significant interest as the appearance of information carriers seems to be a prerequisite to the emergence of cellular life. Black et al. [74] have reported that canonical and some closely related nucleobases will prevent the salt disruption of MCFA vesicles, likely through direct interactions with the bilayers.

The stabilizing interactions with hydrophilic molecules, e.g. guanidine [81, 82], choline [42], or tetrabutylammonium [43] have been reported, which underline the impact of the medium composition on the stability of fatty acid aggregates. LCFAs (e.g. myristic and palmitic acid, C_{14} and C_{16}, respectively) will crystallize as potassium and sodium salts or form waxes below their Krafft point (41 and 61 °C, respectively) that is well above ambient temperatures. However, mixing LCFAs (C_{14} and C_{16}) with guanidine hydrochloride prevents their crystallization at high temperatures, and leads to self-assemblies, such as flat bilayers and elongated micelles, depending on the temperature, which are stable upon cooling of the suspensions.

The recent report on the stabilization of decanoic acid bilayers by sugars is very exciting [74] – in particular, the fact that ribose, the sugar building block of the RNA backbone, exhibited a better protection against flocculation than glucose and xylose. When the effect of nucleobases and ribose were investigated simultaneously, their stabilization effects were shown to be additive. This type of result highlights that simple membranes such as those discussed in this chapter could indeed have been a step in the evolution of the first cells. From the point of view of the selection of specific chemicals during the emergence of biochemistry, the reciprocal support of the development of cellular components may be very significant.

4.4.4 Self-Assembly on Surfaces

The same physical properties of single-hydrocarbon-chain amphiphiles that govern the formation of supramolecular structures in solution also are responsible for how such molecules assemble at a liquid–solid interface. For example, the ability of fatty acids to interact by hydrogen bonding also influences the structure of fatty acids at a solid surface. For hydrophobic surfaces such as graphite, fatty acid physisorption is spontaneous and the resulting monolayer is ordered with rows of dimers formed through hydrogen bonding of the head groups [83]. Accordingly, alkyl amines will form a hydrogen-bonded network at the hydrophobic surface. For charged and metallic surfaces, the interaction and orientation of the amphiphiles is different. It is usually the polar head group of the amphiphile that adsorbs directly to the solid surface leaving the hydrophobic tail free in an air-solid film system [84]. The situation where fatty acids are adsorbed to a solid hydrophilic surface in a liquid–solid system is unclear but likely occurs through covalent bonds, hydrogen bonds, ion exchange, van der Waals forces, and hydrophobic effects [85].

It has been shown that the presence of a mineral surface accelerates the formation of fatty acid vesicles from a solution of micelles. The rate enhancement is drastic and depends little on the type of surface present. Rather the rate enhancement is reliant upon the adsorption of the amphiphiles to a surface. For example, fatty acids tend not to adsorb to the surface of Teflon particles and therefore no vesicle formation rate enhancement is seen. However if the Teflon particles are precoated with a perfluorinated amphiphile, the rate enhancement is present [86].

The interaction of SCAs with solid surfaces has implications for protocell construction and origin of life studies. It has been shown that certain preparations of the mineral montmorillonite can promote the polymerization of RNA monomers in solution into long oligomers of RNA [87]. The same mineral montmorillonite can promote vesicle formation. When the mineral is precoated with RNA, the mineral not only promotes the formation of fatty acid vesicles but also becomes encapsulated inside the resulting vesicles. This system is comprised of a mineral as a catalytic surface that can polymerize RNA as a potential information-carrying polymer and also promote the formation of a fatty acid membrane container around itself. This simple system therefore represents the three main components of life as described previously: a compartment, a reaction network, and an information component. The fatty acid vesicle membrane can certainly be considered a compartment. However this protocell system falls short on the other two criteria. The RNA polymerized likely is mostly random and contains no functional information and the catalytic mineral surface alone may not suffice as the hub of a reaction network. Nevertheless, the basic architecture for a protocell, constructed largely through self-assembly, is apparent.

Another protocell architecture is based on spontaneous formation of coacervates from very slowly diffusing mixtures of molecules [88]. These spheroid assemblies can encapsulate specific organic molecules and functional enzymes, thereby having some lifelike properties [89]. Recently, it was shown that a coacervate formed from cationic peptides/polyelectrolytes and adenosine triphosphate (ATP) or oligo/polyribonucleotides can act as a surface to assemble multilamellar fatty acid membranes. The assembled membrane acts selectively with regard to the uptake of different types of molecules. The hybrid protocell consisting of a fatty acid vesicle membrane enclosing a preformed coacervate can undergo fusion and growth in high ionic strength environments [90].

The propensity for fatty acids and other SCAs to assemble on surfaces not only provides an insight into amphiphile dynamics and self-assembly into membranes, but also shows potential paths forward to creating protocells. In Section 4.6 on dynamism we will discuss how fatty acids on a surface can lead to successive generations of protocellular compartments.

4.5 Catalysis Compartmentalization with Single-Hydrocarbon-Chain Amphiphiles

Amphiphile structures provide a rare opportunity to co-localize or compartmentalize molecules. However, they also provide a unique set of environments for chemical reactions to occur: in the internal volume, in the membrane, and at the lipid/water interface (Figure 4.3). In addition, direct interaction of the

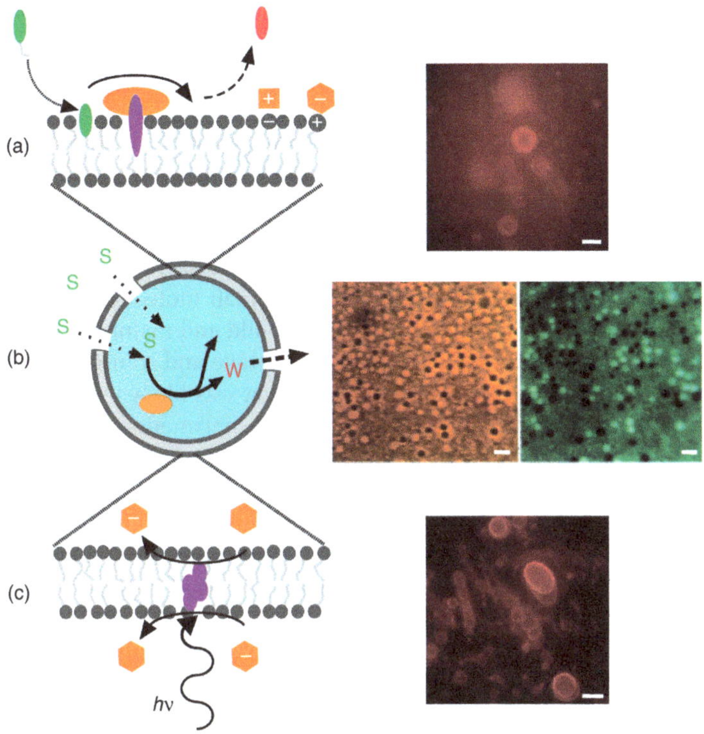

Figure 4.3 Protocell designs: A vesicle can provide a variety of environments for protocell construction and function. On the left side: (a) Metabolic complexes are anchored directly in bilayers with a lipophilic tail or associate with the structures through electrostatic interactions. The resulting protocell system can be called interfacial. (b) Metabolic networks are completely encapsulated into the internal aqueous volume of the vesicle. This is the typical enclosed model. (c) The membrane contains parts of the metabolic system, but the presence of a distinct aqueous lumen is necessary, as is the case for the energy harvesting molecular systems. On the right side: epifluorescence micrographs illustrate the different designs (bar = 10 μm). (a) The interactions between ruthenium trisbipyridine chloride with decanoic acid vesicles. The bilayer interactions with the soluble metal complex allow their observation using the accumulation of the intrinsically fluorescent metal complex. However, the background fluorescence remains high at any concentration. (b) These micrographs show the result of the encapsulation of double stranded DNA fragments (approximately 600 bp in length) according to the dehydration/rehydration method. The bilayers are composed of a mixture of n-dodecanoic acid, n-dodecanol (5 : 1 M ratio) at pH 8. The comparison of the two micrograph types (left picture: phase contrast/right picture: epifluorescence using acridine orange to label the DNA) allows us to distinguish between two vesicle populations: the vesicles with encapsulated DNA have a dark-shaded aspect in the left micrographs and fluorescent (light-shaded) on the right one, while the reverse situation is observed for the empty compartments *Source*: (Micrographs from [70]. Copyright by 2002 Wiley-Liss, Inc). (c) Epifluorescence micrograph shows the intercalation of a hydrophobic dye, Nile Red, added to preformed decanoic acid vesicles. This dye inserts efficiently into the amphiphile bilayers as any hydrophobic molecules, e.g. PAH, will do.

chemical reactions with the container itself may kinetically enhance [91] or inhibit [92] certain chemical reactions.

Protocell development focuses on three types of reactions, which are of paramount importance for the emergence of cells: the synthesis of biopolymers, the synthesis of other protocell building blocks, such as amphiphiles, and the uptake of energy from a primary source, such as light. Biopolymer synthesis is considered central to the origin of life due to the ability of biopolymers to catalyze reactions under mild conditions as well as to encode for genetic information, while the synthesis of other building blocks is necessary for self-maintenance and self-reproduction. In principle, small molecules in the environment can be used by a metabolic protocell to build and maintain itself through an associated metabolism. Protocell energy uptake and transduction is essential for attaining a certain level of autonomy that was likely necessary for the evolution toward more complex systems. Some experimental evidence exists to support scenarios involving membranes as a central participant in these three processes. We will deal here only with the concepts and their embodiment in single-chain amphiphile structures without attempting an in-depth survey of all implementations of phospholipid liposomes (see for review [93–95]), even though the experimental data gathered in these later systems are more numerous.

Two approaches to make a functional compartment have been proposed: the classical enclosed and the interfacial models. In the former, the aqueous lumen of e.g. a vesicle or an emulsion compartment, contains the protocell's chemistry, as shown in Figure 4.3. In the latter, a surface, e.g. a bilayer, is the actual container, co-localizing all of the protocellular components. As proposed by Blobel [96], these two designs might well represent different stages in the development of SCA compartmentalization. Indeed, the interfacial models might have preceded the enclosed models. It should only require association of molecules with the amphiphile structures to achieve co-localization of the components needed to perform reactions, whereas the enclosed model requires an encapsulation process in relatively stable bilayers, and results in limited access to charged substrates due to their low permeability across bilayers.

4.5.1 Enclosed Protocell Models

The enclosed model resembles the basic morphology of all cells (as far as the basic design is considered: an aqueous volume enclosed by a lipid-based boundary is shown in Figure 4.3b), which clearly explains its early investigation. The encapsulation of hydrophilic solutes that would have been necessary to build up a protometabolism could have occurred on the early Earth following perhaps two procedures, dehydration-rehydration and pH vesiculation, that are now commonly used for the self-assembly of single-hydrocarbon-chain vesicles in laboratories.

The dehydration–rehydration method is based on drying of vesicles in the presence of solutes to be encapsulated in the external medium. During drying, solutes will be retained between fragments of bilayers. The molecular arrangement of the bilayers seems to be preserved when vesicles are dehydrated [97]. In the following rehydration step, solutes will be encapsulated within the lumen of the reforming vesicles.

The pH-vesiculation takes advantages of the morphological changes of SCA structures upon formation/breaking of H-bonds [31]. The increase of the pH will disrupt the existing vesicles, while its subsequent lowering will allow for vesicle reformation that also results in the encapsulation of solutes.

Both processes could have occurred in a fluctuating early Earth environment. However, this amphiphile self-assembly, as with any spontaneous process, results in the formation of vesicle suspensions with large variability in terms of the vesicle properties, such as size, multilamellarity, as well as amphiphile composition. More significantly, this compartmentalization leads to a random encapsulation with an overall relatively low efficiency and a large fraction of empty vesicles (Figure 4.3). These factors may not have been so detrimental to the emergence of protocells on the early Earth as they are for the development of efficient and controllable biotechnologies.

The compartmentalization within an aqueous lumen surrounded by bilayers will also reduce the access to charged substrates by passive diffusion. This would have limited the possible protometabolism, even though SCAs build bilayers that are more amenable to the passive diffusion of small molecules [47, 59, 65, 98, 99] than their phospholipid counterparts. This higher permeability might have been advantageous for simple compartmentalized catalytic systems. But in clear contrast to phospholipids, where ATP will cross bilayers and ApUp (adenyl(3′-5′) uridine 3′-monophosphate) will not [100], the enhanced permeability is such that it could have prevented the accumulation of small products with low charge density, and thereby the possibility of their further conversion in more complex substances. Moreover, the relative stability of MCFA vesicles [47] due to continuous fragmentation of their bilayers or exchanges of large fractions of amphiphiles between vesicles could also have constrained the time available for reactions.

Nevertheless, several studies using SCAs as compartment building blocks have shown that chemical reactions that are essential for the emergence of life can be carried out within vesicles. In line with the idea of the RNA-based protocell [101], the synthesis of RNA via nonenzymatic template-directed oligomerization was successfully demonstrated within mixed MCFA vesicles (decanoic acid/decanol/glycerolmonodecanoate) [59, 99]. This is particularly interesting only because the reaction could be carried out within the vesicles, but also because this type of reaction requires magnesium ions as catalysts. In the Mansy et al. study [99], the elongation of a primer-template system (13 and 32 nucleotides long, respectively) could be achieved using the passive diffusion of activated RNA monomers, possibly chelated by magnesium ions. That is, these simple vesicles exhibit a relative selective permeability toward the chemicals that form the reaction system, even though the authors noticed that the MCFA membranes slowed the interaction between the primer-template and activated nucleotides in the vesicle interior. In the second study, the magnesium concentration could be significantly increased (10 mM) using a chelating agent, citrate, without inhibiting the polymerization. That is, the presence of small solutes could have contributed to the stabilization of vesicles beyond the threshold concentration (approximately 4 mM for 30 mM glycerolmonoalkanoate/fatty acid 1:2) that can be reached using mixed bilayers [39, 65].

Chen et al. [65] demonstrated the feasibility of the obvious next step in the development of a protometabolism within the hypothesis of the RNA world. Using again mixed vesicles (myristoleic acid and its glycerol monomyristoleate), they showed that an encapsulated hammerhead ribozyme could be activated by letting its metal cofactor, magnesium, diffuse across the vesicle behavior.

At this point, one can ask whether more complex reaction networks could be implemented within SCA vesicles. The absence of information about plausible protometabolism obliges the researchers to use reconstituted biological systems [70] at the risk of encountering chemical incompatibilities between enzymatic systems and fatty acid vesicle systems with relatively high CVCs. The polymerization of random RNA by polynucleotidyl phosphorylase [98] was performed in oleic acid vesicles supplying the encapsulated enzyme with its ADP substrates by passive diffusion. However, the same reaction was completely inhibited in decanoic acid vesicles (at pH 7.3, unpublished observations). It is however clear from the results obtained in liposomal systems (vesicles composed of phospholipids) that the complexity of the biochemistry that can be carried out in self-assembled systems can reach the level of coupled transcription-translation of proteins. That is, a complex biological network contains about 80 molecular components can be encapsulated. Remarkably, it can be observed that liposomes with a radius of about 100 nm harbor protein expression, and the amount of fluorescent protein in the vesicles can be, on average, about six times greater than in a bulk solution [102].

4.5.2 Interfacial Protocell Models

The difficulties inherent to the usage of an enclosed system for protocellular metabolisms (encapsulation of functions, permeability issue – reduced access to charged substrates and lack of selectivity) have led researchers to propose an alternate approach to co-localization of reaction components (Figure 4.3a). The idea here is to co-localization all chemicals on the interface of a bilayer structure by physically anchoring the entire network to the structure as opposed to encapsulate them inside a semipermeable barrier [103, 104]. In several instances, the anchoring of catalytic molecules and in part their cofactors have been conducive to improved reactions, for example the template-directed RNA polymerase ribozyme exhibited higher activity once anchored to micelles [105]. The advantages of this anchored system are numerous. Nutrients and waste products can freely diffuse to and from the metabolic/informational core of the network. It also eliminates the issue of vesicle lysis and the associated chemical release, as only the bilayer itself needs to stay intact. This system is more flexible in terms of structure type as well, where oil droplets or reverse micelles can also be used. However, it should be noted that the type and complexity of the catalytic assemblies anchored in vesicle bilayers should be limited compared to the complexity of a transcription-translation machinery sometimes encapsulated in enclosed models.

The anchoring of chemicals to SCA structures can be obtained by either hydrophobic derivatization or simply electrostatic association. The former has been explored by chemical ligating the chemical parts to hydrophobic moieties, e.g. aliphatic chains (see Figure 4.4) [61], cholesterol [105], or amphiphilic molecules that can even allow for the spontaneous and relative stable anchoring of polyelectrolytes to decanoic acid vesicles [106]. The use of hydrophobic molecules will spontaneously insert within bilayers [80, 107] can also help here. The latter (anchoring) is based on the spontaneous interactions between species of opposite charges, which inherently limits its applicability as the intended anchoring should not lead to bilayer disruption. Moreover, any such association will depend on the charge type and density on the structure interfaces, the association/dissociation equilibrium of the associated molecules (by extension their concentration in the medium), and their number of charges. This type of anchoring will therefore be defined by a delicate balance of factors, and should be more easily altered by fluctuations in the environment, hereby compromising the catalytic functions.

One such "interfacial" system has recently been proposed [60, 61], which investigates the photochemical production of amphiphiles with its catalytic molecules all anchored in bilayers by aliphatic hydrocarbon chain moieties (Figure 4.4). The production is catalyzed by two parts catalytic system: a photosensitizer/redox metal complex, a ruthenium trisbipyridine derivative, and an electron donor molecule, an 8-oxo guanine derivative. The catalysis is induced by the light excitation of the ruthenium complex, which results in the formation of a long-lived metal-to-ligand-charge-transfer (MLCT) state of the ruthenium complex. This excited state can decay via a number of pathways (including luminescence and various nonradiative pathways), but in the presence of a suitable electron donor (here 8-oxo guanine) it is reduced to a new excited state that can cleave lipid precursors into a waste radical (picolyl radical) and a charged amphiphile. The regeneration catalyst and the scavenging of radicals are obtained through interactions with a hydrogen atom source, 2,5-dihydrophenylglycine.

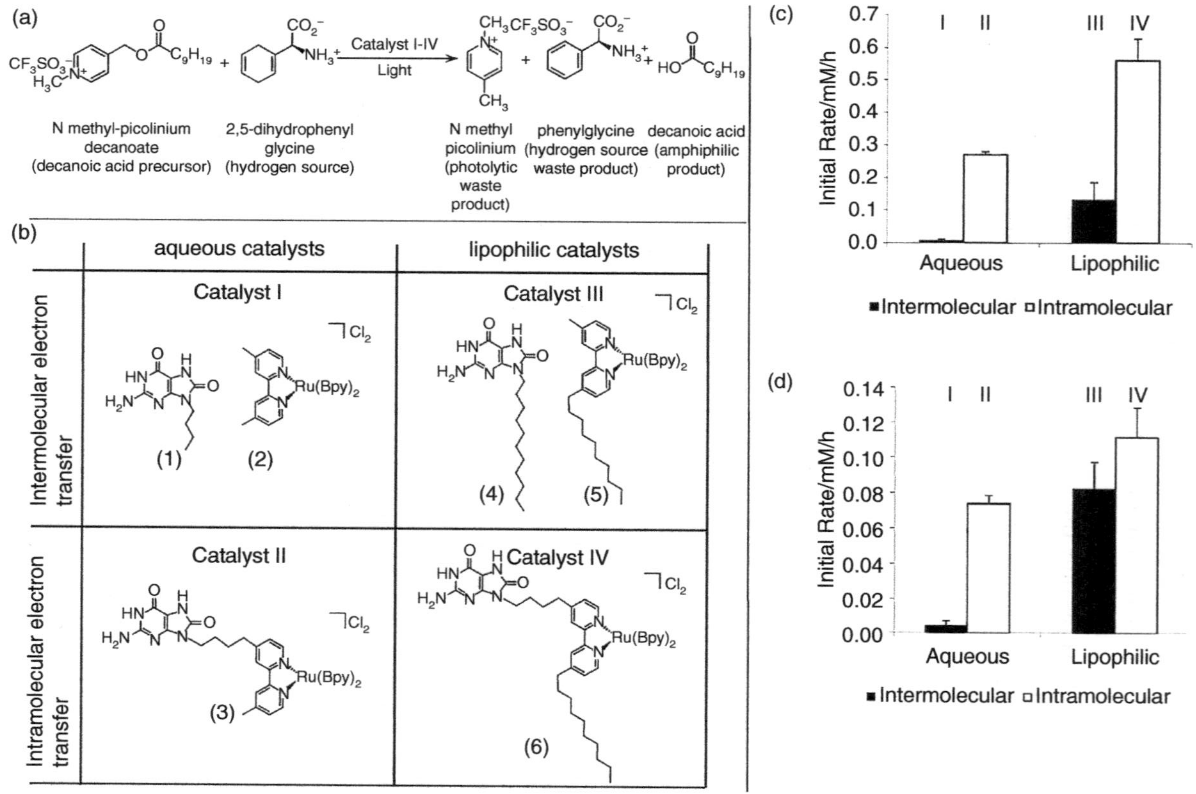
(a)
Catalyst I-IV
Light
N methyl-picolinium decanoate (decanoic acid precursor)
2,5-dihydrophenyl glycine (hydrogen source)
N methyl picolinium (photolytic waste product)
phenylglycine (hydrogen source waste product)
decanoic acid (amphiphilic product)
(b)
aqueous catalysts
lipophilic catalysts
Intermolecular electron transfer
Intramolecular electron transfer
Catalyst I
Catalyst II
Catalyst III
Catalyst IV
(1)
(2)
(3)
(4)
(5)
(6)
(c)
(d)
Initial Rate/mM/h
Aqueous
Lipophilic
Intermolecular
Intramolecular

The role of the decanoic acid bilayers in co-localizing the reaction components is crucial in determining the rate of precursor conversion. The initial rate of precursor conversion into fatty acid is highly dependent on the interaction between the information, the energy harvester, and the bilayer itself. In fact, the conversion rate became less dependent on the ruthenium complex-nucleobase catalyst configuration (bound or unbound, which results in intramolecular or intermolecular electron transfer, respectively) as long as both the ruthenium and the oxoguanine were tightly associated with the bilayers. At the relatively high local concentrations of catalysts, the advantage of the intramolecular electron transfer between the nucleobase and the ruthenium complex almost completely disappears (Figure 4.4c,d). However, with the increasing concentrations of product, a rate reduction can be seen. Contrary to traditional product inhibition, this reduction in rate is caused by the formation/growth of the structures.

This weakening of metabolism was therefore caused by a dilution of the reactant and catalysts in the bilayer, where they are localized. These results demonstrate that the bilayer, i.e. the container, can be sufficient for the co-location needed for efficient electron transfer [61]. That is, the interactions between the container and the other two components of the protocell can alter the expected chemistry.

This embodiment of an interfacial protocell model is still a simplified version of protocells, as it lacks replicable information (the nucleic acid polymer was replaced by a single nucleobase, 8-oxoguanine [60]), but it has already allowed the investigation of several processes essential for the life cycle of a protocell, such as growth and division [108]. Co-localizing information, metabolism, and compartments, the bilayers, can therefore result in significant changes in the expected chemistry, and such a systemic approach, based on the use of chemical systems, might be essential for understanding the way to explain how modern biochemistry emerged through the co-development of its main components.

Figure 4.4 Various configurations of an amphiphile production catalyst. *Source*: (Collage of Figures from [61]. Copyright by WILEY-VCH Verlag). (a) The photolytic cleavage reaction using visible light and its components. The catalyst systems are described below. (b) Catalyst configurations I to IV: (I) aqueous intermolecular catalyst, 9-butyl-8-oxoguanine (1) and [Ru(II)(bpy)$_2$(4,4′-dimethyl-bpy)] chloride (2); (II) aqueous intramolecular catalyst, bis-(2,2′-bipyridyl)-(9-N-[4-(4′-methyl-[2,2′]bipyridinyl-5-yl)-butyl]-8-oxoguanine)-ruthenium chloride (3); (III) lipophilic intermolecular catalyst, 9-decyl-8-oxoguanine (4) and [Ru(II)(bpy)$_2$(4-decyl,4′-methyl-bpy)] chloride (5); (IV) lipophilic intramolecular catalyst, bis-(2,2′-bipyridyl)-(9-N-[4-(4′-decyl-[2,2′]bipyridinyl-5-yl)-butyl]-8-oxoguanine) ruthenium chloride (6). (c) Rate dependence on catalyst configuration in the absence of a bilayer structure. (d) Rate dependence on catalyst configuration in the presence of decanoic acid/decanoate bilayer structures (vesicles). The numbers on top of the bars correspond to the catalyst number in part B.

4.5.3 Membranes as Energy Transduction Systems

In both autotroph and heterotroph contemporary cells, energy harvesting is performed by sophisticated (i.e.: proteinaceous) catalysis using complex lipid membranes and organic redox active chemicals that are the products of a long evolution process, thus could not have been present during the early phases of precellular development. However, energy uptake and transduction must have been essential to permit the emergence of a truly (semi-) autonomous protocell [109, 110]. The issue is then whether SCA bilayers could play this crucial role, as their counterparts in cells do. The development of energy harvesting and conversion into usable chemical energy can be deconvoluted into a series of processes/functions of increasing complexity: (i) directly linked energy harvesting, mostly prominently light (other energy sources, such as thermal and geological electrochemical ones, could be used, but the role of bilayers as reaction environments is less pronounced), and chemical conversion, (ii) creation of chemical gradients, and (iii) energy harvesting and its conversion into high-energy bonds in phosphate chemicals. Some experimental evidence exists to support scenarios involving SCA bilayers as central participants.

4.5.3.1 Linking Light Energy Harvesting and Chemical Conversion

The inventory of chemicals present on the early Earth includes potential pigments, such as PAHs [79], which have been proposed as primitive light-harvesting systems [77]. The formation of mixed systems from lipid aggregates with a large molar fraction of inserted PAHs has been demonstrated to occur by self-assembly (see above).

Other potential photocatalytic systems could have involved photosensitizers that were based on either minerals, e.g. titan oxide particles [111] or metal complexes, e.g. ruthenium complexes [60, 61], either encapsulated within the aqueous lumen of vesicles or incorporated into the bilayers, provided they possess a suited hydrophobic moiety due to the nature of their ligands.

These simple light-harvesting systems could have been used to convert chemical precursors, present from their environment, into their own building blocks or into more complex molecules, and to some extent started a primitive catalytic network. Obviously, the breadth of catalytic processes would likely have been limited to the transformation of simple molecules activated by abiotic processes. Indeed, the formation of carbon–carbon bonds would have required the conversion of these one-electron photochemistry into the net-two-electron chemistry. It would be, however, wrong to underestimate the importance of these simple designs, as complex thermodynamically downhill reactions can promote important processes, e.g. the photochemical production of amphiphiles [60, 61], or the promotion of the division in vesicular protocells [80].

4.5.3.2 Formation of Chemical Gradients

The formation of chemical (either proton or electric) gradients should likely represent the next step toward cellular energetics, as they intrinsically possess the potential to realize a truly up-hill reaction, by a two-electron transfer reaction. Gradients during their build-up have to be maintained until their amplitude is large enough to allow it to drive uphill reactions but simultaneously allow for the free diffusion of other molecules.

Although properties of single hydrocarbon chain compartment systems seem at first sight to preclude the formation of molecular gradients, this requirement of selective permeability can actually be met. Comparison of various solute diffusion rates show that the permeability of charged small polymers [59], such as nucleic acids, as well as small anionic solutes with high charge density, such as potassium ferricyanide, pyranine [64], is low, whereas other charged molecules, such as that of carboxyfluorescein, a model molecule, can readily cross the same bilayers. Thus, molecular gradients of specific molecules can be in principle maintained for periods of time compatible with further use of the stored energy for chemistry, as demonstrated by the complete reduction of the encapsulated ferricyanide to ferrocyanide using naphtho[2,3a]pyrene incorporated in mixed fatty acid membrane. As a sacrificial reductant, bilayer impermeable EDTA was used as a suitable soluble electron donor in the external medium. Thus, simple compartments can harbor a directional charge transfer that is induced by light harvesting.

The formation of proton gradients upon the irradiation of bilayer incorporated PAHs has been reported [77, 78]. The build-up of the proton gradient detected by fluorescence underlines the importance of the compartmentalization in an enclosed volume. Indeed, the proton release upon irradiation of PAHs is not directional, and thus only 50% of the proton generated will enter the lumen of the structures. However, the ensuing local concentration can result in transient pH gradient as large as three units of pH, which could be large enough to couple a proton gradient to a reaction network, provided the reaction rates would outpace gradient dissipation, which is rapid.

Addition of co-surfactants, such as fatty alcohols [56] or glycerol-monoesters [39, 47], have been shown to extend the robustness of these systems in terms of aggregation equilibria, pH, temperature sensitivity, and solute encapsulation. As primitive compartments that were composed of a single amphiphile type seem to be improbable, mixed compartment boundaries composed of (at this point unknown) co-surfactants might have conferred additional selectivity in solute permeability and the production of chemical gradients.

4.5.3.3 Energy Harvesting and Its Conversion into High-Energy Bonds of Phosphate-Chemicals

So far, the energy harvesting and conversion into high-energy bonds of phosphate-chemicals in protocell models composed of fatty acid has, to the best of our knowledge, not been successfully attempted. This might be due to the fact that this bioenergetics is intimately linked to the presence of sophisticated protein machinery for harvesting of light itself, and its conversion to a proton gradient, as well as its dissipation by the formation of ATP.

Experiments to design artificial photosynthetic systems have been carried out that are based on the reconstitution of a photosynthetic machinery in phospholipid liposomes [112, 113] and polymersomes [114]. It is, however, interesting to note that the light-harvesting step can be implemented using photosensitizer triads with hydrogen carriers, such as hydrophobically derivatized quinones [112] instead of protein-based machinery. Triads are rather complex organic molecules. These molecular machineries have been extensively optimized to improve charge separation on the photosensitizers upon their irradiation and induce a directional transfer of hydrogen. But that being said, from the chemical point of view the resulting processes are not truly different from those achievable with PAHs.

4.6 Dynamism

As discussed, the formation of fatty acids and similar amphiphiles into supramolecular structures is a function of the molecular and chemical properties of the monomer, the concentration of the amphiphile in the system, and the environmental conditions. Also, it is important to understand that once the supramolecular structure is formed, the overall structure may be kinetically stable but the individual amphiphiles that make up the structure are continuously exchanging with the amphiphiles in solution and in other structures. Therefore, at equilibrium a certain set and number of supramolecular structures may stably persist, but any perturbation of the molecules due to chemical transformation, change in concentration, or change in environmental parameter such as pH can make the system dynamic. Such perturbations are useful to consider when trying to push a supramolecular system from a stable equilibrium state to a dynamic state that could lead to self-replicating structures.

The pioneering work of Luisi et al. showed that fatty acids micelles [115, 116], reverse micelles [117], and vesicles [98, 118] could transform fatty acid anhydride precursors with autocatalytic kinetics into more supramolecular structures. In addition, it has been shown that fatty acids vesicles can grow through the slow addition of fatty acid micelles to a buffered solution of preformed fatty acid vesicles [119] and by evaporation of solvent [120].

Division of the supramolecular structures could be either spontaneous or controlled through external physical shear forces by extrusion through pores [121] or in fluid flow [120]. An alternative pathway for fatty acid vesicle division was implemented using either an encapsulated fluorescent dye or membrane-associated hydroxypyrene. Upon irradiation, the vesicles rapidly change shape and produce a pearling configuration, and subsequent division occurs in the presence of thiols [80]. This photochemical system induces a change in surface tension in the vesicles and causes controlled division.

Similarly, by exploiting the phase transition from micelle fatty acid structures to vesicles by manipulating local concentrations, Budin et al. [122] used a borosilicate microchannel in a temperature gradient to mimic gradients observed in deep-sea thermal vents due to fluid flow dynamics. They were able to show a local concentration effect of up to 800-fold and the formation of fatty acid vesicles in such a system.

As discussed, individual fatty acids constantly and rapidly exchange between supramolecular structures. This effect was exploited by introducing a small amount of phospholipids into fatty acid vesicles (see Section 4.4.2). The mixed-membrane vesicles were able to competitively retain incoming fatty acids, resulting in their growth at the expense of the pure fatty acid vesicles [62].

So far, we have discussed the self-formation of vesicles and how the dynamics of the amphiphiles can be exploited to show vesicle growth and division, but there is a concern about how clearly such systems present a true replication cycle. For example in Hanczyc et al. [119], several (five) total generations of growth and division were performed with an internal encapsulated fluorescent dye to evaluate the continuity of the replication process. However, in these studies the vesicles were too small to be directly observed during their growth and division events.

Finally, another issue concerns the correlation between vesicle growth and division and the replication of the catalytic network of protocells. Indeed, if the replication rates of the vesicle structures were to outpace that of the catalytic network, this could lead to the formation of a nonfunctional protocell [101]. This is especially true for the enclosed model (see Section 4.5.1), as replication can occur in a disjointed fashion [98]. Recently, the transmission of a simple catalytic function (the photochemical production of amphiphiles [55]) has been carried out using glass microsphere supported decanoic acid bilayers with the tethered ruthenium complex (photocatalyst). The advantage of the interfacial construct was to allow for a physical separation between the original protocell (containing large microspheres) and any newly formed second protocell generation (vesicles lacking microspheres) by physical separation using centrifugation. With such a chemical system, it was found that the capacity to photochemically synthesize amphiphiles was transmitted to the second protocell generation. Interestingly, this phenomenon could be considered as a form of identity inheritance, without truly replicable information polymers.

4.7 Conclusion

The study of SCA self-assembly is still in progress as the delicate nature of the resulting structures and their polymorphism render the investigations more challenging compared to studying their larger and more stable phospholipid counterparts. However, the delicate balance between stability and dynamism, and the easy functionalization of these structures, offer an opportunity for the construction of out-of-equilibrium protocell systems. Such dynamic systems are not only essential to gain insights in the origins of cellular systems, but could also be applied to the development of biotechnologies requiring adaptability or on-demand responsive behavior, such as environmentally controlled release of cargo.

In terms of the origins of life research, single-hydrocarbon-chain amphiphiles produce cell-like compartments that could help to delineate an evolutionary path from these prebiotically plausible amphiphiles to the building blocks of contemporary cellular membranes. Obviously, the investigation of this route still does not truly account for the effects of peptides and other molecules that might have formed or participated to the formation of primitive membranes. In the future, it will be imperative to consider the contribution of the non-coded peptides, whose spontaneous synthesis under plausible prebiotic conditions is certain. In addition, other small molecules associated with a primitive membrane could not only enhance amphiphile structure stabilization but also increase the catalytic repertoire of the structure.

The application of SCA self-assembly to study models of primitive membranes highlight the importance of composition and context in the formation and persistence of the structures. While simply composed pure fatty acid membranes are capable of encapsulation, simple metabolism, and growth and division, they are not realistic when considering the chemical complexity likely present in a real-world origins context. In addition such membrane structures are not stable in many common real-life contexts like salty water. By increasing the chemical complexity of the membranes, stability and robustness are affected, usually positively, and now more realistic environmental contexts are considered compatible with protocell models. However in these models, dynamics are still present that intimately link the protocell model to the environment. This complex interplay partially governed by the dynamic properties of amphiphile structures (reconfigurability, selective permeability, and potentially catalytic power) sets the stage for future studies that attempt to explain the origin and evolution of early life. All important characteristics of living systems can be considered with SCA protocell models, from information to energy harvesting to chemical conversion as well as others phenomena, such as reproduction. These systems are, however, simple enough to allow for the implementation and investigation of protocell systems as a whole, thus approaching the transition from chemical to biological evolution.

Acknowledgments

The authors are grateful for many fruitful discussion and suggestions from colleagues at the University of Southern Denmark and Los Alamos National Laboratory, in particular Drs. J. M. Boncella, H. Fellermann, G. Goranovic, S.E. Maurer, H.-J. Ziock, and S. Rasmussen. The Universities of Southern Denmark and of Trento, Italy, supported the preparation of this article.

References

1 Overton, E. (1900). Studien über die Aufnahme der Anilinfarben durch die lebenden Zelle. *Jahrb. Wiss. Bot.* 34: 669–701.
2 Singer, S.J. and Nicolson, G.L. (1972). The fluid mosaic model of the structure of cell membranes. *Science* 175 (4023): 720–731.
3 Luisi, P.L. (1998). About various definitions of life. *Origins Life Evol. Biosphere* 28: 613–622.
4 Cleland, C.E. and Chyba, C.F. (2002). Defining life. *Origins Life Evol. Biosphere* 32: 387–393.
5 Chen, I.A. and Walde, P. (2010). From self-assembled vesicles to protocells. *Cold Spring Harbor Perspect. Biol.* 2: a002170.
6 Morigaki, K. and Walde, P. (2007). Fatty acid vesicles. *Curr. Opin. Colloid Interface Sci.* 12: 75–80.
7 Oparin, A.I., Orlovskii, A.F., Bukhlaeva, V.Y. et al. (1976). Influence of the enzymatic synthesis of polyadenylic acid on a coacervate system. *Dokl. Akad. Nauk. SSSR* 226: 972–794.
8 Russell, M.J. and Hall, A.J.J. (1997). The emergence of life from iron monosulphide bubbles at a submarine hydrothermal redox and pH front. *Geol. Soc.* 154: 377–402.
9 Williams, D.S., Koga, S., Hak, C.R.C. et al. (2012). Polymer/nucleotide droplets as bio-inspired functional micro-compartments. *Soft Matter* 8: 6004–6014.
10 Zhang, J., Song, Y.-F., Cronin, L. et al. (2008). Self-assembly of organic-inorganic hybrid amphiphilic surfactants with large polyoxometalates as polar head groups. *J. Am. Chem. Soc.* 130: 14408–14409.
11 Li, M., Huang, X., and Mann, S. (2014). Spontaneous growth and division in self-reproducing inorganic colloidosomes. *Small* 10: 3291–3298.
12 Epps, D.E., Sherwood, E., Eichberg, J. et al. (1978). Cyanamide mediated syntheses under plausible primitive earth conditions. *Mol. Evol.* 11: 279–292.
13 Yuen, G.U. and Kvenvolden, K.A. (1973). Monocarboxylic acids in Murray and Murchison carbonaceous meteorites. *Nature* 246: 301–303.

14 Lawless, J.G. and Yuen, G.U. (1979). Quantitation of monocarboxylic acids in the Murchison carbonaceous meteorite. *Nature* 282: 431–454.

15 Yuen, G., Blair, N., Des Marais, D.J. et al. (1984). Carbon isotope composition of low molecular weight hydrocarbons and monocarboxylic acids from Murchison meteorite. *Nature* 307: 252–254.

16 Epstein, S., Krishnamurthy, R.V., Cronin, J.R. et al. (1987). Unusual stable isotope ratios in amino acid and carboxylic acid extracts from the Murchison meteorite. *Nature* 326: 477–479.

17 Krishnamurthy, R.V., Epstein, S., Cronin, J.R. et al. (1992). Isotopic and molecular analyses of hydrocarbons and monocarboxylic acids of the Murchison meteorite. *Geochim. Cosmochim. Acta* 56: 4045–4058.

18 Mautner, M., Leonard, R.L., and Deamer, D.W. (1995). Meteorite organics in planetary environments: hydrothermal release, surface activity and microbial utilization. *Planet. Space Sci.* 43: 139–147.

19 Naraoka, H., Shimoyama, A., and Harada, K. (1999). Molecular distribution of monocarboxylic acids in Asuka carbonaceous chondrites from Antarctica. *Origins Life Evol. Biosphere* 29: 187–201.

20 Dworkin, J.P., Deamer, D.W., Sandford, S.A. et al. (2001). Self-assembling amphiphilic molecules: synthesis in simulated interstellar/precometary ices. *Proc. Natl. Acad. Sci. U.S.A.* 98: 815–819.

21 Cooper, G.W., Onwo, W.M., and Cronin, J.R. (1992). Alkyl phosphonic acids and sulfonic acids in the Murchison meteorite. *Geochim. Cosmochim. Acta* 56: 4109–4115.

22 Degraaf, R.M., Visscher, J., and Schwartz, A.W. (1995). A plausibly prebiotic synthesis of phosphonic acids. *Nature* 378: 474–477.

23 Pizzarello, S., Williams, L.B., Lehman, J. et al. (2011). Abundant ammonia in primitive asteroids and the case for a possible exobiology. *Proc. Natl. Acad. Sci. U.S.A.* 108: 4303–4306.

24 McCollom, T.M., Ritter, G., and Simoneit, B.R.T. (1999). Lipid synthesis under hydrothermal conditions by Fischer-Tropsch-type reactions. *Origins Life Evol. Biosphere* 29: 153–166.

25 Rushdi, A.I. and Simoneit, B.R.T. (2001). Lipid formation by aqueous Fischer-Tropsch-type synthesis over a temperature range of 100 to 400 °C. *Origins Life Evol. Biosphere* 31: 103–118.

26 Powner, M.W. and Sutherland, J.D. (2011). Prebiotic chemistry: a new modus operandi. *Phil. Trans. R. Soc. B* 366: 2870–2877.

27 Rastogi, A., Ghosh, A.K., and Suresh, S.J. (2011). Hydrogen bond interactions between water molecules in bulk liquid, near electrode surfaces and around ions. In: *Thermodynamics – Physical Chemistry of Aqueous Systems* (ed. J.C. Moreno-Piraján), 351–364. InTech. ISBN: 978-953-307-979-0.

28 Israelachvili, J.N., Mitchell, D.J., and Ninham, B.W.J. (1976). Intermolecular and surface forces. *Chem. Soc., Faraday Trans. II* 72: 1525–1568.

29 Kanicky, J.R., Poniatowski, A.F., Mehta, N.R. et al. (2000). Cooperativity among molecules at interfaces in relation to various technological processes: effect of chain length on the pK(a) of fatty acid salt solutions. *Langmuir* 16: 172–177.

30 Hargreaves, W.R. and Deamer, D.W. (1978). Liposomes from ionic, single-chain amphiphiles. *Biochemistry* 17: 3759–3768.

31 Cistola, D.P., Atkinson, D., Hamilton, J.A. et al. (1986). Phase-behavior and bilayer properties of fatty-acids – hydrated 1-1 acid soaps. *Biochemistry* 25: 2804–2812.

32 Walde, P., Wessicken, M., Rädler, U. et al. (1997). Autopoietic self-reproduction of fatty acid vesicles. *Phys. Chem. B* 101: 7390–7397.

33 Sakai, T., Miyaki, M., Tajima, H. et al. (2012). Precipitate deposition around CMC and vesicle-to-micelle transition of monopotassium monododecyl phosphate in water. *Phys. Chem. B* 116: 11225–11233.

34 Albertsen, A.N., Duffy, C.D., Sutherland, J.D. et al. (2014). Self-assembly of phosphate amphiphiles in mixtures of prebiotically plausible surfactants. *Astrobiology* 14: 462–447.

35 Maurer, E., Belloni, L., Zemb, T. et al. (2007). Ion exchange in catanionic mixtures: from ion pair amphiphiles to surfactant mixtures. *Langmuir* 23: 6554–6560.

36 Namani, T. and Deamer, D.W. (2008). Stability of model membranes in extreme environments. *Origins Life Evol. Biosphere* 38: 329–341.

37 Michina, Y., Carriere, D., Mariet, C. et al. (2009). Ripening of catanionic aggregates upon dialysis. *Langmuir* 25: 698–706.

38 Caschera, F., Bernadino de la Serna, J., Löffler, P.M.G. et al. (2011). Production of RNA by polymerase protein encapsulated within phospholipid vesicles. *Langmuir* 27: 14078–14090.

39 Monnard, P.-A., Apel, C.L., Kanavarioti, A. et al. (2002). Current ideas about prebiological compartmentalization. *Astrobiology* 2: 139–152.

40 Xu, W., Wang, X., Zhong, Z. et al. (2012). Influence of counterions on lauric acid vesicles and theoretical consideration of vesicle stability. *Phys. Chem. B* 117: 242–251.

41 Douliez, J.-P. (2004). Phase behavior of the palmitic acid/palmitin system. A 2H NMR study. *Langmuir* 20: 1543–1550.

42 Douliez, J.-P. and Gaillard, C. (2014). Self-assembly of fatty acids: from foams to protocell vesicles. *New J. Chem.* 38: 5142–5148.

43 Zana, R. (2004). Partial phase behavior and micellar properties of tetrabutylammonium salts of fatty acids: unusual solubility in water and formation of unexpectedly small micelles. *Langmuir* 20: 5666–5668.

44 Klein, R., Kellermeier, M., Drechsler, M. et al. (2009). Self assembly of anastomosis-like superstructures in fatty acid/guanidine hydrochloride aqueous dispersions. *Colloids Surf., A* 338: 129–134.

45 Mouritsen, O.G. (2005). *Life-As a Matter of Fat: The Emerging Science of Lipidomics*. Springer Science & Business Media.

46 Nagarajan, R. (2002). Molecular packing parameter and surfactant self-assembly: the neglected role of the surfactant tail. *Langmuir* 18: 31–38.

47 Maurer, S.E., Deamer, D.W., Boncella, J.M. et al. (2009). Impact of glycerol monoacyl on the stability of plausible prebiotic fatty acid membranes. *Astrobiology* 9: 979–987.

48 Lewis, R.N., Mannock, D.A., and McElhaney, R.N. (1997). Membrane lipid molecular structure and polymorphism. *Curr. Top. Membr.* 44: 25–102.

49 Israelachvili, J.N. (2011). *Intermolecular and Surface Forces*: Revised third edition. Academic press.

50 Stano, P. and Luisi, P.L. (2010). Reactions in liposomes. In: *Molecular Encapsulation: Organic Reactions in Constrained Systems* (ed. U. Brinker and J. Mieusset), 455. Chichester: Wiley.

51 Stano, P. and Luisi, P.L. (2010). Achievements and open questions in the self-reproduction of vesicles and synthetic minimal cells. *Chem. Commun.* 46: 3639–3653.

52 Svetina, S. and Žekš, B. (1989). Membrane bending energy and shape determination of phospholipid vesicles and red blood cells. *Eur. Biophys. J.* 17: 101–111.

53 Svetina, S. and Žekš, B. (2002). Shape behavior of lipid vesicles as the basis of some cellular processes. *Anat. Rec.* 268: 215–225.

54 Huo, C., Li, M., Huang, X. et al. (2014). Membrane engineering of colloidosome microcompartments using partially hydrophobic mesoporous silica nanoparticles. *Langmuir* 30: 15047–15052.

55 Albertsen, A.N., Maurer, S.E., Nielsen, K.A. et al. (2014). Transmission of photo-catalytic function in a self-replicating chemical system: in-situ amphiphile production over two protocell generations. *Chem. Commun.* 50: 8989–8992.

56 Apel, C.L., Mautner, M.N., and Deamer, D.W. (2002). Self-assembled vesicles of monocarboxylic acids and alcohols: conditions for stability and for encapsulation of biopolymers. *Biochim. Biophys. Acta* 1559: 1–9.

57 Budin, I., Prywes, N., Zhang, N. et al. (2014). Chain-length heterogeneity allows for the assembly of fatty acid vesicles in dilute solutions. *Biophys. J.* 107: 1582–1590.

58 Schuler, L.D., Walde, P., Luisi, P.L. et al. (2001). Molecular dynamics simulation of n-dodecyl phosphate aggregate structures. *Eur. Biophys. J. Biophys. Lett.* 30: 330–343.

59 Adamala, K. and Szostak, J.W. (2013). Competition between model protocells driven by an encapsulated catalyst. *Science* 342: 1098–1100.

60 DeClue, M., Monnard, P.-A., Bailey, J. et al. (2009). Nucleobase mediated, photocatalytic vesicle formation from an ester precursor. *Am. Chem. Soc.* 131: 931–933.

61 Maurer, S.E., DeClue, M.S., Albertsen, A.N. et al. (2011). Interactions between catalysts and amphiphile structures and their implications for a protocell model. *ChemPhysChem* 12: 828–835.

62 Budin, I. and Szostak, J.W. (2011). Physical effects underlying the transition from primitive to modern cell membranes. *Proc. Natl. Acad. Sci. U.S.A.* 108: 5249–5254.

63 Monnard, P.-A. and Walde, P. (2015). Current ideas about prebiological compartmentalization. *Life* 5: 1239–1263.

64 Cape, J., Monnard, P.-A., and Boncella, J.M. (2011). Prebiotically relevant mixed fatty acid vesicles support anionic solute encapsulation and photochemically catalyzed trans-membrane charge transport. *Chem. Sci.* 2: 661–667.

65 Chen, I.A., Salehi-Ashtiani, K., and Szostak, J.W. (2005). RNA catalysis in model protocell vesicles. *J. Am. Chem. Soc.* 127: 13213–13219.

66 Simoneit, B.R.T., Rushdi, A.I., and Deamer, D.W. (2007). Abiotic formation of acylglycerols under simulated hydrothermal conditions and self-assembly properties of such lipid products. *Adv. Space Res.* 40: 1649–1656.

67 Mansy, S.S., Schrum, J.P., Krishnamurthy, M. et al. (2008). Thermostability of model protocell membranes. *Nature* 454: 122–125.

68 Bales, B.L. and Zana, R.J. (2002). Characterization of micelles of quaternary ammonium surfactants as reaction media I: dodeclytrimethylammonium bromide and chloride. *Phys. Chem. B* 106: 1926–1939.

69 Boiteau, L. and Pascal, R. (2011). Energy sources, self-organization, and the Origin of Life. *Origins Life Evol. Biosphere* 41: 23–33.

70 Monnard, P.-A. and Deamer, D.W. (2002). Membrane self-assembly processes: steps towards the first cellular life. *Anat. Rec.* 268: 196–207.

71 Chakrabarti, A.C., Breaker, R.R., Joyce, G.F. et al. (1994). Production of RNA by polymerase protein encapsulated within phospholipid vesicles. *Mol. Evol.* 39: 555–559.

72 Vist, M.R. and Davis, J.H. (1990). Phase equilibria of cholesterol/DPPC mixtures: 2H-NMR and DSC. *Biochemistry* 29: 451–464.

73 Groen, J., Deamer, D.W., Kros, A. et al. (2012). Polycyclic aromatic hydrocarbons as plausible prebiotic membrane components. *Origins Life Evol. Biosphere* 42: 295–306.

74 Black, R.A., Blosser, M.C., Stottrup, B.L. et al. (2013). Nucleobases bind to and stabilize aggregates of a prebiotic amphiphile, providing a viable mechanism for the emergence of protocells. *Proc. Natl. Acad. Sci. U.S.A.* 110: 13272–13276.

75 Whittet, D.C. (1997). Is extraterrestrial organic matter relevant to the origin of life on earth? *Origins Life Evol. Biosphere* 27: 249–262.

76 Pasek, M. and Lauretta, D. (2008). Extraterrestrial flux of potentially prebiotic C, N, and P to the early earth. *Origins Life Evol. Biosphere* 38: 5–21.

77 Deamer, D.W. (1992). Polycyclic aromatic hydrocarbons: primitive pigment systems in the prebiotic environment. *Adv. Space Res.* 12: 183–189.

78 Volkov, A.G., Gugeshashvili, M.I., and Deamer, D.W. (1985). Energy conversion at liquid/liquid interfaces: artificial photosynthetic systems. *Electrochim. Acta* 40: 2849–2868.

79 Briggs, M.H. and Mamikunian, G. (1962). Organic constituents of the carbonaceous chondrites. *Space Sci. Rev.* 1: 647–682.

80 Zhu, T.F., Adamala, K., Zhang, N. et al. (2012). Coupled growth and division of model protocell membranes. *Proc. Natl. Acad. Sci. U.S.A.* 109: 9828–9832.

81 Fameau, A.-L., Houinsou-Houssou, B., Ventureira, J.B. et al. (2011). Self-assembly, foaming, and emulsifying properties of sodium alkyl carboxylate/guanidine hydrochloride aqueous mixtures. *Langmuir* 27: 4505–4513.

82 Douliez, J.-P., Houinsou-Houssou, B., Fameau, A.-L. et al. (2010). Self assembly of anastomosis-like superstructures in fatty acid/guanidine hydrochloride aqueous dispersions. *J. Colloid Interface Sci.* 341: 386–389.

83 De Feyter, S. and De Schryver, F.C. (2003). Two-dimensional supramolecular self-assembly probed by scanning tunneling microscopy. *Chem. Soc. Rev.* 32: 139–150.

84 Ulman, A. (1996). Formation and structure of self-assembled monolayers. *Chem. Rev.* 96: 1533–1554.

85 Giles, C.H. (1982). *Forces Operating in Adsorption of Surfactants and Other Solutes at Solid Surfaces: A Survey*. New York: Plenum.

86 Hanczyc, M.M., Mansy, S.S., and Szostak, J.W. (2007). Experimental models of primitive cellular compartments: encapsulation, growth, and division. *Origins Life Evol. Biosphere* 37: 67–82.

87 Ferris, J.P., Hill, A.R. Jr.,, Liu, R. et al. (1996). Synthesis of long prebiotic oligomers on mineral surfaces. *Nature* 381: 59–61.

88 Bungenberg de Jong, V.H.G. (1932). Die koazervation und ihre bedeutung für die biologie. *Protoplasma* 15: 110–176.

89 Oparin, A.I. (1966). *The Origin and Initial Development of Life*. Moscow, Springfield: Meditsina Publishing House. English translation: NASA TT F-488 (1968), Clearinghouse for Federal Scientific and Technical Information.

90 Dora Tang, T.Y., Rohaida Che Hak, C., Thompson, A.J. et al. (2014). Fatty acid membrane assembly on coacervate microdroplets as a step towards a hybrid protocell model. *Nat. Chem.* 6: 527–533.

91 Fallah-Araghi, A., Meguellati, K., Baret, J.-C. et al. (2014). Enhanced chemical synthesis at soft interfaces: a universal reaction-adsorption mechanism in microcompartments. *Phys. Rev. Lett.* 112: 028301.
92 Caschera, F., Sunami, T., Matsuura, T. et al. (2011). Programmed vesicle fusion triggers gene expression. *Langmuir* 27: 13082–13090.
93 Walde, P. and Ichikawa, S. (2001). Review. Enzyme inside lipid vesicles: preparation, reactivity and applications. *Biomol. Eng.* 18: 143–177.
94 Luisi, P.L., Ferri, F., and Stano, P. (2006). Approaches to semi-synthetic minimal cells: a review. *Naturwissenschaften* 93: 1–13.
95 Monnard, P.A., DeClue, M.S., and Ziock, H. (2008). Organic nano-compartments as biomimetic reactors, and protocells. *J. Curr. Nanosci.* 4: 71–87.
96 Blobel, G. (1980). Intracellular protein topogenesis. *Proc. Natl. Acad. Sci. U.S.A.* 77: 1496–1500.
97 Shew, R.L. and Deamer, D.W. (1985). A novel method for encapsulation of macromolecules in liposomes. *Biochim. Biophys. Acta* 816: 1–8.
98 Walde, P., Goto, A., Monnard, P.-A. et al. (1994). Oparin's reaction revisited: enzymatic synthesis of poly(adenyl acid) in micelles and self-reproducing vesicles. *J. Am. Chem. Soc.* 116: 7541–7547.
99 Mansy, S.S. and Szostak, J.W. (2008). Thermostability of model protocell membranes. *Proc. Natl. Acad. Sci. U.S.A.* 105: 13351–13355.
100 Monnard, P.-A. and Deamer, D.W. (2001). Nutrient uptake by protocells: a liposome model system. *Origins Life Evol. Biosphere* 31: 147–155.
101 Szostak, J.W., Bartel, D.P., and Luisi, P.L. (2001). Synthesizing life. *Nature* 409: 387–390.
102 Kuruma, Y., Stano, P., Ueda, T. et al. (2009). A synthetic biology approach to the construction of membrane proteins in semi-synthetic minimal cells. *Biochim. Biophys. Acta* 1788: 567–574.
103 Rasmussen, S., Chen, L., Nilsson, M. et al. (2003). Bridging nonliving and living matter. *Artif. Life* 9: 269–316.
104 Wieczorek, R., Wamberg, M.C., Albertsen, A.N. et al. (2013). Bottom-up protocell design: gaining insights in the emergence of complex function(s). In: *Evolutionary Biology*: : 16th Meeting 2012 (ed. P. Pontarottis), 81–94. Springer-Verlag.
105 Müller, U.F. and Bartel, D.P. (2008). Improved polymerase ribozyme efficiency on hydrophobic assemblies. *RNA* 14: 552–562.
106 Wamberg, M.C., Wieczorek, R., Bo Brier, S. et al. (2014). Synthesis and characterization of bolaamphiphilic-DNA conjugates for anchoring in fatty acid membranes. *Bioconjugate Chem.* 25: 1678–1688.
107 Adamala, K. and Szostak, J.W. (2013). Nonenzymatic template-directed RNA synthesis inside model protocells. *Nat. Chem.* 5: 634–634.
108 Hanczyc, M.M. and Szostak, J.W. (2004). Replicating vesicles as models of primitive cell growth and division. *Curr. Opin. Chem. Biol.* 8: 660–664.

109 Deamer, D.W. (1997). The first living systems: a bioenergetic perspective. *Microbiol. Mol. Biol. Rev.* 61: 230–261.

110 Deamer, D.W. and Weber, A.L. (2010). Bioenergetics and life's origins. *Cold Spring Harbor Perspect. Biol.* 2: a004929.

111 Summers, D.P., Noveron, J., and Basa, R.C.B. (2009). Energy transduction inside of amphiphilic vesicles: encapsulation of photochemically active semiconducting particles. *Origins Life Evol. Biosphere* 39: 127–140.

112 Steinberg-Yfrach, G., Liddell, P.A., Hung, S.-C. et al. (1997). Conversion of light energy to proton potential in liposomes by artificial photosythetic reaction centres. *Nature* 385: 239–241.

113 Steinberg-Yfrach, G., Rigaud, J.-L., Durantini, E.N. et al. (1998). Light-driven production of ATP catalyzed by F0F1-ATP synthase in an artificial photosynthetic membrane. *Nature* 392: 479–482.

114 Choi, H.-J. and Montemagno, C.D. (2005). Artificial organelle: ATP synthesis from cellular mimetic polymersomes. *Nano Lett.* 5: 2538–2542.

115 Bachmann, P.A., Walde, P., Luisi, P.L. et al. (1991). Self-replicating micelles: aqueous micelles and enzymatically driven reactions in reverse micelles. *J. Am. Chem. Soc.* 113: 8204–8209.

116 Bachmann, P.A., Luisi, P.L., and Lang, J. (1992). Autocatalytic self-replicating micelles as models for prebiotic structures. *Nature* 357: 57–59.

117 Bachmann, P.A., Walde, P., Luisi, P.L. et al. (1990). Self-replicating reverse micelles and chemical autopoiesis. *Am. Chem. Soc.* 112: 8200–8201.

118 Walde, P., Wick, R., Fresta, M. et al. (1994). Autopoietic self-reproduction of fatty acid vesicles. *J. Am. Chem. Soc.* 116: 11649–11654.

119 Hanczyc, M.M., Fujikawa, S.M., and Szostak, J.W. (2003). Experimental models of primitive cellular compartments: encapsulation, growth, and division. *Science* 302: 618–622.

120 Budin, I., Debnath, A., and Szostak, J.W. (2012). Concentration-driven growth of model protocell membranes. *J. Am. Chem. Soc.* 134: 20812–20819.

121 Zhu, T.F. and Szostak, J.W. (2009). Coupled growth and division of model protocell membranes. *J. Am. Chem. Soc.* 131: 5705–5713.

122 Budin, I., Bruckner, R.J., and Szostak, J.W. (2009). Formation of protocell-like vesicles in a thermal diffusion column. *J. Am. Chem. Soc.* 131: 9628–9629.

5

Programming Micelles with Biomolecules

Matthew P. Thompson and Nathan C. Gianneschi

Northwestern University, Department of Chemistry, 633 Clark Street, Evanston, IL, 60208, USA

5.1 Introduction

The term biomolecule is used to describe organic molecules, more frequently meaning macromolecules, which are found in and produced by living organisms. These molecules are responsible for both the diverse and specific function of biological processes. Therefore, it is no surprise that the use of biomolecules in synthetic systems is garnering increased attention as chemists look to take advantage of the unique and specific properties imparted by these molecules that have been refined through the evolutionary process for specific and selective function. Semi-synthetic, biohybrid materials incorporating biomolecules in their design have the potential to respond to the biochemical signals that trigger responses from the native biomolecules themselves, including changes in morphology, chemical structure, and/or function. In biological systems there are numerous examples of materials that are programmed to undergo shape and size changes from allosteric enzymes to pseudopodium formation during chemotaxis, and endosomes during cell-uptake. Semi-synthetic, biohybrid supramolecular systems that are programmed for these types of changes are just beginning to be exploited [1–4]. Understanding how these nanoscale assemblies undergo size and shape changes is expected to have a broad impact in a number of applications. One of the major hurdles in the synthesis of complex, well-defined nanomaterials have been controlling the assembly of small individual components at this length scale. Small molecule or (co)polymer amphiphiles are well suited as building blocks for functional supramolecular systems as small changes in the hydrophilic segment of an amphiphile can lead to the formation, destruction, or a change in morphology of the assemblies. As the concentration of an amphiphile in solution increases they begin to form different assemblies. The simplest of these assemblies

Self-Assembly: From Surfactants to Nanoparticles, First Edition. Edited by Ramanathan Nagarajan.

is the spherical micelle, where the hydrophilic units occupy the periphery (corona) of the structure and are in contact with the solvent (water) while the hydrophobic units are sequestered in the center of the assembly shielded from contact with the solvent. To successfully program a system, one is reliant on the ability of the biomolecule to encode specific information. Two classes of biomolecules capable of this are composed of either amino acids (peptides and proteins) or nucleotides (nucleic acids in DNA or RNA). Conceivably, this type of approach could be extended to complex sugars, although examples of this type of system are rare and/or remain unexplored. In this chapter we will focus on the use of biomolecules as programmable units in micellar assemblies and how they influence and direct the transformation of these assemblies in response to stimuli.

5.2 Peptide-Containing Micelles

Natural and non-natural amino acid containing peptides offer broad chemical diversity and have the ability to bind and recognize other molecules. They are also susceptible to optimization via molecular evolution strategies [5, 6], and are capable of selective targeting [7, 8], signaling [9–11], receptor binding [12, 13] and behaving as substrates for specific enzymes [14–17]. Therefore, the incorporation of amino acids and peptides into amphiphilic systems as functional and structural building blocks is of great interest and has tremendous potential utility by virtue of making synthetic particles susceptible to the inherent biological characteristics of the peptide sequence. Amphiphiles that contain peptides fall into three general categories, (i) amphiphilic peptides; these are peptides where the amphiphilicity arises from the sequences themselves, i.e. they include both hydrophilic and hydrophobic residues. (ii) Small molecule peptide amphiphiles (PAs), these generally consist of a peptide sequence that is hydrophilic in nature covalently linked to a hydrophobic small molecule and, (iii) Peptide-polymer amphiphiles (PPAs), these systems are composed of a polymer unit and a segment containing one or more peptides either covalently attached to the polymer through a post polymerization conjugation strategy or polymerized as a block itself. Due to the large number of systems comprised of amphiphiles containing peptides we will only consider systems that meet the following criteria: (i) the amphiphiles must form discrete spherical micelles in solution, (ii) the peptide sequence has been designed as a substrate for a specific enzyme which upon interaction, results in a change in function and/or morphology of the material. Thus, amphiphiles where the peptide segment is used purely as a targeting moiety or systems where the stimulus is a bulk property of the solution (temperature, salt concentration, and pH) are excluded from further examination here. In addition, systems with different starting nanostructures (fibers, aggregates, hydrogels, etc.) will not be covered in this chapter. While these systems are of great interest, they are beyond the scope of this discussion and are either covered in other chapters of this book or have recently been the subject of numerous excellent reviews [18–25].

Enzymes play a critical role in biology through a myriad of natural processes involved in the manipulation of nanoscale self-assemblies including the replication of nucleic acids, decomposition of biomaterials including extracellular matrices, and in the assembly of viruses. To date, the majority of research into organic enzyme-responsive systems has focused on the use of phosphorylation, dephosphorylation, or other enzymatic reactions to control the formation of peptide-only nanostructures. In the following sections we highlight research where enzymatic reactions are used to control the structural characteristics of peptide containing micelles.

5.2.1 Peptide Amphiphiles

Matrix metalloproteinases are a family of zinc containing endopeptidases that are responsible for tissue remodeling and degradation of the extracellular matrix. They have been associated with a number of diseases including cancer and as a result they have been utilized as the triggering enzyme for a number of peptide-based substrates. With this in mind, Goto et al. [26] designed a PA that would undergo a micelle-to-nanofiber transition resulting in gelation after an MMP-7 catalyzed reaction as shown in Figure 5.1. The PA consisted of a palmitoyl group conjugated to the N-terminus of a peptide with the sequence NH_2-Gly-Gly-Gly-His-Gly-**Pro-Leu-Gly-Leu**-Ala-Arg-Lys where the -**Pro-Leu-Gly-Leu**- sequence was the substrate for MMP-7. The inclusion of the Proline residue imparts a curved structure to the amphiphile promoting formation of a micellar assembly rather than gelation. Addition of MMP-7 to the solution resulted in cleavage of the substrate sequence between the Gly-Leu residues, which was confirmed through matrix-assisted laser desorption/ionization (MALDI) and high-performance liquid chromatography (HPLC) analysis. The cleavage event reduced the curved structure of the amphiphile allowing for the formation of a translucent hydrogel. Addition of either bovine serum albumin (BSA) or thrombin to the micelle solution rather than MMP-7 did not result in hydrogel formation. Gelation was also prevented when an inhibitor of MMP-7 was present in the micelle solution.

Phosphatases are a class of enzymes that remove phosphoryl groups attached to serine, threonine, or tyrosine residues. Theses enzymes play a vital role in turning off signaling pathways that are activated by kinases. Ulijn et al. [27] have demonstrated a micelle-to-fiber morphology change driven by the dephosphorylation of a simple PA composed of the dipeptide Fmoc-Phe-*p*Tyr, as shown in Figure 5.2. In buffer at concentrations above 6 mM, the dipeptide formed micellar assemblies with a diameter of 5 nm, where the hydrophobic Fmoc groups were packed in the core of the micelle while the

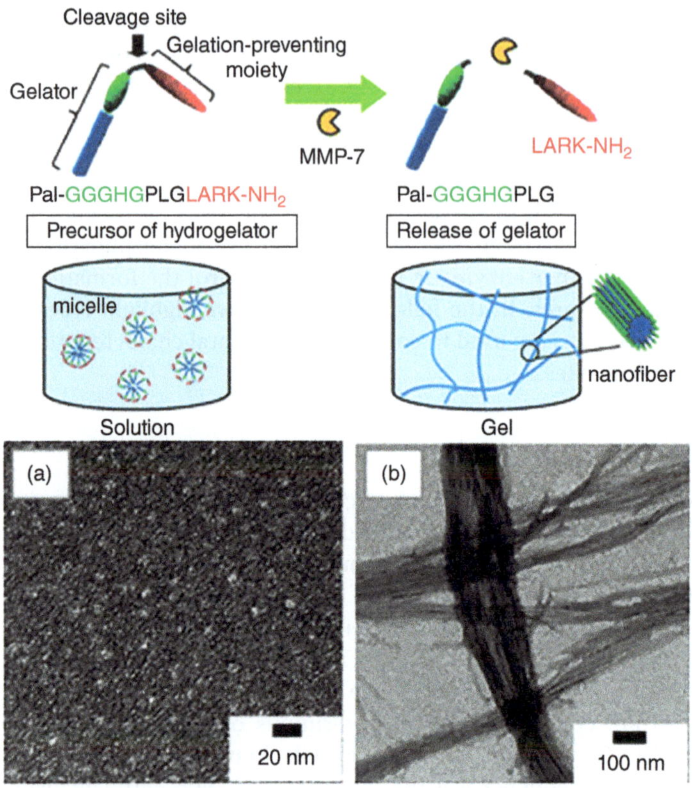

Figure 5.1 Schematic illustration of MMP-7-mediated hydrogelation of a precursor solution and TEM images of the cryo-dried 0.2 wt% precursor solution (a) and hydrogel (b) obtained after adding MMP-7 (1 mg ml^{-1}) to the solution. Both were stained by 2% uranyl acetate. *Source*: Reproduced from ref. [26] with permission of The Royal Society of Chemistry.

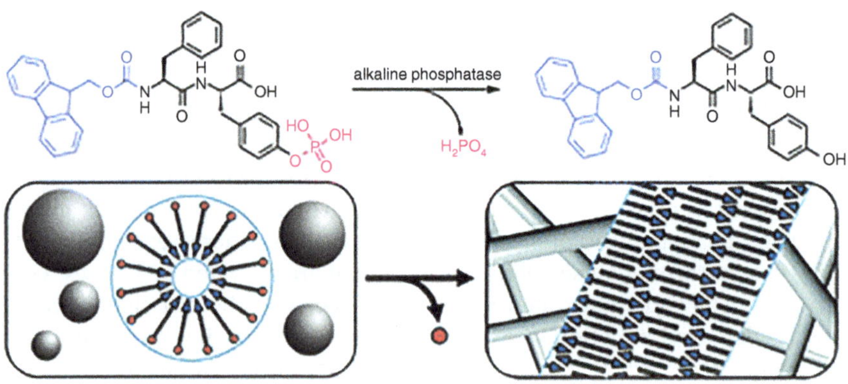

Figure 5.2 Enzyme catalyzed dephosphorylation reaction of Fmoc-FpY and a schematic representation of the transformation from micelles to fibers. *Source*: Reproduced from ref. [27] with permission of The Royal Society of Chemistry.

hydrophilic phosphate groups were in contact with the solvent. Subsequent treatment of the micellar assemblies with alkaline phosphatase resulted in the formation of extended fiber structures and gelation after 90 minutes. The dephosphorylation reaction was monitored by HPLC while morphology transitions were monitored by both fluorescence and circular dichroism spectroscopy. At 90 minutes, approximately 70% of the dipeptide had undergone dephosphorylation while the onset of gelation occurs at approximately 40 minutes.

5.2.2 Peptide–Polymer Amphiphiles (PPAs)

This class of amphiphile combines the structure and functionality inherent to peptides with the stability and processability of synthetic polymers. The synthesis of this type of amphiphile is divided into two categories: (i) conjugation of peptide(s) to polymers via a coupling reaction between two complementary chemical handles (graft to) or (ii) direct polymerization of the hydrophilic and hydrophobic units (graft-through).

Cenker et al. [28] designed a peptide–polymer conjugate capable of hydrolysis upon treatment with α-chymotrypsin (Figure 5.3). α-Chymotrypsin is a serine protease responsible for cleavage on the C-terminal side of tyrosine, phenylalanine, and tryptophan [29]. Conjugation of PEG_{3000} to the peptide βAβAKLVFF led to the formation of spherical micelles approximately 10 nm in diameter, where the peptide unit formed the hydrophobic core and the polyethylene glycol (PEG) unit formed the hydrophilic shell. After treatment with α-chymotrypsin, which cleaves between the F–F residues of the peptide, two fragments, F–PEG_{3000} and βAβAKLVF were generated and confirmed by mass spectrometry. Both the released peptide fragment and the F–PEG_{3000} unit showed no noticeable secondary nanostructures in solution.

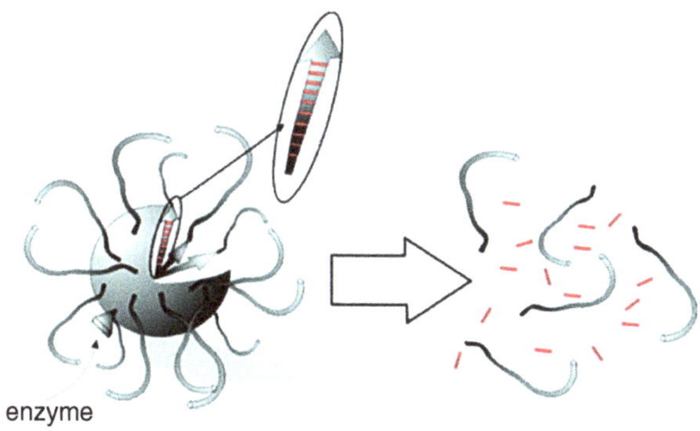

Figure 5.3 Enzyme-triggered degradation of a PEG-peptide micelle. The peptides within the β-strands (red lines) in the core of the micelle (left-hand side) are released in unassociated form (right-hand side) after treatment with enzyme. *Source*: Reproduced from ref. [28]. Copyright 2010 American Chemical Society.

The systems highlighted to this point have all consisted of amphiphiles that contain only single copies of the hydrophobic and hydrophilic moieties. By contrast, block-copolymer amphiphiles contain multiple copies of each monomer allowing one to assess the effects this has on the performance of the material. The first example of an amphiphilic peptide–polymer micellar system, capable of undergoing an enzyme triggered morphology change was reported by Gianneschi et al. [30]. The PPAs were designed to respond to four different cancer-associated enzymes: protein kinase A (PKA), protein phosphatase-1 (PP1) and matrix-metalloproteinases 2 and 9, (MMP-2 and MMP-9). Two amphiphilic brush block-copolymers were synthesized from a block copolymer containing a hydrophobic phenyl block and a N-hydroxysuccinimide ester (NHS) block via conjugation of the N-terminus of the peptide to the activated NHS ester. The two PPAs differed only in the relative ordering of the peptide sequence. The resulting PPAs were characterized by size-exclusion chromatography with a multiangle light scattering detector (SEC-MALS) and revealed a coupling efficiency of approximately 33% (6 and 5 peptides for PPA-1 and PPA-2, respectively). Well-defined micellar aggregates were formed upon transitioning the PPAs from a 1 : 1 dimethyl sulfoxide (DMSO): dimethyl formamide (DMF) solution to buffered water (Figure 5.4).

The resulting micelles were treated with MMP-2 and MMP-9, which resulted in cleavage of the peptide substrate between the G-L residues of the recognition sequence –L-G-L-. The morphology of the systems was monitored by dynamic light scattering (DLS), transmission electron microscopy (TEM) and scanning electron microscopy (SEM) as shown in Figure 5.5. No change in morphology was observed by TEM or SEM for **M1** while DLS indicated the presence of some larger aggregates in solution. By contrast a large change in morphology was observed upon treatment of **M2** with MMP. TEM and SEM showed the formation of an amorphous network post cleavage, while DLS measurements indicated complete conversion of the particles to large aggregates in solution. These results indicated that the position of the cleavage site in the amphiphile plays a critical role in how the morphology of the micelle is affected by proteolysis, with cleavage sites closer to the backbone giving rise to greater changes in morphology. Interestingly the efficiency of the enzyme cleavage reaction was estimated to be only 21% by HPLC analysis (one peptide per PPA) suggesting that small changes in the PPA structure can lead to significant changes in morphology.

These PPAs were also designed to be responsive to phosphorylation/dephosphorylation reactions. Treatment of the micelles with PKA resulted in phosphorylation of the serine residue which led to a dramatic change in morphology (Figure 5.6). TEM and SEM of **M2** showed the formation of amorphous structures while DLS indicated a 50-fold increase in the hydrodynamic radius. Subsequent treatment of the phosphorylated micelles with PP1 for 24 hours at 30 °C followed by dialysis, resulted in conversion of the amorphous

	Peptide substrate[a]	m^b	n^b	Polymer M_n, g/mol[c]	M_w/M_n[c]	D_h[d]	PDI[d]
M1	N-H- LRRASLGKGPLGLAG	34	6	19470	1.01	24	0.27
M2	N-H- KKPLGLAGLRRASLG	34	5	19430	1.17	33	0.19

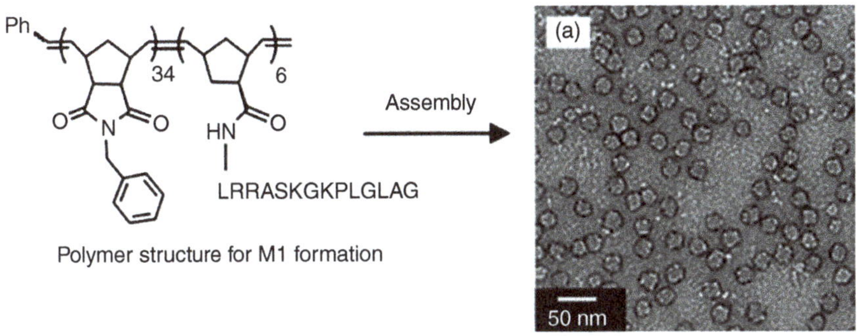

Figure 5.4 Polymer properties, structure, and characterization of PPA micelles. Top – [a] PKA/PP1 and MMP substrates are shown in red and blue, respectively, with phosphorylation or cleavage sites boldfaced and underlined. Peptides are conjugated to the polymer through the amino termini. [b] Block size of m ("phenyl block") was determined by SEC-MALS. ($M_n = 8553\ g\ mol^{-1}$), and n was estimated via SEC-MALS and UV–vis. [c] Polymer M_n and M_w/M_n determined by SEC-MALS. [d] Hydrodynamic diameter and micelle PDI (polydispersity) were determined by DLS. Bottom - TEM characterization of **M1**. (a) Micrograph of **M1** sample, stained with 1% uranyl acetate. *Source*: Reprinted with permission from ref. [30]. Copyright 2011 American Chemical Society.

structures to particles of the same size and morphology as the initial micelles. To confirm that the phosphorylation/dephosphorylation reaction occurred through a cycling process, radioactive adenosine triphosphate (ATP) was used to label the micelles. The phosphorylation reaction with PKA proceeded with a greater than 95% yield, while the dephosphorylation reaction with PP1 proceeded nearly quantitatively. The system showed no significant degradation over three phosphorylation/dephosphorylation cycles.

Utilizing information obtained in the previous example Gianneschi et al. designed a PPA that would act as a substrate for the cancer-associated enzyme MMP-9 [31]. The PPAs were generated from polymers labeled with peptides and dyes via post-polymerization conjugation reactions. The peptide sequence was chosen to position the MMP-9 cleavage site near the polymer backbone thus maximizing changes in both the shape and hydrophilicity of the peptide (hydrophilic) block, leading to changes in the morphology of the particles. Dye molecules fluorescein and rhodamine, capable of undergoing efficient Förster Resonance Energy Transfer (FRET) were chosen to allow for the detection of structural and aggregation states of the particles. Three sets of well-defined

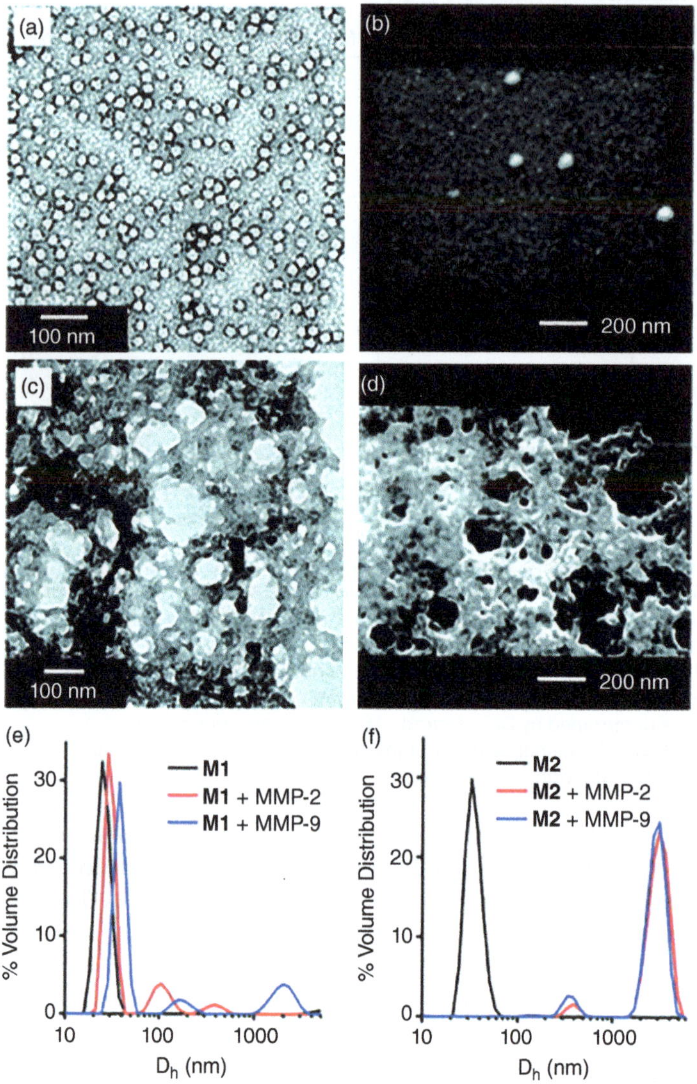

Figure 5.5 Response of micelles **M1** and **M2** to treatment with MMPs. (a) TEM **M1** + MMP-2. (b) SEM **M1** + MMP-2. (c) TEM **M2** + MMP-2. (d) SEM **M2** + MMP-2. (e-f) DLS for particles indicating changes in particle size upon MMP treatment. Micelles at 20 μM (with respect to polymer-peptide amphiphile) were incubated with MMPs (100 μM) at 37 °C for 24 h (Tris–HCl, 50 mM, pH 7.4). Reprinted with permission from ref. [30]. Copyright 2011 American Chemical Society.

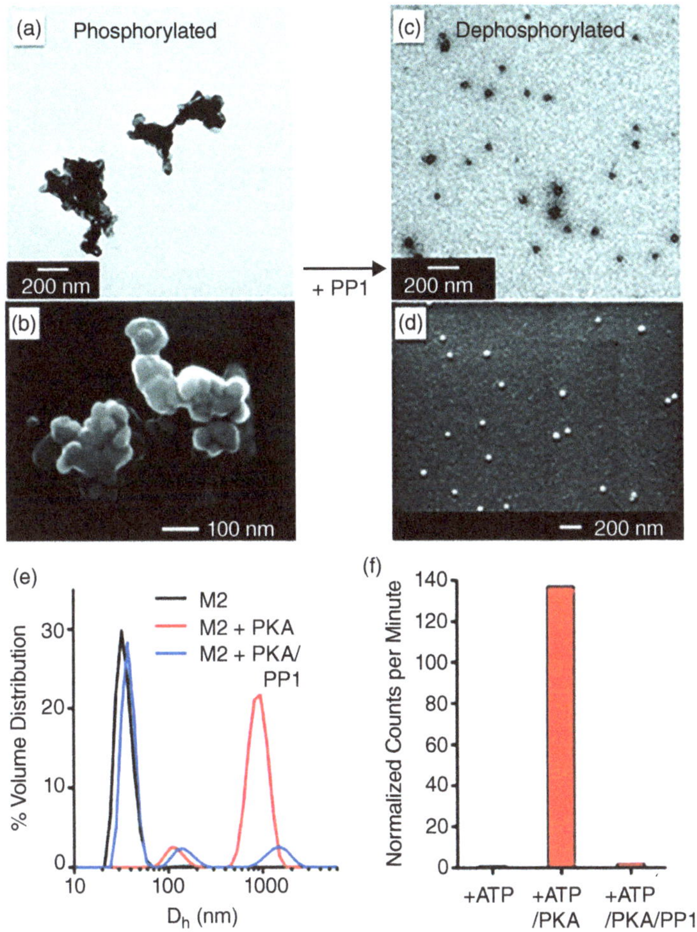

Figure 5.6 Response of **M2** to sequential additions of PKA and PP1. (a) TEM and (b) SEM: **M2** (20 μM with respect to polymer–peptide amphiphile) treated with PKA (2500 U) plus ATP (2 mM) and incubated at 30 °C for 24 h followed by dialysis. (c) TEM and (d) SEM: Phosphorylated particles subjected to either dialysis or heat denaturation of PKA (20 min, 65 °C) prior to treatment with PP1 (2.5 U) at 30 °C for 24 h. (e) DLS confirms increase and decrease of aggregate size in solution via phosphorylation and dephosphorylation respectively. (f) Phosphorylation and dephosphorylation were confirmed by radiolabelling the particles using [γ-^{32}P] ATP. Heat denaturation and extended dialysis had no effect on **M2** micelles alone. No change is observed with ATP (2 mM) without addition of PKA. *Source*: Reprinted with permission from ref. [30]. Copyright 2011 American Chemical Society.

spherical micelles **M1** (fluorescein only), **M2** (rhodamine only) and **M3** (fluorescein and rhodamine) were obtained following transition of the PPAs from 1 : 1 DMSO/DMF solutions to buffered water as shown in Figure 5.7. While the single dye particles showed the expected spectroscopic properties, a fluorescence signal at 512 nm for **M1** (fluorescein) and no observable peak for

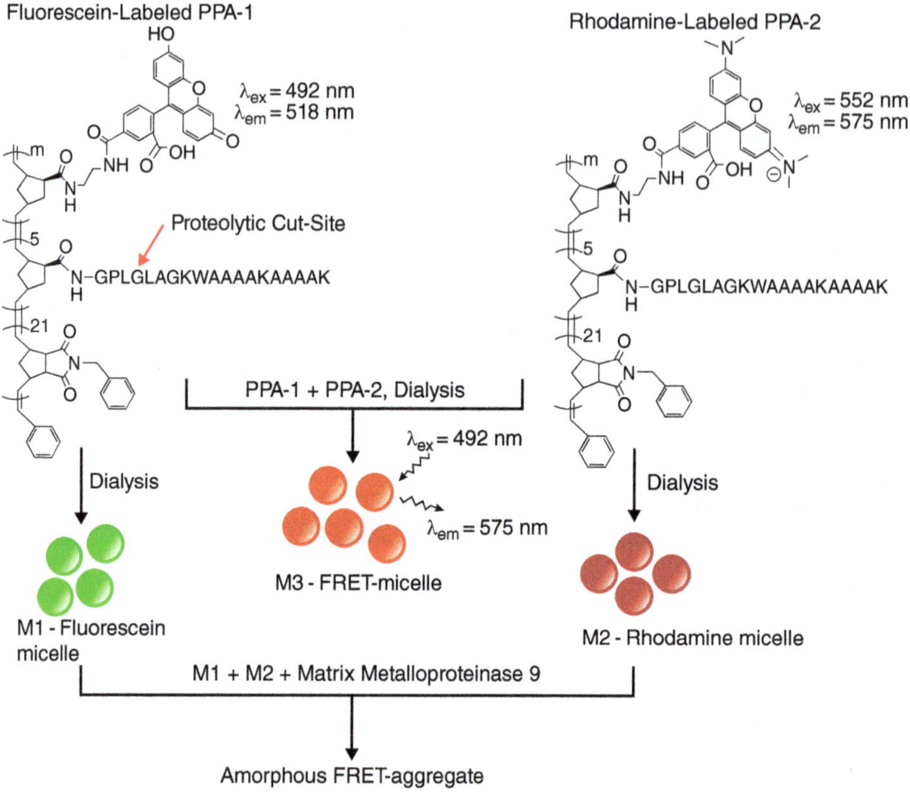

Figure 5.7 Assembly of peptide–polymer amphiphiles (PPAs) to generate fluorogenic micellar nanoparticles. Polymers were labeled with peptides and dyes, post-polymerization with block sizes determined by SEC-MALS analysis and spectroscopy. Degree of dye incorporation (m), was between 1 and 2 for both **PPA-1** and **PPA-2**. *Source*: From ref. [31]. Reproduced by permission of The Royal Society of Chemistry.

M2 (rhodamine) when excited at 470 nm, the mixed dye **M3** particles showed a fluorescence signal due to rhodamine at 563 nm, indicative of a FRET pair within the Förster radius.

A solution containing both **M1** and **M2** was treated with MMP-9, which resulted in the cleavage of the peptide substrate from the shell of the micelle. The cleavage event resulted in the formation of a new FRET-active species in solution as the PPAs rearrange into aggregates containing both dyes. The process was monitored by fluorescence spectroscopy and DLS with the final structures also being characterized by TEM (Figure 5.8). The micelles were also shown to undergo enzyme-induced aggregation in response to MMPs expressed from WEP1-NA45 cells as well as in blood serum doped cell media samples treated with MMP-9. These studies demonstrated the viability of

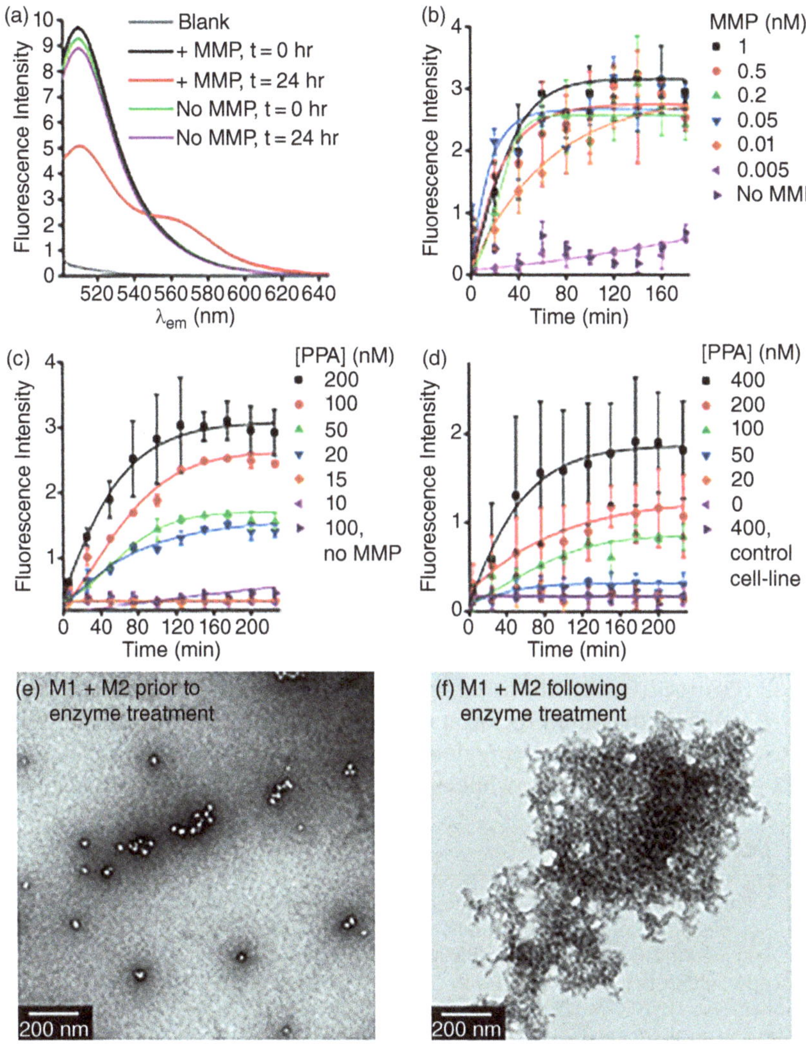

Figure 5.8 Response of mixtures of **M1** and **M2** to MMPs. (a) Fluorescence spectra of **M1** and **M2** (0.5 mM each with respect to PPA) with and without MMP-9 (10 nM) at times indicated following enzyme addition; λ_{ex}=470 nm. (b–d) Fluorescence intensity vs. time plots via plate reader analysis, to monitor rearrangement of **PPA-1** and **PPA-2** into new FRET active aggregates; λ_{ex}=490 nm and λ_{em}=590 nm. (b) Detection of MMP-9 down to 10 pM of enzyme with **M1** and **M2** (at 0.5 mM, [PPA]). (c) Detection of MMP-9 at 10 nM with varying concentrations of **M1** and **M2** shown with respect to [PPA], detectable down to 20 nM of polymer. (d) Detection of cell-secreted (WPE1-NA45 cells) MMP-2 and -9 with varying concentrations of **M1** and **M2** shown with respect to [PPA]. Cells were seeded at 1.6×10^4 cells/well in a clear bottom 96-well plate in Dulbecco Modified Eagle's Medium (DMEM). After 24 hours, cell medium was added to solutions of **M1** and **M2**. MMP-2 and -9 were at 0.048 and 0.005 nM respectively as quantified by an enzyme-linked immunosorbent assay (ELISA). Control was the non-MMP expressing MCF-7 cell-line cultured in the same manner. All reactions run in PBS, unless otherwise noted. (e–f) TEM of **M1** and **M2** before and after 24 hours following MMP-9 treatment. *Source*: From ref. [31]. Reproduced by permission of The Royal Society of Chemistry.

this method for detecting the presence of specific enzymes under biologically relevant conditions.

Building on the success of the previous example Gianneschi et al. designed a similar PPA to explore the viability of using disease associated enzymes to induce a micelle to aggregate morphology change in tumor tissue [32]. The PPAs were generated from block copolymers terminated with either fluorescein or rhodamine dyes, via a post-polymerization conjugation reaction of the peptide with an activated NHS ester moiety. Six sets of well-defined spherical micelles **M1** (fluorescein only), **M2** (rhodamine only), **M3** (fluorescein and rhodamine) and their D-amino acid equivalent, $\mathbf{M1_D}$-$\mathbf{M3_D}$ were obtained following transition of the PPAs from a 1 : 1 DMSO/DMF solution to buffered water as shown in Figure 5.9.

Initial *in vitro* experiments showed the expected morphology change and induction of a FRET signal for samples containing both **M1** and **M2** in the presence of MMP-9. By contrast, micelles containing D-amino acids ($\mathbf{M1_D}$ and $\mathbf{M2_D}$) maintained both their original structures and spectroscopic properties in the presence of active enzyme.

These particles were then used to assess if enzymatic signals within tumor tissue could trigger the micelle to aggregate assembly process. Three groups of HT-1080 xenograph mice were treated via tail vein injection with either a mixture of **M1** and **M2** (group 1), a mixture of $\mathbf{M1_D}$ and $\mathbf{M2_D}$ (group 2) or **M3** (group 3) (Figure 5.10 a–c). The assembly process was monitored via FRET signal development in the tumor region at various time points. A noticeable FRET signal was observed in the tumor region of mice injected with a mixture of **M1** and **M2** particles after one day, while no signal was observed in mice injected

Figure 5.9 Preparation of MMP-responsive fluorescent micellar nanoparticles. Peptide–polymer amphiphiles (**PPAs**) with fluorescent tags were generated utilizing ring-opening metathesis polymerization. Block sizes were determined by SEC-MALS and spectroscopy. Synthesis: (i) **1** was mixed with Grubbs' modified initiator for 45 minutes and an aliquot analyzed by SEC-MALS to confirm degree of polymerization. (ii) **2** was added for 45 minutes and an aliquot analyzed for confirmation of the 21 : 8 block copolymer ratio by SEC-MALS. (iii) The resulting polymer was split into two pots and mixed with novel fluorescent termination agents **3** and **4**, for two hours followed by addition of ethyl vinyl ether to quench the catalyst. This procedure generates the fluorescently labeled polymers as shown (green circle – fluorescein; red circle – rhodamine). (iv) Peptides were added to the block copolymers to form PPAs via amide-bond formation through the amino terminus of the peptide and the polymeric NHS-ester, with degree of incorporation determined by SEC-MALS. (v) Dialysis of **PPAs** from DMSO to PBS (pH 7.4) buffered water over 24 hours to generate micellar nanoparticles. **M1** contains **PPA—F**, **M2** contains **PPA-R**, and **M3** contains **PPA-F** and **PPA-R** mixed in a 1 : 1 ratio. Peptide sequences: L-amino acid sequence: GPLGLAGGWGERDGS. D-amino acid sequence: gplglagGWGERDGS (lowercase indicates portion prepared from D-amino acids). *Source*: From ref. [32]. Copyright © 2000 by John Wiley & Sons, Inc. Reprinted by permission of John Wiley & Sons, Inc.

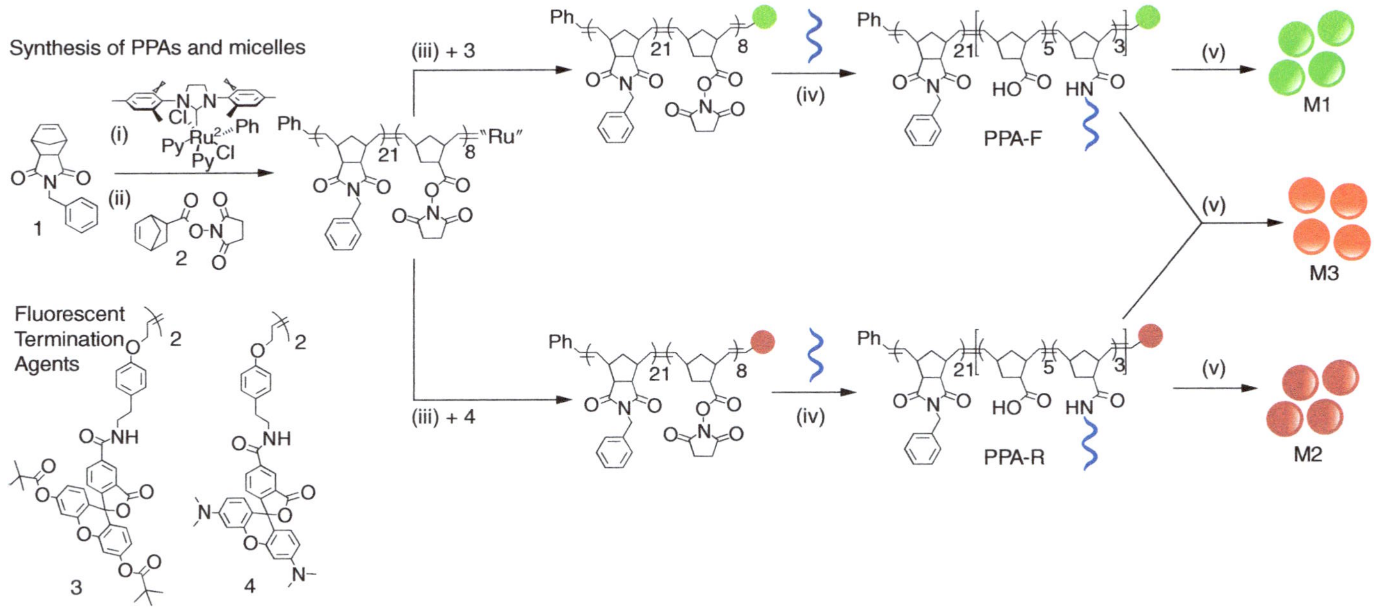
Synthesis of PPAs and micelles
(i)
(ii)
1
2
(iii) + 3
(iii) + 4
(iv)
(v)
PPA-F
PPA-R
M1
M2
M3
Fluorescent
Termination
Agents
3
4

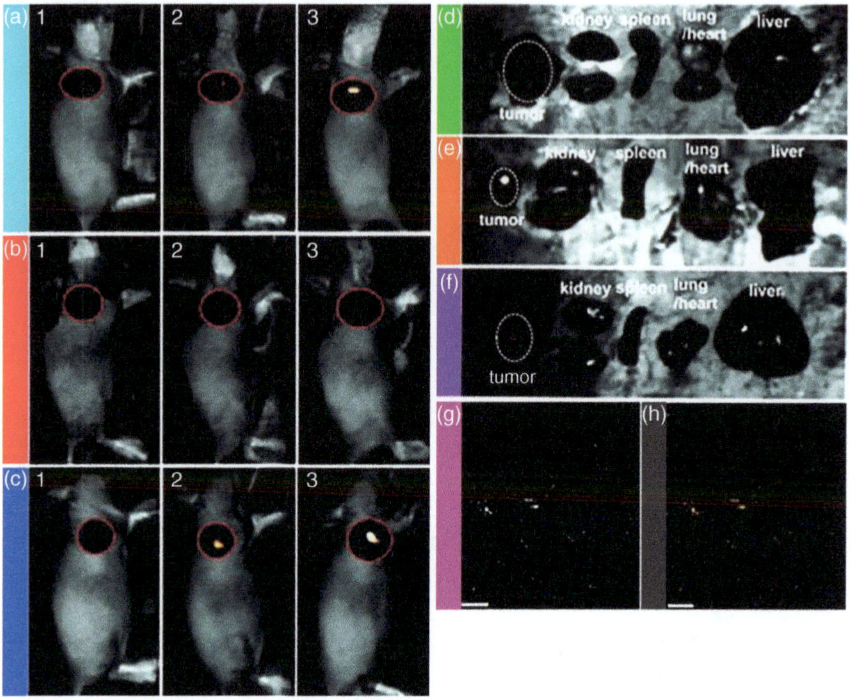

Figure 5.10 Intravenously injected mice imaged to show the time course of accumulation of enzyme-responsive particles in the region of interest (tumor). (a) **M1** + **M2** co-injected. (b) $\mathbf{M1_D}$ + $\mathbf{M2_D}$ co-injected. (c) **M3** injected. (1) Background prior to injection. (2) t = one day. (3) t = two days. (d–f) Ex vivo organ analyses: (d) Background; no injection. (e) **M1** + **M2** co-injected, removed after two days. (f) $\mathbf{M1_D}$ + $\mathbf{M2_D}$ co-injected, removed after two days. General conditions: Micellar nanoparticles were injected via tail vein at 40 nmoles (100 μl volume) as a solution in PBS buffer. HT-1080 xenograft nude mice with tumor sizes approximately 150 mm^3 were utilized for this study. Red circles indicate tumor location identified visually and scanned. Fluorescence images are overlayed on the bright field images of the animals for point of reference as to tumor location. *Source*: From ref. [32]. Copyright © 2000 by John Wiley & Sons, Inc. Reprinted by permission of John Wiley & Sons, Inc.

with the D-amino acid containing peptide substrate, consistent with the results observed in the *in vitro* study. The injection of **M3** particles also resulted in the development of a FRET signal in the tumor region as expected for particles containing both donor and acceptor dyes and active enzyme substrate. To assess if the accumulation event was localized to the tumor, selected organs in addition to the tumor were excised and the FRET signals measured. For mice injected with a mixture of **M1** and **M2** a significantly larger FRET signal was observed in the tumor compared to the other organs. Mice injected with a mixture of $\mathbf{M1_D}$ and $\mathbf{M2_D}$ particles showed a very weak FRET signal in the tumor tissue indicating the possibility of low level enzymatic cleavage which is consistent with other studies of D-amino acid substrates *in vivo*.

To determine if the programmed morphology change from micelle to aggregate resulted in a slow clearance of the materials from the tumor, FRET active micelles **M3** and **M3*D*** were injected directly into the tumor tissue and the loss of FRET signal was monitored over time. A strong FRET signal is observed for both sets of micelles immediately following injection; however after one hour the signal is no longer visible in the tumor tissue injected with $\mathbf{M3_D}$ particles. By contrast a strong signal is observed in the tumor tissue injected with **M3** particles. These studies demonstrated that it is in fact possible to program PPAs to use endogenous, enzymatic signals to guide and control their accumulation *in vivo* via a micelle to aggregate morphology change.

The peptide polymer amphiphilic systems described above were all synthesized via conjugation reactions between the polymer and the peptide unit. While this strategy proved successful in the development of enzyme responsive micellar systems it inherently suffers from a lack of quantitative modification of the polymer backbone which inhibits one's ability to determine and regulate the density of peptide in the corona of the micelle, which may play a role in the programming and response of these systems to enzymes.

To address this, Gianneschi et al. [33] synthesized two PPAs by directly polymerizing both the hydrophobic block, comprised of a phenyl norbornene unit, and the hydrophilic block, comprised of a norbornenyl-peptide unit, via ring-opening metathesis polymerization (ROMP) using a modified Grubbs' second-generation catalyst (Scheme 5.1). The two peptide sequences used in the study are known substrates for enzymes MMP-2 and -9. This method generated well-defined PPAs where the density and location of the peptide moieties were known. These PPAs formed micellar nanoparticles of uniform size and shape, as shown in Figure 5.11. Interestingly, when the micellar nanoparticles were treated with MMP-2, no cleavage of the peptide substrate was observed. This is in contrast to the micellar systems composed of PPAs synthesized via a post-polymerization strategy highlighted above, indicating that the spatial arrangement and density of the peptide may allow for the programming of a micelles susceptibility to proteolytic cleavage and thus the ability to undergo a change in morphology and or function.

5.3 DNA-Programmed Micelle Systems

Nucleic acids are unique informational molecules capable of highly predictable sequence-specific recognition by other synthetic or natural nucleic acids, enzymes, and proteins. Additionally, nucleic acids are capable of multiple modes of selective binding with small molecules, as observed for aptameric systems [34–37] and via intercalation or DNA sequence-specific

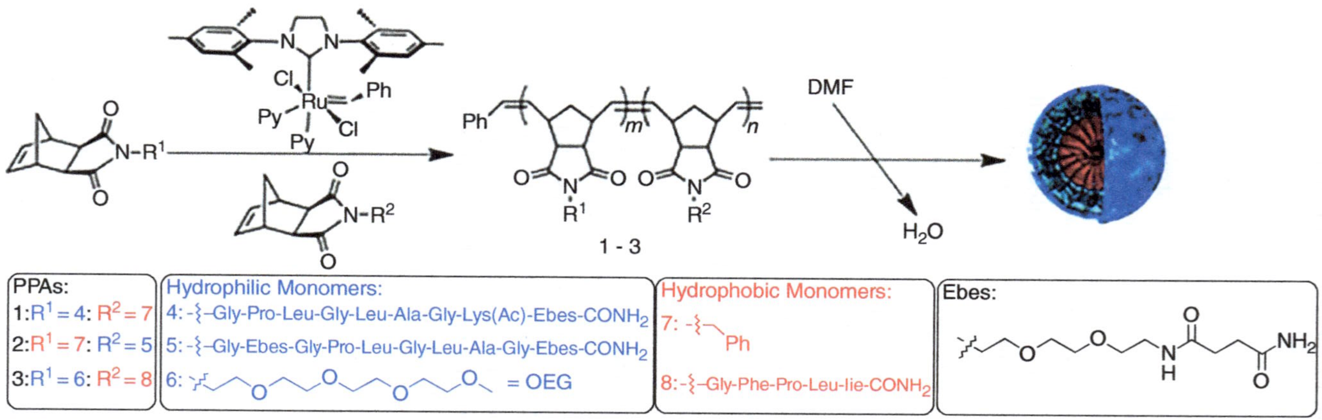

Scheme 5.1 Synthesis of peptide–polymer amphiphiles (PPAs) and formulation into nanoparticles. Hydrophobic components of the PPAs are color-coded in red. Hydrophilic components of the PPAs are color-coded with blue. *Source*: From ref. [33]. Reproduced by permission of The Royal Society of Chemistry.

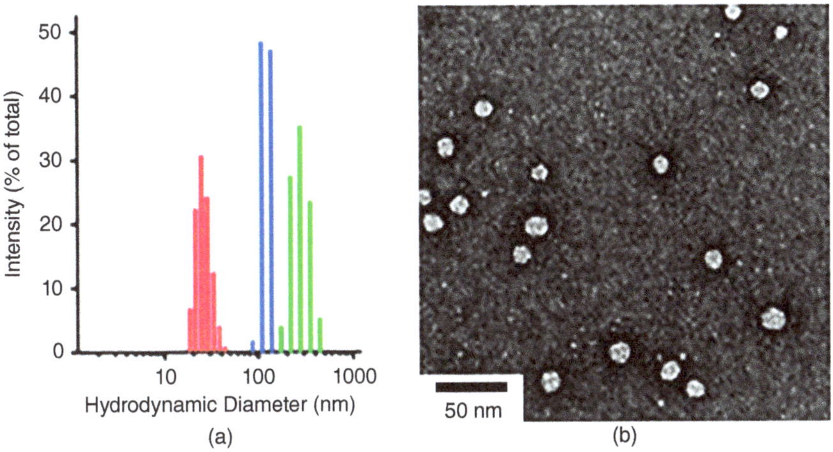

Figure 5.11 (a) DLS size intensity distribution of nanoparticles derived from **PPA 1** (red), **PPA 2** (blue), and **PPA 3** (green). (b) TEM of nanoparticles derived from **PPA 1**. *Source*: From ref. [33]. Reproduced by permission of The Royal Society of Chemistry.

binding [38, 39]. In addition, nucleic acids are known to occur naturally as catalysts [40] and can be evolved *in vitro* [41–44] to behave as such. This functional diversity arises because of natural evolution of function in combination with non-natural function facilitated by modern synthetic molecular evolution and selection processes [45]. At the heart of these capabilities is the information content of nucleic acids. This is the cornerstone of the genetic code and is increasingly of value in the construction of synthetic nanoscale materials. The field of DNA nanotechnology has blossomed in recent years, with DNA increasingly employed in the construction of complex nanoscale architectures made entirely from nucleic acids [46–52] or for the decoration of both organic and inorganic nanoparticles [53–57]. DNA, with its highly negative charged backbone has been used as the hydrophilic portion of DNA-hybrid-amphiphiles where the hydrophobic portion is comprised of either a lipid-like small molecule or a polymer chain [58–82]. In this section we highlight systems where the DNA is used to program spherical micelles for a specific function. In most of these examples, the sequence selective binding capabilities of the DNA unit(s) are used to bind complementary target strands as a method to incorporate additional functionality such as dye or drug molecules or to induce a change in the function of the material. This is in contrast to most of the peptide-based micelle systems where the programming element was used to trigger a change in morphology. While there are a number of publications that report the use of single nucleic acid bases or very short nucleotides in the synthesis of amphiphilic systems these will not be discussed here as these materials lack the programming element imparted by longer

DNA sequences, however, these systems have been the subject of recent reviews [83–85]. In addition, we have excluded synthetic mimics of DNA such as peptide nucleic acids, which are garnering increased attention [86–89].

5.3.1 Lipid-Like DNA Amphiphiles

The simplest method for synthesizing DNA-amphiphiles capable of forming micellar aggregates is to conjugate an oligonucleotide to a lipid like molecule. Gosse et al. [59] and Tan et al. [90] demonstrated that well-defined micelles ranging between 8 and 35 nm in diameter could be assembled in aqueous solution when oligonucleotides of various lengths were conjugated to lipid-like molecules containing either one or two alkyl chains respectively (Figure 5.12). The DNA micelles retained their ability to undergo hybridization with complementary DNA strands, demonstrating the potential use of these types of materials as delivery vehicles/carriers for moieties that can be attached to complementary DNA strands.

In addition to their ability to bind complementary DNA sequences via traditional Watson-Crick base pairings, single-stranded DNA sequences can form complex tertiary structures that recognize proteins, small molecule targets, and cancer cells through noncovalent interactions. This type of

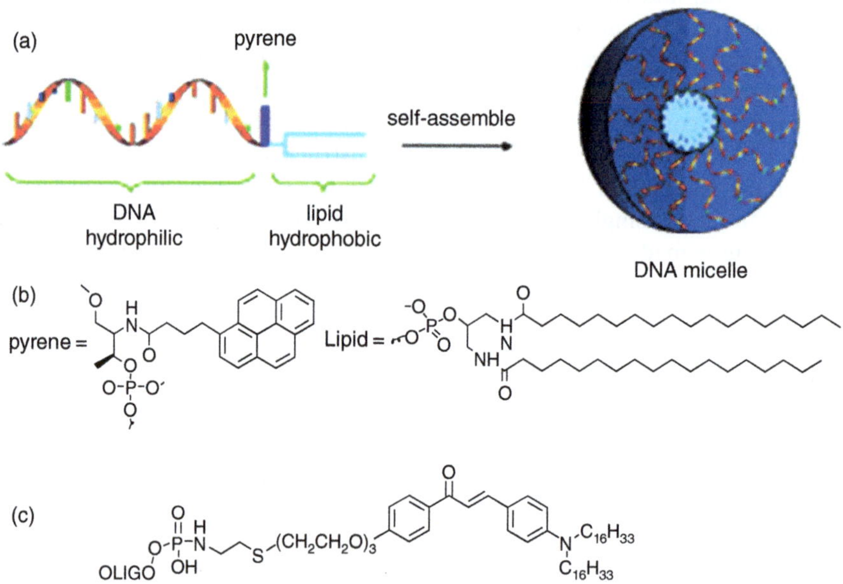

Figure 5.12 Schematic of the design and assembly of DNA micelles. (a) Oligonucleotide micelles contain a DNA corona, a pyrene unit (fluorescence reporter), and a lipid core. (b) Molecular structure of pyrene and the lipid unit. *Source*: From ref. [59]. Copyright © 2000 by John Wiley & Sons, Inc. Reprinted by permission of John Wiley & Sons, Inc. (c) Molecular structure of the oligonucleotide amphiphile synthesized by Gosse et al. See ref. [59].

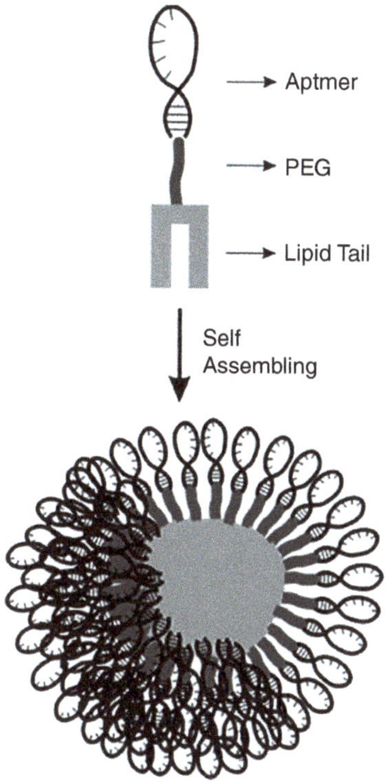

Scheme 5.2 Illustration of DNA aptamer amphiphile and micelle formation. *Source*: Redrawn from ref. [61].

oligonucleotide is referred to as an aptamer. In 2010, Tan et al. [61] synthesized a number of DNA–amphiphile systems by conjugating a lipid phosphomidite containing two C11 alkyl chains to the end of various aptamers (Scheme 5.2). These amphiphiles formed micelles 68 nm in diameter, as determined by TEM and DLS. Interestingly, two unexpected outcomes arose as a result of the DNA aptamer-amphiphiles, forming micellar nanostructures. First, the binding capability of the aptamers was found to increase by up to 80 times at physiological temperatures, and second, aptamers that did not undergo internalization themselves were taken up in cells when formulated as a micelle, as demonstrated by optical imaging with z-axis scanning.

Expanding on the use of aptamers as the hydrophilic moiety in DNA-amphiphiles, Tan et al. [62] developed DNA-micelle flares for imaging and gene therapy, where the DNA unit is an aptamer beacon. The sequence of the aptamer results in a loop-stem structure where a fluorophore and quencher at the 3′ and 5′ ends of the sequence are brought in close proximity, resulting in quenching of the fluorescent signal. Upon binding the target molecule (ATP) the aptamer undergoes a conformational change that results in the fluorescence signal being restored (Figure 5.13). Conjugation of the aptamer

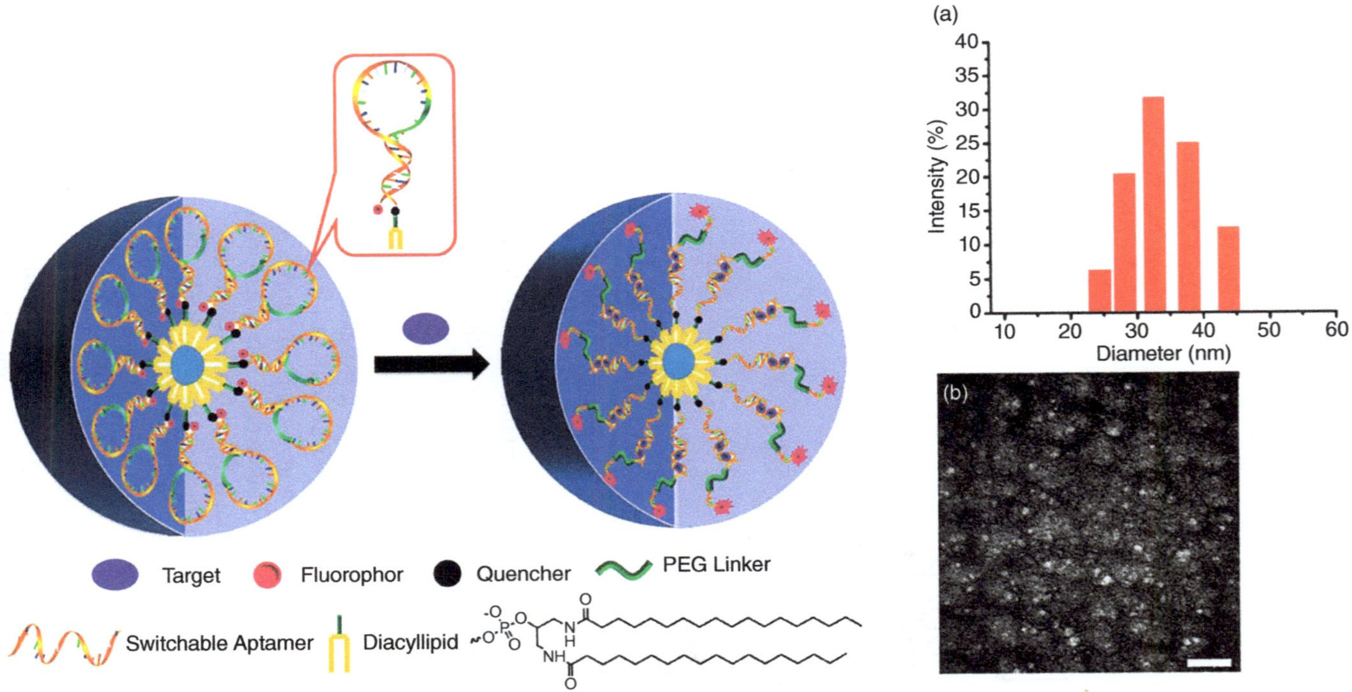

Figure 5.13 Working principle of switchable aptamer micelle flares (SAMFs) (left) and their characterization (right). Aptamer switch probe and diacyllipid conjugates are self-assembled to form the nanostructure of micelle flares. In the absence of target, the aptamer switch probe maintains a loopstem structure, and fluorescence is quenched because of the close proximity between fluorophore and quencher. Upon target binding, the conformation of switchable aptamer is altered, resulting in the restoration of fluorescence signal. (a) Dynamic light scattering of SAMFs. (b) TEM image of SAMFs after negative staining by 2% aqueous uranyl acetate. Scale bar: 200 nm. *Source*: Reprinted with permission from ref. [62]. Copyright 2011 American Chemical Society.

to a lipid phosphoramidite containing two C11 alkyl chains yielded a DNA aptamer-amphiphile that formed micelles 35 nm in diameter in aqueous solution, as determined by DLS and TEM (Figure 5.13).

Upon treatment of the aptamer-micelles (1 μM) with a series of ATP concentrations, a 3.9-fold increase in the fluorescence signal was observed, signifying successful binding of the ATP to the aptamer. The selectivity of the aptamer-micelles for ATP was demonstrated by treating the micelles with structural analogues thymidine triphosphate (TTP), guanidine triphosphate (GTP), cytosine triphosphate (CTP) and uridine triphosphate (UTP) (Figure 5.14). Finally, the aptamer micelles were shown to effectively internalize into HeLa cells compared to only the aptamer.

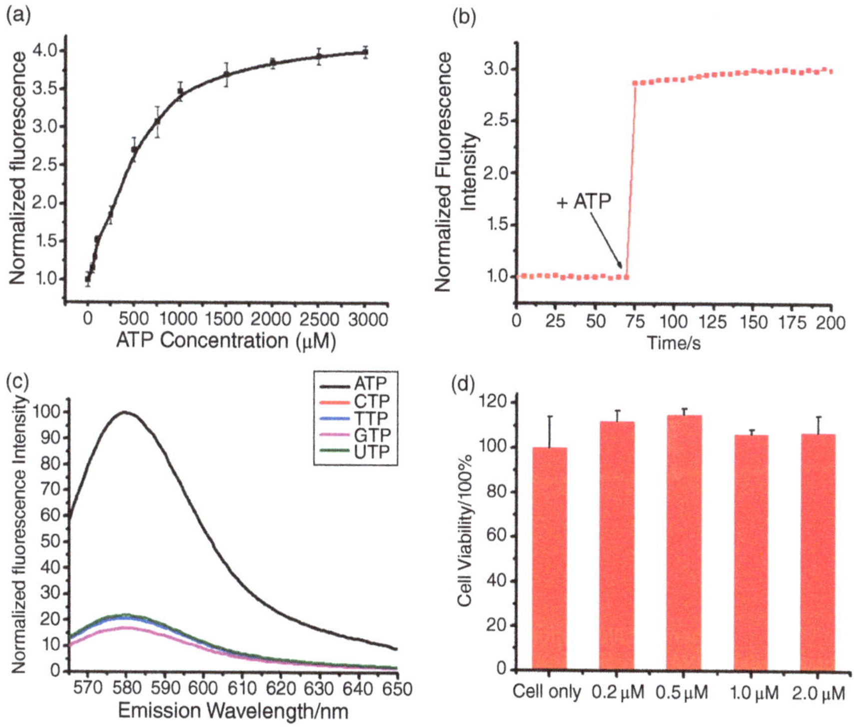

Figure 5.14 Switchable aptamer micelle flares respond to target ATP molecules *in vitro*. (a) Fluorescence titration of SAMF (1 μM) against increased concentrations of ATP (0.1–3.0 mM). (b) Time course of fluorescence associated with the binding of target molecules. The concentration of ATP was 1 mM. (c) Switchable aptamer micelle flares distinguish ATP from its analogues: TTP, GTP, CTP, and UTP. The concentration of ATP and its analogues was 2 mM. (d) Cytotoxicity assay of HeLa cells treated with different concentrations of SAMFs (0, 0.2, 0.5, 1, 2 μM). *Source*: Reprinted with permission from [62]. Copyright 2011 American Chemical Society.

Having successfully demonstrated the effectiveness of this method, Tan et al. [63] applied these principles to design a system capable of intracellular mRNA imaging and gene therapy (Figure 5.15). Conjugation of a DNA aptamer beacon to a dialkyl lipid moiety resulted in the spontaneous formation of molecular beacon micelle flares (MBMFs) in aqueous solution. The diameter of the micelles increased from 20 to 30 nm upon binding of the complementary DNA sequence, which coincided with an increase in the fluorescence signal

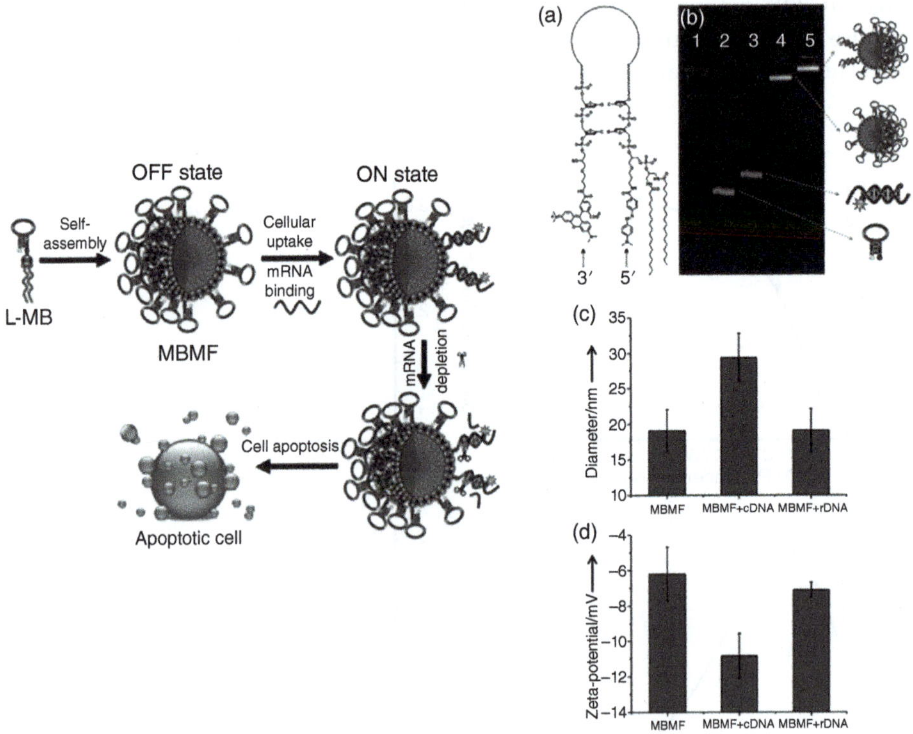

Figure 5.15 Schematic illustration of molecular beacon micelle flares (MBMFs) for intracellular mRNA detection and gene therapy (left) and their characterization (right). Diacyllipid–molecular-beacon conjugates (L-MBs) self-assemble into MBMFs and enter living cells. Before binding their target mRNA, the fluorophore and the quencher of the MBMFs are in close proximity (OFF state). Hybridization between the loop region and the target mRNA separates the fluorophore and the quencher, producing a fluorescence signal (ON state) and DNA/RNA heteroduplex for RNase H action. Note: not all MBs are shown on the MBMF. Characterization of MBMFs. a) Structure of MBMFs. Note: not all the bases are shown. b) Agarose gel electrophoresis of DNA marker (lane 1), MBs (lane 2), MBs with synthetic complementary target (cDNA; lane 3), MBMFs (lane 4), and MBMFs with cDNA (lane 5). c) DLS and d) zeta-potential measurements of MBMFs, MBMFs with cDNA, and MBMFs with a synthetic non-complementary target (rDNA). *Source*: From ref. [63]. Copyright © 2013 by John Wiley & Sons, Inc. Reprinted by permission of John Wiley & Sons, Inc.

(Figure 5.15). Treatment of A549 cells that express the mRNA sequence (c-*raf-1*) with complementary MBMFs showed a 2.5× increase in fluorescence signal compared to noncomplementary MBMF's. Finally, it was shown that these types of systems were also effective at inhibiting cell proliferation.

5.3.2 DNA–Polymer Amphiphiles

There are two general categories of DNA–polymer amphiphiles: linear systems, where one end of an oligonucleotide is coupled to the end of a polymer chain, and brush systems, where oligonucleotides are conjugated to the side chains of a polymer. This is most often accomplished via a post-polymerization reaction between complementary reactive groups on the DNA and polymer side chains. However, recently there have been reports of the polymerization of DNA functionalized monomers [60, 89, 91]. Amphiphilic DNA-based biohybrid polymers have the ability to be manipulated via two key approaches: (i) isothermal and thermal DNA hybridization, and (ii) response to enzymatic activity.

In 2001, Park and Jeong [64] demonstrated that a DNA–polymer amphiphile could assemble into micelles upon addition to aqueous solution. The amphiphile was synthesized by conjugation of the free terminal carboxylic group of Poly(D,L-lactic-*co*-glycolic acid) to the primary amine group at the 5′ end of the antisense oligonucleotide *c-myc*. Micelles approximately 65 nm in diameter as determined by DLS and atomic force microscopy (AFM) were obtained after transition of the system from DMSO to water (Figure 5.16). In order to visualize the endocytosis process of the micelles, a complementary rhodamine-labeled sense oligonucleotide was hybridized to the antisense

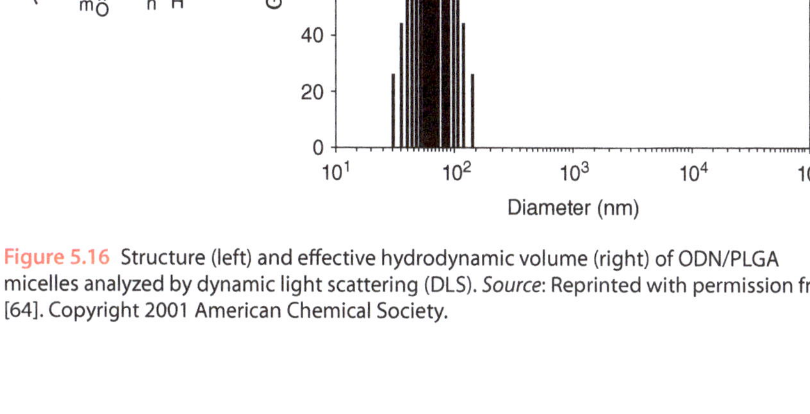

Figure 5.16 Structure (left) and effective hydrodynamic volume (right) of ODN/PLGA micelles analyzed by dynamic light scattering (DLS). *Source*: Reprinted with permission from [64]. Copyright 2001 American Chemical Society.

oligonucleotide in the corona of the micelle. Treatment of NIH3T3 mouse fibroblast cells showed even distribution of the micelles in the cytoplasm, confirming that this type of system could be used to deliver DNA into a cell.

In 2004 Mirkin et al. [65] described a micellar system assembled from DNA–polymer amphiphiles where the DNA was used as a programming moiety. The amphiphiles were synthesized by coupling a polystyrene polymer containing a phosphoramidite end group to alcohol terminate oligonucleotides on controlled pore glass beads (CPG). Well-defined micellar structures with diameters of 13–18 nm were obtained as characterized by tapping mode AFM and DLS, following slow addition of water to a solution of the amphiphiles in DMF (Figure 5.17). The selective programming nature of these micelles was demonstrated by adding them to a solution of 13 nm gold nanoparticles modified with the complementary DNA sequence. Upon addition, sequence-selective assembly takes place due to hybridization of the complementary DNA strands, which results in aggregate formation. The aggregate could be disassembled by heating the system above the melting temperature of the duplex DNA, as monitored by the surface plasmon band of the gold nanoparticles. This example demonstrated that DNA–polymer micelles are capable of recognition properties as defined by the DNA sequence and that they can be used to build higher-order structures through sequence-selective

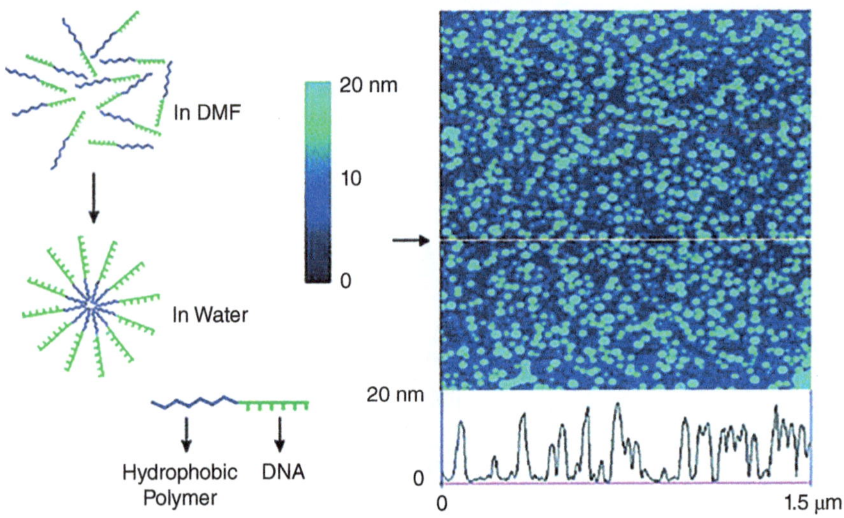

Figure 5.17 Schematic illustration of DNA micelle formation (left) and their characterization (right). Tapping mode AFM image showing the spherical micelle structures constructed from polymer-DNA amphiphiles. Typically, a drop of micelle solution (5 μl) was placed on an aminopropyltrimethoxysilane functionalized mica surface, sprayed with dry N_2, washed with deionized water, and dried again with flowing N_2. *Source*: Reprinted with permission from [65]. Copyright 2004 American Chemical Society.

hybridization with materials functionalized with the complementary DNA sequence.

In 2007, the Herrmann group developed a DNA–polymer micellar system capable of undergoing a programmed response to complementary DNA sequences (Figure 5.18) [66]. Amphiphilic polymers comprising a 22-base DNA hydrophilic block and a polypropyleneoxide (PPO) hydrophobic block were used to generate well-defined micelles with a diameter of approximately 5 nm. Micelles were treated with a short, complementary DNA sequence (22 bases), which resulted in double-helix formation within the corona of the micelles (Figure 5.18). This hybridization event had no noticeable effect on the structure of the micelles as determined by both scanning force microscopy and fluorescence correlation spectroscopy. The effect of DNA length on the morphology of the DNA–polymer assemblies was probed by the addition of either a sequence containing 110 bases (T110) or 88 bases (T88) to solutions of the micelles. These sequences encoded the complementary sequence of the DNA–polymer multiple times (five for T110 and four for T88), allowing for multiple polymers to be hybridized to a single complement strand. The addition of either sequence to a solution of the ssDNA micelles resulted in the formation of rodlike assemblies (Figure 5.18). SFM analysis of the assemblies on a mica surface showed that the length of the rods were 29.1 ± 6.5 and 22.7 ± 5.1 nm with heights of 1.95 ± 0.1 and 1.72 ± 0.2 nm for the T110 and T88 sequences, respectively. This system demonstrates the ability of DNA–polymer micelles to undergo a morphology change as a result of hybridization to long complementary DNA strands and that the assembly of DNA–polymer nanostructures can be mediated via interactions of the hydrophobic polymer block.

In addition to controlling the morphology of DNA–polymer micelles Herrmann et al. have also shown that the enzyme, terminal deoxynucleotidyl transferase (TdT) can be used to grow (extend) the DNA shell of a DNA-b-PPO micelle (Scheme 5.3) [67, 92]. The growth of the micelles was monitored *in situ* by SFM in soft-tapping mode on a freshly cleaved mica surface. Adding 2′deoxythymidine-5′-triphosphate (dTTP) mononucleotide and TdT to the micelles resulted in a 40% increase in size (4.5–6.3 nm) after two hours, while longer reaction times did not result in further DNA chain growth. Although the DNA chain extension via TdT is not sequence dependent, one can imagine using the initial DNA sequence to program the micelle for multiple purposes.

The systems just outlined illustrate the two strategies used to manipulate DNA-programmed micelles – namely, hybridization to complementary ssDNA and enzymes. In 2010, Gianneschi et al. developed a DNA–polymer system that was responsive to both of these strategies [68]. DNA–polymer amphiphiles were synthesized via a post-polymerization conjugation reaction between the CPG-bound DNA and the activated NHS ester moieties (hydrophilic block) of a block copolymer. The resulting DNA-polymer amphiphiles assembled into spherical micelles approximately 25 nm in diameter, as shown by TEM,

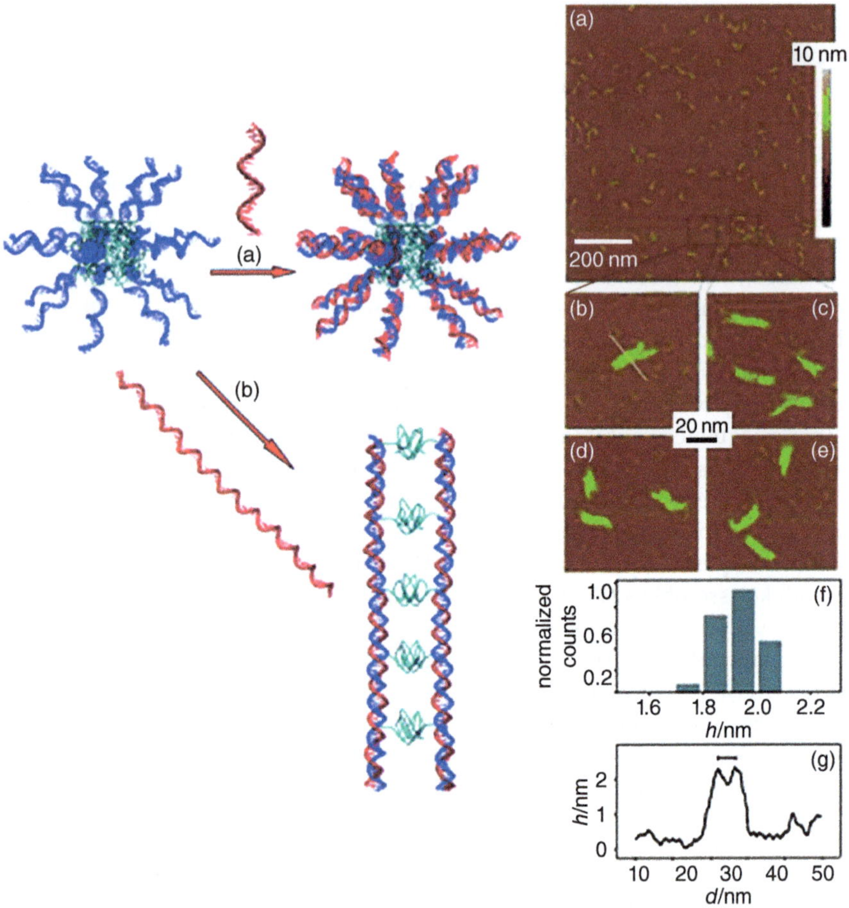

Figure 5.18 Schematic representation of hybridization of ss-DNA-b-PPO micelles with different DNA molecules (left) and their characterization (right). Left – (a) Base pairing with a short complementary sequence yields micelles with a ds corona, which maintains the overall shape of the aggregates. (b) Hybridization with long DNA templates results in rodlike micelles consisting of two parallel aligned double helices. Right(a) SFM topography image of the hybridization products of DNA-b-PPO and T110. (b–e) Close-up images of rodlike micelles consisting of two DNA helices arranged parallel to each other forming a dimer. (b and c) Enlarged images recorded at higher resolution than in (a). (d and e) Topography images taken from other experiments obtained under similar conditions (f) Histogram showing the height of the rodlike aggregates. (g) The line profile along the indicated line in (b). h = height, d = distance. *Source*: From ref. [66]. Copyright © 2013 by John Wiley & Sons, Inc. Reprinted by permission of John Wiley & Sons, Inc.

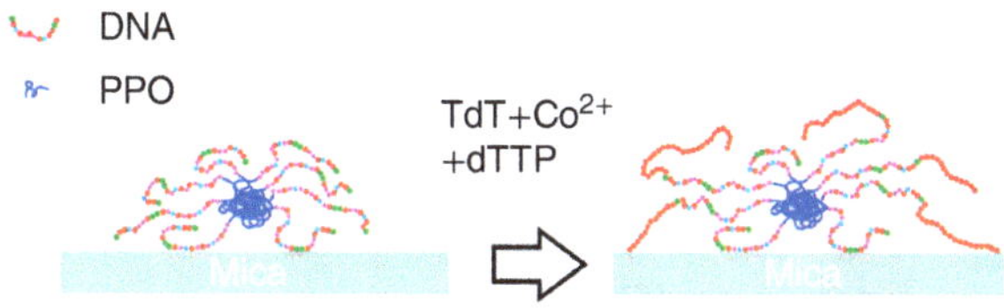

Scheme 5.3 Diagram showing the growth of the DNA-b-PPO diblock copolymer micelles on mica in a buffer solution in the presence of TdT, Co^{2+}, and dTTP. Reprinted with permission from *Source*: [67]. Copyright 2008 American Chemical Society.

AFM, and DLS (Figure 5.19). The DNA sequence contained an RNA base as the enzymatic cleavage site and two 18-member ethylene glycol moieties to increase the steric bulk of the hydrophilic block. Treatment of the micelles with a DNA-based phosphodiesterase (DNAzyme) resulted in complete and rapid cleavage of the DNA in the shell of the micelle, leaving a short ssDNA sequence (Figure 5.19). This cleavage event resulted in a sphere-to-cylinder morphology change as the "new" DNA amphiphiles reorganized to minimize packing parameters. Interestingly, the catalytic efficiency of the DNAzyme was shown to be greatly enhanced in the micelle substrate compared to the analogous ssDNA substrate.

The morphology change was shown to be reversible upon addition of a 19-base DNA sequence that could form a 9-base double-helix with the truncated DNA in the shell of the cylinder. The resulting spherical micelle contained a shell composed of a noncovalent DNA duplex that could be disrupted by strand invasion. Addition of a perfectly complementary 19-base DNA strand to the micelles resulted in strand invasion and hybridization, which regenerated the truncated DNA amphiphile and hence the reformation of the cylinder morphology. In addition to using the sequence-selective nature of DNA to isothermally modulate the morphology of the system, these morphology changes could also be triggered via thermally controlling the DNA hybridization by heating the system above the melting temperature of the DNA duplex.

As a demonstration of the selectivity and programmability of these DNA–polymer micelles, fluorescence studies were carried out on a micelle solution containing two different micelles, each with a unique DNA sequence in the shell of the particle and labeled with either fluorescein or rhodamine dyes. The mixed micelle solution was treated with two different DNAzymes, each one encoded to cleave only one of the dye-labeled micelles. Addition of the DNAzyme encoded for the fluorescein particle resulted in the formation of only green cylinders, while addition of the DNAzyme encoded for the rhodamine particle resulted in the formation of only red cylinders (Figure 5.20). When both DNAzymes were added to the solution, cylinders containing both fluorescein and rhodamine were formed. This study demonstrated how one can manipulate specific micelles selectively even in the presence of structurally

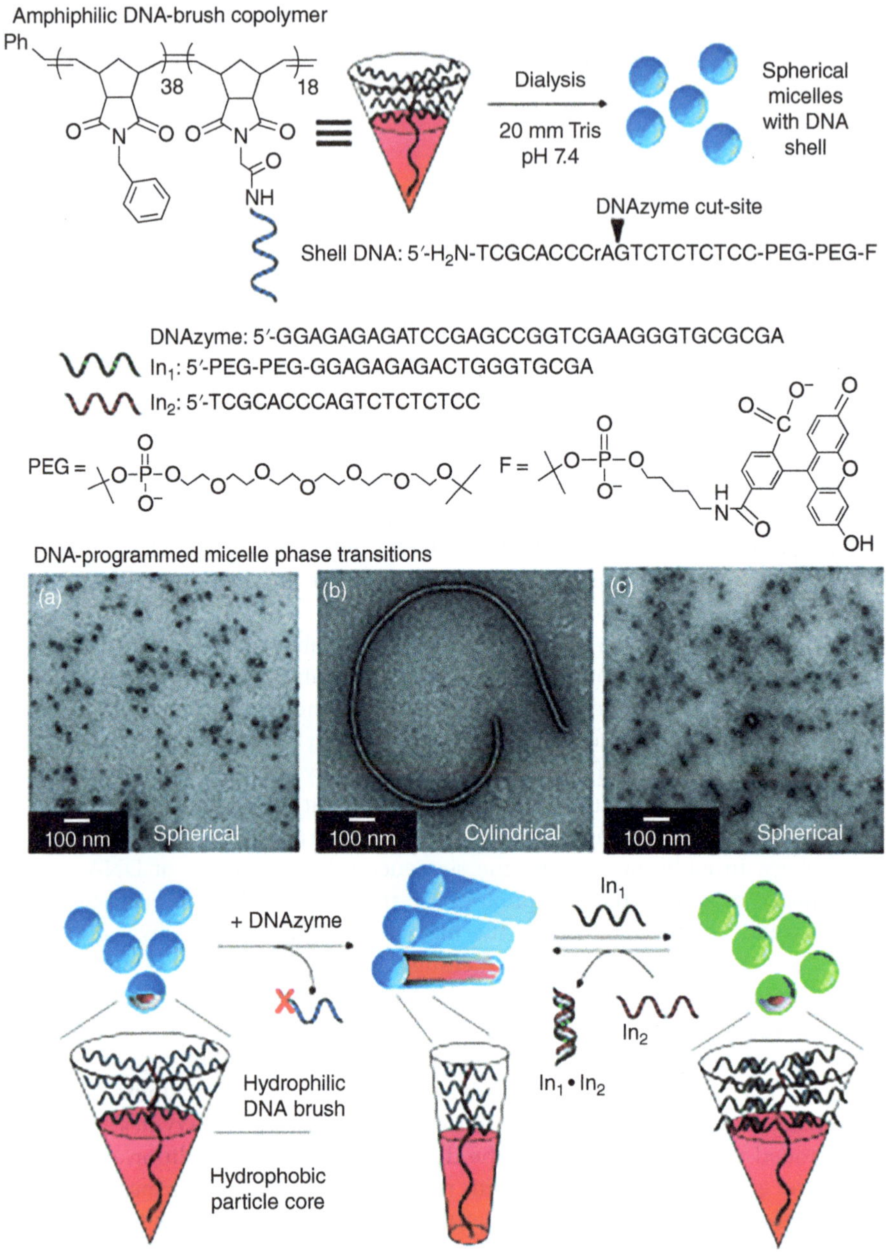

Figure 5.19 Assembly of DNA-brush copolymers into micelles with spherical or cylindrical morphologies. Amphiphile structures are represented as cones for each respective morphology, with the hydrophobic domain highlighted in red. TEM images of (a) 25 nm spherical micelles assembled from initial DNA-brush copolymers; (b) cylindrical morphology formed following DNAzyme addition to spheres; (c) spherical micelles (green) formed after addition of $\mathbf{In_1}$ to cylinders. *Source*: From ref. [67]. Copyright © 2010 by John Wiley & Sons, Inc. Reprinted by permission of John Wiley & Sons, Inc.

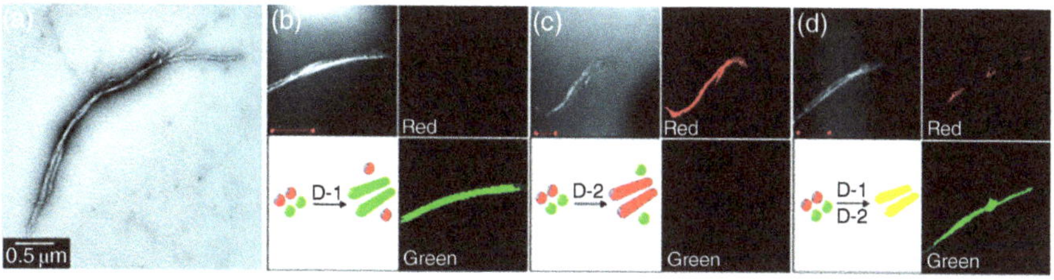

Figure 5.20 Sequence-selective phase shifting observed by fluorescence microscopy. (a) TEM image showing bundled cylinder structures analogous to optical images. Optical microscopy images show bright field, green and red fluorescence images taken after treatment with DNAzyme. (b) **D-1** recognizes only fluorescein labeled particles. (c) **D-2** recognizes only rhodamine labeled particles. (d) **D-1** and **D-2** together cause fiber formation containing both labels. Optical image scale bars = 10 mm. Conditions: Micelles (0.14 gl^{-1}), DNAzyme (5 nm), Tris (20 mm, pH 7.4), $MgCl_2$ (50 mm). Shell DNA sequences are analogous to that shown in Figure 5.19, with an extra three bases at the 5′-terminus. The third base from the 5′-amide linkage to the polymer backbone is a dye-labeled thymidine residue. *Source*: From ref. [67]. Copyright © 2010 by John Wiley & Sons, Inc. Reprinted by permission of John Wiley & Sons, Inc.

and chemically similar particles through encoding specific information in the respective DNA sequences.

Recently, Gianneschi et al. [70] reported the synthesis of DNA–polymer amphiphiles resulting from the conjugation of a hydrophobic polymer terminally modified with a carboxylic acid moiety to a 5′-amino-modified oligonucleotide on solid support. The DNA sequences were chosen as substrates for a sequence-specific endonuclease and incorporated a dye-quencher pair for monitoring nuclease activity. Well-defined particles 20 nm in diameter were obtained following dialysis into deionized water (Figure 5.21). Free ssDNA was separated from the DNA–polymer micelles via size-exclusion chromatography (SEC-FPLC) yielding pure particles as shown in Figure 5.21.

The DNA–polymer micelles were allowed to hybridize to complementary ssDNA and subsequently treated with the sequence-selective nicking endonuclease Nt.CviPll, which recognizes double-stranded DNA (dsDNA) and introduces a single-strand break on the 5′ side of the recognition site. The cleavage reaction was monitored by fluorescence and by the melting temperature (T_m) of the DNA. After hybridization, the DNA-micelles showed almost no nuclease activity via fluorescence which was in stark contrast to free dsDNA, which was readily cleaved as shown in both the fluorescence and T_m data (Figure 5.22). The inhibition of nuclease activity on the micelles was attributed to the high density of DNA on the micelle surface. These results suggest that it is possible to program DNA-micelles to protect complementary DNA from nuclease degradation, which could have significant impacts on the delivery of DNA in a number of *in vivo* applications.

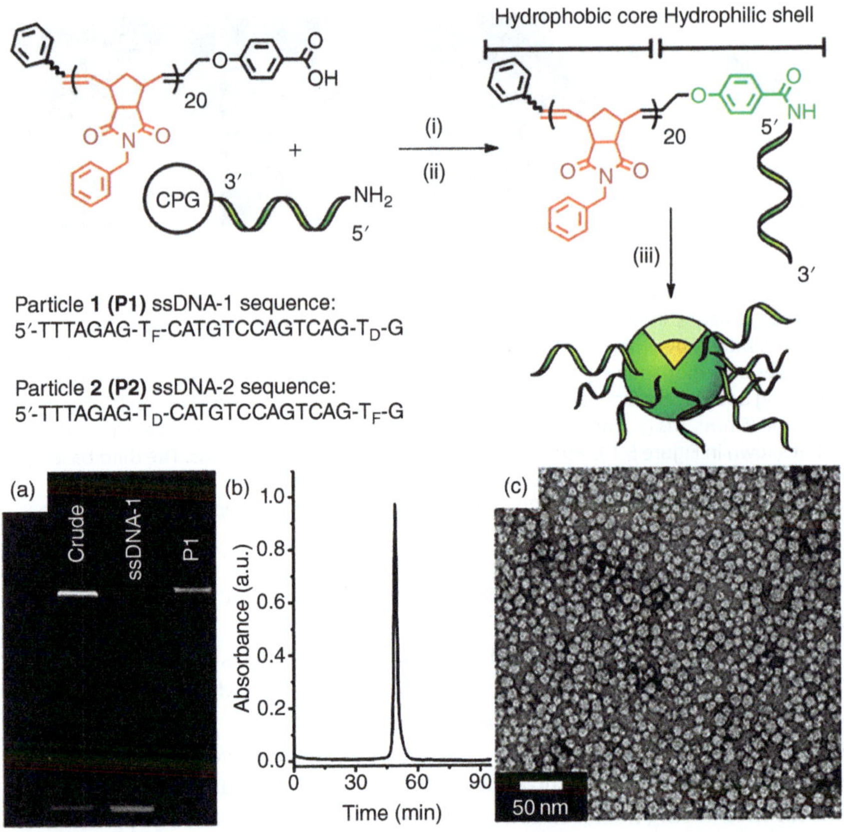

Figure 5.21 Preparation of DNA–polymer amphiphiles (DPAs) and assembly of micelles. Synthesis: (i) a hydrophobic polymer, terminally modified with a carboxylic acid moiety was mixed with a coupling agent and reacted with a 50-amino-modified oligonucleotide on solid support (CPG); (ii) deprotection and cleavage of the resulting DNA–polymer conjugate from solid support; (iii) dialysis of cleaved DPA into deionized water to form a mixture of micelles and free, nonconjugated nucleic acid. T_F and T_D correspond to fluorescein- and dabcyl-modified thymidine phosphoramidites. (a) PAGE analysis. Lane 1: Crude material post micelle (**P1**) formation showing conjugate (top band) and free ssDNA (lower band). Lane 2: HPLC purified sample of ssDNA-1. Lane 3: Purified P1, isolated via size-exclusion chromatography (SEC). (b) SEC trace of purified **P1** (λ_{abs} = 260 nm). (c) Transmission electron micrograph of **P1**. *Source*: Reprinted with permission from ref. [70]. Copyright 2013 American Chemical Society.

In fact, Gianneschi et al. recently used this approach to show that mRNA could be regulated in HeLa cells using locked nucleic acid (LNA)-polymer-amphiphile-based micelles (LPA) [69]. The amphiphilic LNA-polymer conjugates were synthesized as outlined above and formed micelles approximately 20 nm in diameter as determined by TEM and DLS (Figure 5.23). Three

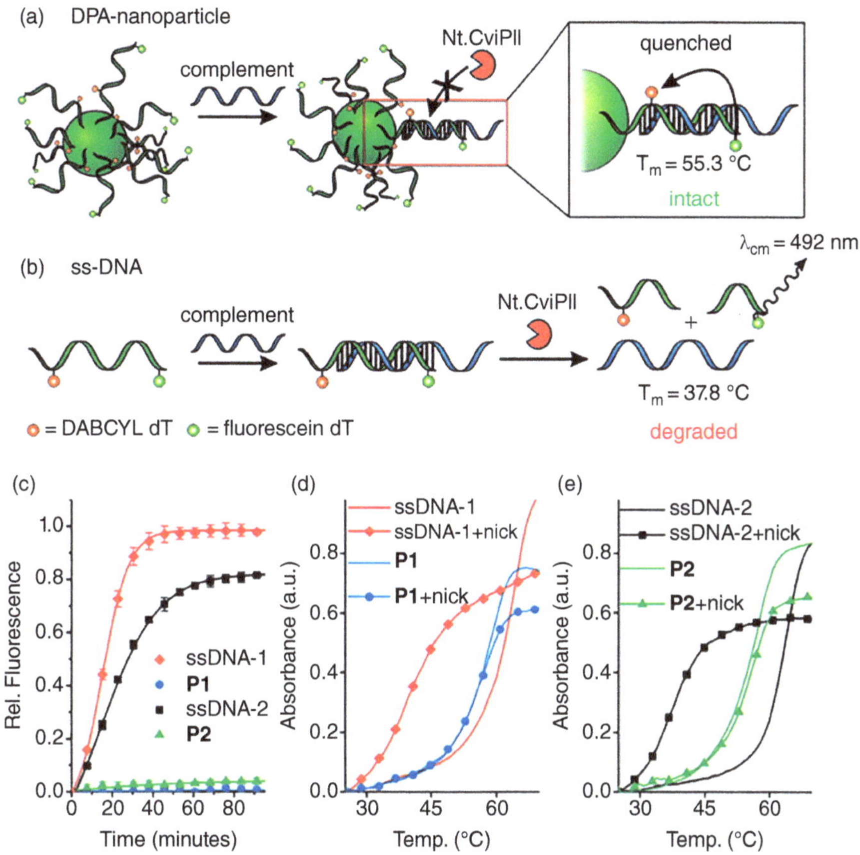

Figure 5.22 Endonuclease resistance of DPA nanoparticles. (a) Scheme depicting DPA nanoparticle (**P2**) resistance to nicking endonuclease (Nt.CviPII) and consequently intact, quenched duplex DNA at the particle surface. (b) Scheme depicting dsDNA degradation by Nt.CviPII and consequently a decrease in duplex melting temperature and increase in fluorescein fluorescence. (c) Nt.CviPII activity over time, monitored via fluorescein fluorescence dequenching (λ_{ex} = 485 nm, λ_{em} = 535 nm). (d) Thermal denaturation analysis with and without Nt.CviPII treatment for **P1** and ssDNA-1; λ_{abs} = 260 nm. Sample subjected to enzyme for 100 min at 37 °C. ssDNA-1 + complement: T_m = 63.9 °C; ssDNA-1 + Nt.CviPII + complement: T_m = 37.8 °C; P1 + complement: T_m = 58.8 °C; P1 + Nt.CviPII + complement: T_m = 58.3 °C. (e) Thermal denaturation analysis with and without Nt.CviPII treatment for **P2** and ssDNA-2; λ_{abs} = 260 nm. Sample subjected to enzyme for 100 min at 37 °C. ssDNA-2 + complement: T_m = 63.9 °C; ssDNA-2 + Nt.CviPII + complement: T_m = 37.8 °C; **P2** + complement: T_m = 56.9 °C; **P2** + Nt.CviPII + complement: T_m = 55.3 °C. Complement: 5′-TATTATATCTTTAGACACTGACTGGACATGACTCT-3′. Reprinted with permission from Ref. [70]. Copyright 2013 American Chemical Society.

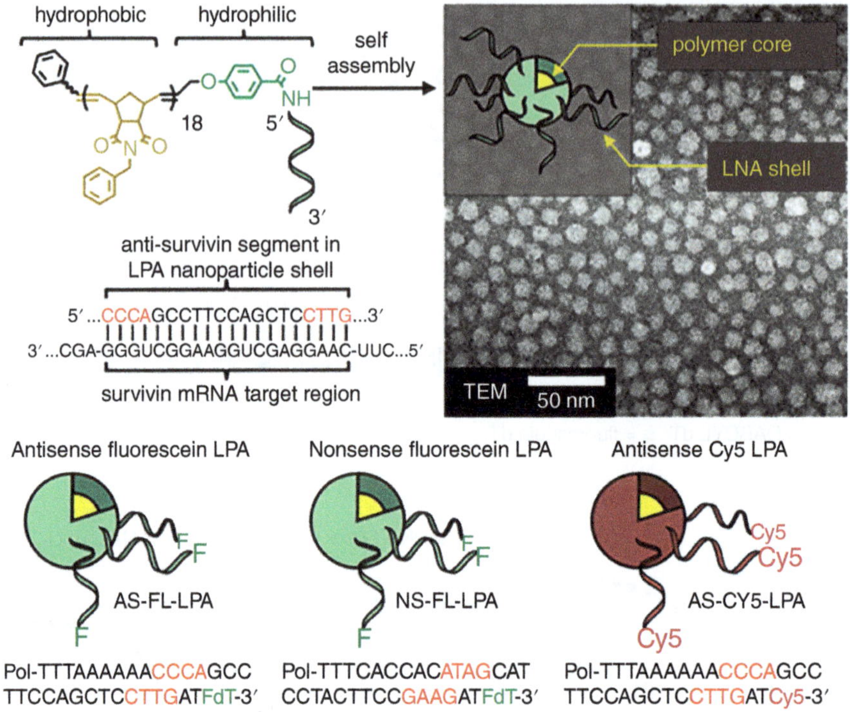

Figure 5.23 LPA composition and characterization by electron microscopy. LPAs assemble into spherical micellar nanoparticles as they are released from solid support and dispersed into aqueous solution. The resulting nanoparticles are roughly 20 nm in diameter as evidenced by negative stain TEM and DLS. LPA nanoparticles consist of a hydrophobic polynorbornyl core and a fluorescently labeled hydrophilic LNA shell designed to be complementary (antisense) or noncomplementary (nonsense) to a 20-base region of mRNA responsible for synthesizing the protein survivin. LNA bases are indicated in orange, "Pol" indicates the norbornene polymer conjugated to the 5′ end of the LNA sequence, F and FdT represent fluorescein-modified thymidine, and Cy5 represents an incorporated cyanine 5 phosphoramidite. *Source*: Reprinted with permission from ref. [70]. Copyright 2014 American Chemical Society.

different LPA micelles were synthesized to study the ability of LPA micelles to undergo cellular uptake and mRNA regulation. The first, termed antisense fluorescein labeled LPA (**AS-FL-LPA**) contained a fluorescein labeled LNA sequence complementary to a 20 base region of surviving mRNA in HeLa cells. The second, termed nonsense fluorescein labeled LPA (**NS-FL-LPA**) contained the same nucleotides but as a scrambled sequence for a control micelle. The third, termed antisense cyanine 5 labeled LPA (**NS-Cy5-LPA**) contained the same LNA sequence but incorporated cyanine 5 dye to probe the effect of dye structure on cellular uptake.

Treatment of HeLa cells with **AS-FL-LPA** or **NS-Cy5-LPA** micelles for two hours showed a 10-fold increase in cellular uptake compared to the ssLNA analogue as determined by fluorescence-activated cell sorting (FACS) analysis. Cy5-labeled ssLNA showed an increased association with the HeLa cells compared to the fluorescein labeled ssLNA, suggesting that the dye structure can have an impact on the association of materials with cells. The micelles were shown to be evenly distributed within the cytosol of the HeLa cells via live-cell z-stack confocal fluorescence spectroscopy. The ability of these micelles to modulate intracellular survivin mRNA levels was determined by treating HeLa cells with either **AS-FL-LPA** or **NS-FL-LPA** micelles for two hours and subsequently measuring the total mRNA content for the relative abundance of survivin mRNA (Figure 5.24). Cells treated with **AS-FL-LPA** micelles shown greatly reduced levels of surviving mRNA compared to both untreated and **NS-FL-LPA**–treated micelles, while both cell viability and uptake of the two micelles remained the same. This study demonstrated that it is possible to program micelles with DNA to target specific mRNA sequences in complex biological environments.

In 2012, Sleiman et al. reported that DNA–polymer micelles could be programmed to assemble selectively on DNA nanotubes [93]. A diblock

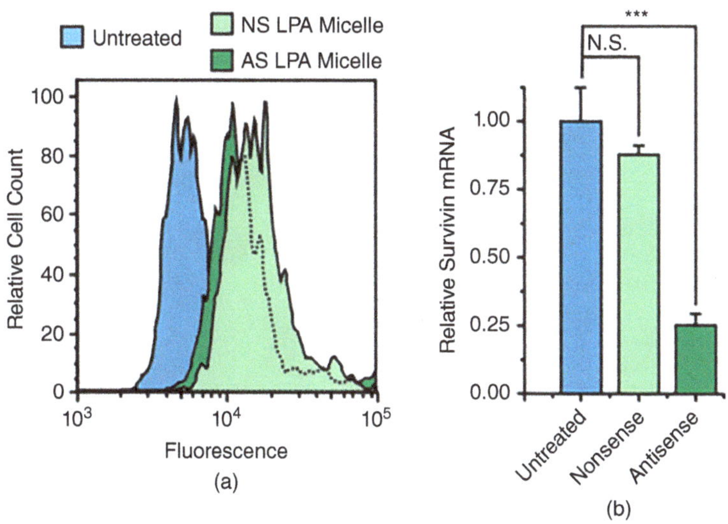

Figure 5.24 LPA nanoparticle uptake and survivin mRNA depletion. (a) FACS data showing similar uptake for both antisense and nonsense LPA nanoparticles after incubation with HeLa cells for 2 h. $\lambda_{ex} = 488$ nm, $\lambda_{em} = 533 \pm 15$ nm. (b) Reverse transcription polymerase chain reaction (RT-PCR) results showing sequence-selective surviving mRNA knockdown due to treatment with antisense LPA nanoparticles (AS-FL-LPA). Reprinted with permission from ref.[69]. Copyright 2014 American Chemical Society.

co-oligomer composed of two PEG_{2000} units and a single NHS-activated carboxylic acid unit was synthesized via ROMP and subsequently conjugated to a 5′-amino-modified oligonucleotide on solid support (Figure 5.25). The PEG-oligomer-DNA conjugates formed spherical micelles in both buffered solution and pure water. When these aggregates were mixed with a DNA nanotube containing the complementary DNA sequence at specific intervals, bundled fibers were formed where the aggregates were positioned longitudinally along the fibers, as shown by the AFM images in Figure 5.25. While aggregates containing a noncomplementary DNA sequence also adhere to the surface of the DNA nanotubes they showed no positional order. DNA nanotube-DNA micelle aggregates could be reversibly disassembled via strand invasion and hybridization by addition of a fully complementary ssDNA strand

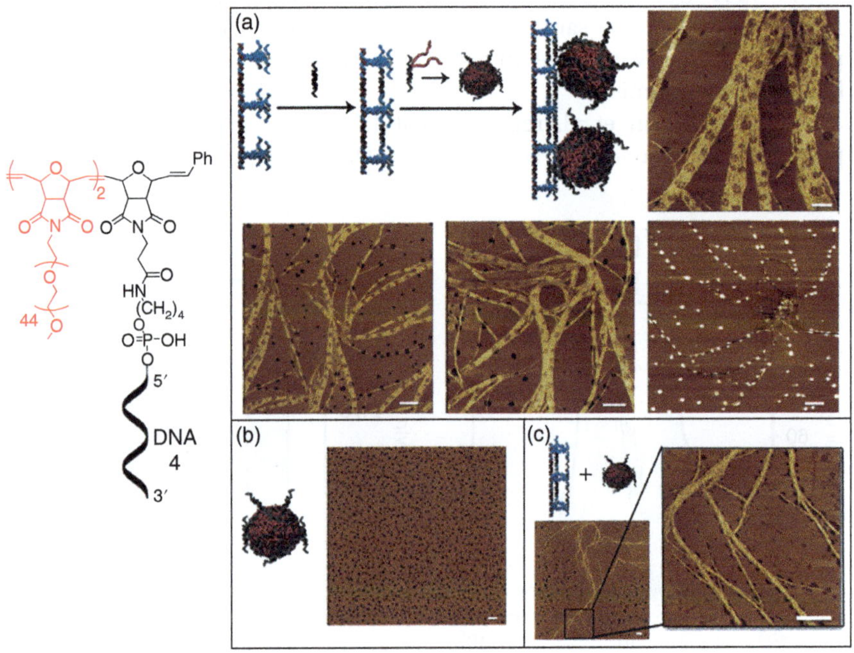

Figure 5.25 Structure of poly(ethyleneglycol)-DNA polymer (left) and (a) A scheme of the incorporation of graft PEG–DNA aggregates **ds4** onto rolling circle amplification (RCA) nanotubes and their AFM characterization (the lower right micrograph is a height image, while the others are phase images, bar is 1 μm). Control AFM experiments showing (b) graft–PEG DNA conjugates **ds4** alone and (c) the addition of non-complementary graft–PEG DNA conjugates to DNA nanotubes (phase images, bar is 1 μm). All air AFM experiments were performed by depositing a 1.5 μM sample solution in $1 \times TA_{Mg}$ on highly oriented pyrolytic graphite (HOPG). *Source*: From ref. [93]. Copyright © 2012 by John Wiley & Sons, Inc. Reprinted by permission of John Wiley & Sons, Inc.

when DNA-PEG conjugates containing a 10-base overhang were used in the assembly process.

The generality of this approach to positional ordering of DNA-micelles on DNA-nanotubes was demonstrated with three other polymer-DNA systems composed of poly styrene (PS_{13}-b-DNA), or one of two ROMP-based polymers containing either a shorter PEG unit ($(PEG660)_2$-b-DNA) or a hydrophobic phenyl unit ($(phenyl)_{10}$-b-DNA). In each case, addition of the polymer-DNA micelles to complementary DNA-nanotubes resulted in the formation of aggregate structures where the polymer-DNA micelles showed positional ordering along the DNA nanotubes (Figure 5.26).

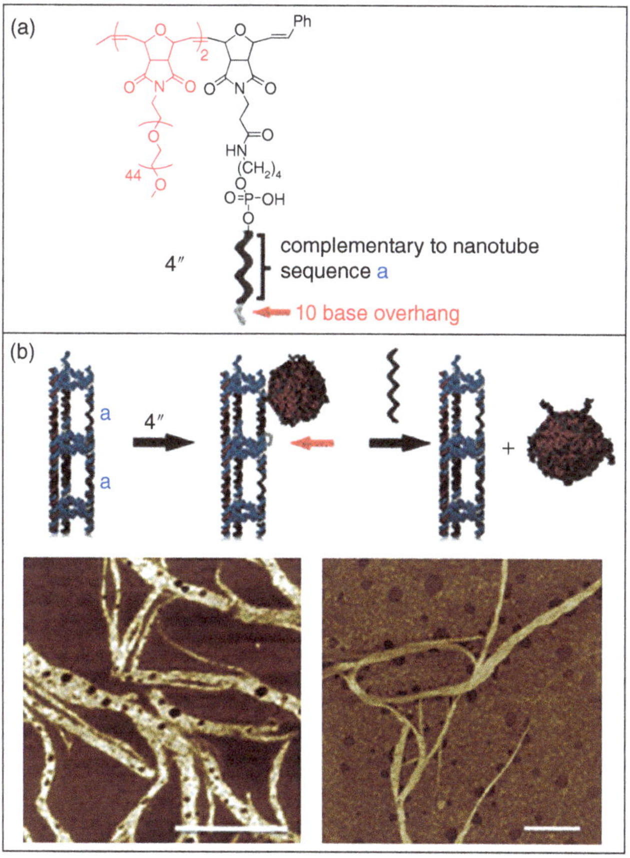

Figure 5.26 (a) The chemical structure of graft-PEG DNA **4″** and (b) a scheme and AFM images of the write–erase of graft-PEG DNA **4″** strand from RCA-nanotubes (phase images, bar is 1 μm). *Source*: From ref. [93]. Copyright © 2012 by John Wiley & Sons, Inc. Reprinted by permission of John Wiley & Sons, Inc.

5.4 Summary

In summary, a number of micellar systems incorporating peptides and DNA as programmable biomolecular units have been described. The sequence-specific properties associated with peptides and DNA has been exploited to program the micellar systems to respond to specific stimuli such as enzymes and complementary DNA/RNA sequences. The responses typically observed involve a change in morphology such as gel formation, rodlike micelle formation, or aggregate formation. The structure of the micelles has been shown to improve the stability of DNA toward degradation via enzymes, ultimately allowing these materials to be used to regulate mRNA levels or as a method for the positional ordering of other materials. Recent developments in this area of research only scrape the surface of the potential that these types of micellar systems possess.

References

1 Hu, J., Zhang, G., and Liu, S. (2012). Enzyme-responsive polymeric assemblies, nanoparticles and hydrogels. *Chemical Society Reviews* 41 (18): 5933–5949.

2 Zelzer, M. et al. (2013). Enzyme responsive materials: design strategies and future developments. *Biomaterials Science* 1 (1): 11–39.

3 Ding, Y., Kang, Y., and Zhang, X. (2015). Enzyme-responsive polymer assemblies constructed through covalent synthesis and supramolecular strategy. *Chemical Communications* 51 (6): 996–1003.

4 Hu, Q., Katti, P.S., and Gu, Z. (2014). Enzyme-responsive nanomaterials for controlled drug delivery. *Nanoscale* 6 (21): 12273–12286.

5 Azzazy, H.M.E. and Highsmith, W.E. Jr., (2002). Phage display technology: clinical applications and recent innovations. *Clinical Biochemistry* 35 (6): 425–445.

6 Bratkovič, T. (2010). Progress in phage display: evolution of the technique and its applications. *Cellular and Molecular Life Sciences* 67 (5): 749–767.

7 Duvshani-Eshet, M., Keren, H., Oz, S. et al. (2008). Effect of peptides bearing nuclear localization signals on therapeutic ultrasound mediated gene delivery. *The Journal of Gene Medicine* 10 (10): 1150–1159.

8 Ruoslahti, E. (2007). Vascular zip codes and nanoparticle targeting. In: *BioMEMS and Biomedical Nanotechnology* (ed. M. Ferrari, T. Desai and S. Bhatia), 127–136. US: Springer.

9 Boonen, K., Creemers, J.W., and Schoofs, L. (2009). Bioactive peptides, networks and systems biology. *BioEssays* 31 (3): 300–314.

10 Kawamura, K., Oishi, J., Kang, J.H. et al. (2004). Intracellular signal-responsive gene carrier for cell-specific gene expression. *Biomacromolecules* 6 (2): 908–913.

11 Molkentin, J.D. (2003). A friend within the heart: natriuretic peptide receptor signaling. *The Journal of Clinical Investigation* 111 (9): 1275–1277.

12 Huang, J. and Ding, J. (2010). Nanostructured interfaces with RGD arrays to control cell-matrix interaction. *Soft Matter* 6 (15): 3395–3401.

13 Ruoslahti, E. (1996). RGD and other recognition sequences for integrins. *Annual Review of Cell and Developmental Biology* 12 (1): 697–715.

14 Galande, A.K., Hilderbrand, S.A., Weissleder, R. et al. (2006). Enzyme-targeted fluorescent imaging probes on a multiple antigenic peptide Core. *Journal of Medicinal Chemistry* 49 (15): 4715–4720.

15 Harris, T.J., von Maltzahn, G., Derfus, A.M. et al. (2006). Proteolytic actuation of nanoparticle self-assembly. *Angewandte Chemie International Edition* 45 (19): 3161–3165.

16 Jiang, T., Olson, E.S., Nyugen, Q.T. et al. (2004). Tumor imaging by means of proteolytic activation of cell-penetrating peptides. *Proceedings of the National Academy of Sciences of the United States of America* 101 (51): 17867–17872.

17 von Maltzahn, G., Harris, T.J., Park, J.H. et al. (2007). Nanoparticle self-assembly gated by logical Proteolytic triggers. *Journal of the American Chemical Society* 129 (19): 6064–6065.

18 Cui, H., Webber, M.J., and Stupp, S.I. (2010). Self-assembly of peptide amphiphiles: from molecules to nanostructures to biomaterials. *Peptide Science* 94 (1): 1–18.

19 Dehsorkhi, A., Castelletto, V., and Hamley, I.W. (2014). Self-assembling amphiphilic peptides. *Journal of Peptide Science* 20 (7): 453–467.

20 Fleming, S. and Ulijn, R.V. (2014). Design of nanostructures based on aromatic peptide amphiphiles. *Chemical Society Reviews* 43 (23): 8150–8177.

21 Hamley, I.W. (2011). Self-assembly of amphiphilic peptides. *Soft Matter* 7 (9): 4122–4138.

22 Mandal, D., Nasrolahi Shirazi, A., and Parang, K. (2014). Self-assembly of peptides to nanostructures. *Organic & Biomolecular Chemistry* 12 (22): 3544–3561.

23 Trent, A., Marullo, R., Lin, B. et al. (2011). Structural properties of soluble peptide amphiphile micelles. *Soft Matter* 7 (20): 9572–9582.

24 Zhang, J., Zhao, Y., Han, S. et al. (2014). Self-assembly of surfactant-like peptides and their applications. *Science China Chemistry* 57 (12): 1634–1645.

25 Zhao, X., Pan, F., Xu, H. et al. (2010). Molecular self-assembly and applications of designer peptide amphiphiles. *Chemical Society Reviews* 39 (9): 3480–3498.

26 Koda, D., Maruyama, T., Minakuchi, N. et al. (2010). Proteinase-mediated drastic morphological change of peptide-amphiphile to induce supramolecular hydrogelation. *Chemical Communications* 46 (6): 979–981.

27 Sadownik, J.W., Leckie, J., and Ulijn, R.V. (2011). Micelle to fibre biocatalytic supramolecular transformation of an aromatic peptide amphiphile. *Chemical Communications* 47 (2): 728–730.
28 Castelletto, V., McKendrick, J.E., Hamley, I.W. et al. (2010). PEGylated amyloid peptide nanocontainer delivery and release system. *Langmuir* 26 (14): 11624–11627.
29 Berg, J.M., Tymoczko, J.L., and Stryer, L. (2007). *Biochemistry*, 6e. New York: W.H. Freeman and Company.
30 Ku, T.-H., Chien, M.-P., Thompson, M.P. et al. (2011). Controlling and switching the morphology of micellar nanoparticles with enzymes. *Journal of the American Chemical Society* 133 (22): 8392–8395.
31 Chien, M.-P., Thompson, M.P., Lin, E.C. et al. (2012). Fluorogenic enzyme-responsive micellar nanoparticles. *Chemical Science* 3 (9): 2690–2694.
32 Chien, M.-P., Thompson, M.P., Barback, C.V. et al. (2013). Enzyme-directed assembly of a nanoparticle probe in tumor tissue. *Advanced Materials* 25 (26): 3599–3604.
33 Hahn, M.E., Randolph, L.M., Adamiak, L. et al. (2013). Polymerization of a peptide-based enzyme substrate. *Chemical Communications* 49 (28): 2873–2875.
34 Brody, E.N. and Gold, L. (2000). Aptamers as therapeutic and diagnostic agents. *Reviews in Molecular Biotechnology* 74 (1): 5–13.
35 Farokhzad, O.C., Jon, S., Khademhosseini, A. et al. (2004). Nanoparticle-aptamer bioconjugates: a new approach for targeting prostate cancer cells. *Cancer Research* 64 (21): 7668–7672.
36 Takeuchi, T. and Matile, S. (2009). DNA aptamers as analyte-responsive cation transporters in fluorogenic vesicles: signal amplification by supramolecular polymerization. *Journal of the American Chemical Society* 131 (50): 18048–18049.
37 Thomson, K., Amin, I., Morales, E. et al. (2007). Preliminary nanopore cheminformatics analysis of aptamer-target binding strength. *BMC Bioinformatics* 8 (Suppl 7): S11.
38 Dervan, P. (1986). Design of sequence-specific DNA-binding molecules. *Science* 232 (4749): 464–471.
39 Persil, Ö. and Hud, N.V. (2007). Harnessing DNA intercalation. *Trends in Biotechnology* 25 (10): 433–436.
40 Baum, D.A. and Silverman, S.K. (2008). Deoxyribozymes: useful DNA catalysts in vitro and in vivo. *Cellular and Molecular Life Sciences* 65 (14): 2156–2174.
41 Breaker, R.R. and Joyce, G.F. (1995). A DNA enzyme with Mg^{2+}–dependent RNA phosphoesterase activity. *Chemistry & Biology* 2 (10): 655–660.

42 Jäschke, A. (2001). Artificial ribozymes and deoxyribozymes. *Current Opinion in Structural Biology* 11 (3): 321–326.
43 Jäschke, A. and Seelig, B. (2000). Evolution of DNA and RNA as catalysts for chemical reactions. *Current Opinion in Chemical Biology* 4 (3): 257–262.
44 Santoro, S.W. and Joyce, G.F. (1997). A general purpose RNA-cleaving DNA enzyme. *Proceedings of the National Academy of Sciences* 94 (9): 4262–4266.
45 Tuerk, C. and Gold, L. (1990). Systematic evolution of ligands by exponential enrichment: RNA ligands to bacteriophage T4 DNA polymerase. *Science* 249 (4968): 505–510.
46 Aldaye, F.A., Palmer, A.L., and Sleiman, H.F. (2008). Assembling materials with DNA as the guide. *Science* 321 (5897): 1795–1799.
47 Dirks, R.M. et al. (2004). Paradigms for computational nucleic acid design. *Nucleic Acids Research* 32 (4): 1392–1403.
48 Han, D., Pal, S., Nangreave, J. et al. (2011). DNA origami with complex curvatures in three-dimensional space. *Science* 332 (6027): 342–346.
49 Kato, T., Goodman, R.P., Erben, C.M. et al. (2009). High-resolution structural analysis of a DNA nanostructure by cryoEM. *Nano Letters* 9 (7): 2747–2750.
50 Rothemund, P.W.K. (2006). Folding DNA to create nanoscale shapes and patterns. *Nature* 440 (7082): 297–302.
51 Seeman, N. (2007). An overview of structural DNA nanotechnology. *Molecular Biotechnology* 37 (3): 246–257.
52 Zhang, C., Su, M., He, Y. et al. (2008). Conformational flexibility facilitates self-assembly of complex DNA nanostructures. *Proceedings of the National Academy of Sciences* 105 (31): 10665–10669.
53 Bertin, P.A., Gibbs, J.M., Shen, C.K.-F. et al. (2006). Multifunctional polymeric nanoparticles from diverse bioactive agents. *Journal of the American Chemical Society* 128 (13): 4168–4169.
54 Farokhzad, O.C., Cheng, J., Teply, B.A. et al. (2006). Targeted nanoparticle-aptamer bioconjugates for cancer chemotherapy in vivo. *Proceedings of the National Academy of Sciences* 103 (16): 6315–6320.
55 Feldkamp, U., Saccà, B., and Niemeyer, C.M. (2009). Dendritic DNA building blocks for amplified detection assays and biomaterials. *Angewandte Chemie International Edition* 48 (33): 5996–6000.
56 Mirkin, C.A., Letsinger, R.L., Mucic, R.C. et al. (1996). A DNA-based method for rationally assembling nanoparticles into macroscopic materials. *Nature* 382 (6592): 607–609.
57 Alivisatos, A.P., Johnsson, K.P., Peng, X. et al. (1996). Organization of 'nanocrystal molecules' using DNA. *Nature* 382 (6592): 609–611.

58 Kwak, M. and Herrmann, A. (2011). Nucleic acid amphiphiles: synthesis and self-assembled nanostructures. *Chemical Society Reviews* 40 (12): 5745–5755.

59 Gosse, C., Boutorine, A., Aujard, I. et al. (2004). Micelles of lipid–oligonucleotide conjugates: implications for membrane anchoring and base pairing. *The Journal of Physical Chemistry B* 108 (20): 6485–6497.

60 Liu, K., Zheng, L., Liu, Q. et al. (2014). Nucleic acid chemistry in the organic phase: from functionalized oligonucleotides to DNA side chain polymers. *Journal of the American Chemical Society* 136 (40): 14255–14262.

61 Wu, Y., Sefah, K., Liu, H. et al. (2010). DNA aptamer–micelle as an efficient detection/delivery vehicle toward cancer cells. *Proceedings of the National Academy of Sciences* 107 (1): 5–10.

62 Wu, C., Chen, T., Han, D. et al. (2013). Engineering of switchable aptamer micelle flares for molecular imaging in living cells. *ACS Nano* 7 (7): 5724–5731.

63 Chen, T., Wu, C.S., Jimenez, E. et al. (2013). DNA micelle flares for intracellular mRNA imaging and gene therapy. *Angewandte Chemie International Edition* 52 (7): 2012–2016.

64 Jeong, J.H. and Park, T.G. (2001). Novel polymer–DNA hybrid polymeric micelles composed of hydrophobic poly(d,l-lactic-co-glycolic acid) and hydrophilic oligonucleotides. *Bioconjugate Chemistry* 12 (6): 917–923.

65 Li, Z., Zhang, Y., Fullhart, P. et al. (2004). Reversible and chemically programmable micelle assembly with DNA block-copolymer amphiphiles. *Nano Letters* 4 (6): 1055–1058.

66 Ding, K., Alemdaroglu, F.E., Börsch, M. et al. (2007). Engineering the structural properties of DNA block copolymer micelles by molecular recognition. *Angewandte Chemie International Edition* 46 (7): 1172–1175.

67 Wang, J., Alemdaroglu, F.E., Prusty, D.K. et al. (2008). In-situ visualization of the enzymatic growth of surface-immobilized DNA block copolymer micelles by scanning force microscopy. *Macromolecules* 41 (8): 2914–2919.

68 Chien, M.P., Rush, A.M., Thompson, M.P. et al. (2010). Programmable shape-shifting micelles. *Angew Chem Int Ed Engl* 49 (30): 5076–5080.

69 Rush, A.M., Nelles, D.A., Blum, A.P. et al. (2014). Intracellular mRNA regulation with self-assembled locked nucleic acid polymer nanoparticles. *Journal of the American Chemical Society* 136 (21): 7615–7618.

70 Rush, A.M., Thompson, M.P., Tatro, E.T. et al. (2013). Nuclease-resistant DNA via high-density packing in polymeric micellar nanoparticle coronas. *ACS Nano* 7 (2): 1379–1387.

71 Thompson, M.P., Chien, M.-P., Ku, T.-H. et al. (2010). Smart lipids for programmable nanomaterials. *Nano Letters* 10 (7): 2690–2693.

72 Chien, M.-P., Thompson, M.P., and Gianneschi, N.C. (2011). DNA-nanoparticle micelles as supramolecular fluorogenic substrates enabling catalytic signal amplification and detection by DNAzyme probes. *Chemical Communications* 47 (1): 167–169.

73 Alemdaroglu, F.E. and Herrmann, A. (2007). DNA meets synthetic polymers-highly versatile hybrid materials. *Organic & Biomolecular Chemistry* 5 (9): 1311–1320.

74 de Vries, J.W., Zhang, F., and Herrmann, A. (2013). Drug delivery systems based on nucleic acid nanostructures. *Journal of Controlled Release* 172 (2): 467–483.

75 Kwak, M., Minten, I.J., Anaya, D.-M. et al. (2010). Virus-like particles templated by DNA micelles: a general method for loading virus nanocarriers. *Journal of the American Chemical Society* 132 (23): 7834–7835.

76 Schnitzler, T. and Herrmann, A. (2012). DNA block copolymers: functional materials for nanoscience and biomedicine. *Accounts of Chemical Research* 45 (9): 1419–1430.

77 Zimmermann, J., Kwak, M., Musser, A. et al. (2011). Amphiphilic DNA block copolymers: nucleic acid-polymer hybrid materials for diagnostics and biomedicine. In: *Bioconjugation Protocols* (ed. S.S. Mark), 239–266. Humana Press.

78 Raouane, M., Desmaële, D., Urbinati, G. et al. (2012). Lipid conjugated oligonucleotides: a useful strategy for delivery. *Bioconjugate Chemistry* 23 (6): 1091–1104.

79 Godeau, G., Staedel, C., and Barthélémy, P. (2008). Lipid-conjugated oligonucleotides via "click chemistry" efficiently inhibit hepatitis C virus translation. *Journal of Medicinal Chemistry* 51 (15): 4374–4376.

80 Gissot, A., Di Primo, C., Bestel, I. et al. (2008). Sensitive liposomes encoded with oligonucleotide amphiphiles: a biocompatible switch. *Chemical Communications* (43): 5550–5552.

81 Patwa, A., Gissot, A., Bestel, I. et al. (2011). Hybrid lipid oligonucleotide conjugates: synthesis, self-assemblies and biomedical applications. *Chemical Society Reviews* 40 (12): 5844–5854.

82 Schade, M., Berti, D., Huster, D. et al. (2014). Lipophilic nucleic acids – a flexible construction kit for organization and functionalization of surfaces. *Advances in Colloid and Interface Science* 208 (0): 235–251.

83 Rosemeyer, H. (2005). Nucleolipids: natural occurrence, synthesis, molecular recognition, and supramolecular assemblies as potential precursors of life and bioorganic materials. *Chemistry & Biodiversity* 2 (8): 977–1063.

84 Allain, V., Bourgaux, C., and Couvreur, P. (2012). Self-assembled nucleolipids: from supramolecular structure to soft nucleic acid and drug delivery devices. *Nucleic Acids Research* 40 (5): 1891–1903.

85 Barthélemy, P. (2009). Nucleoside-based lipids at work: from supramolecular assemblies to biological applications. *Comptes Rendus Chimie* 12 (1–2): 171–179.

86 Vernille, J.P., Kovell, L.C., and Schneider, J.W. (2004). Peptide nucleic acid (PNA) amphiphiles: synthesis, self-assembly, and duplex stability. *Bioconjugate Chemistry* 15 (6): 1314–1321.

87 Liu, L.-H., Li, Z.-Y., Rong, L. et al. (2014). Self-assembly of hybridized peptide nucleic acid amphiphiles. *ACS Macro Letters* 3 (5): 467–471.

88 Lau, C., Bitton, R., Bianco-Peled, H. et al. (2006). Morphological characterization of self-assembled peptide nucleic acid amphiphiles. *The Journal of Physical Chemistry B* 110 (18): 9027–9033.

89 James, C.R., Rush, A.M., Insley, T. et al. (2014). Poly(oligonucleotide). *Journal of the American Chemical Society* 136 (32): 11216–11219.

90 Liu, H., Zhu, Z., Kang, H. et al. (2010). DNA-based micelles: synthesis, micellar properties and size-dependent cell permeability. *Chemistry – A European Journal* 16 (12): 3791–3797.

91 Averick, S.E., Dey, S.K., Grahacharya, D. et al. (2014). Solid-phase incorporation of an ATRP initiator for polymer–DNA biohybrids. *Angewandte Chemie International Edition* 53 (10): 2739–2744.

92 Alemdaroglu, F.E., Wang, J., Börsch, M. et al. (2008). Enzymatic control of the size of DNA block copolymer nanoparticles. *Angewandte Chemie International Edition* 47 (5): 974–976.

93 Carneiro, K.M.M., Hamblin, G.D., Hanni, K.D. et al. (2012). Stimuli-responsive organization of block copolymers on DNA nanotubes. *Chemical Science* 3 (6): 1980–1986.

6

Protein Analogous Micelles

Lorraine Leon[1] *and Matthew Tirrell*[2,3]

[1] *University of Central Florida, Materials Science and Engineering, 12760 Pegasus Drive, Orlando, FL 32816, USA*
[2] *The University of Chicago, Institute for Molecular Engineering, 5640 S. Ellis Ave., Chicago, IL 60637, USA*
[3] *Argonne National Laboratory, 9700 S. Cass Ave., Argonne, IL 60439, USA*

6.1 Introduction

Protein analogous micelles (PAMs) are biomimetic nanostructures that self-assemble by packing hydrophobic moieties in the core, thereby displaying structurally ordered protein fragments at the periphery, which can specifically recognize and bind cellular ligands and other bioactive molecules. A variety of self-assembled micelles, such as those formed using antibodies or proteins conjugated to lipids, or amphiphiles with headgroups consisting only of amino acids, could be classified as PAMs. Here, we choose to focus on PAMs formed using peptide molecules conjugated to lipid tails, called peptide amphiphiles (PAs) [1, 2].

PAs form micelles following the same thermodynamic principles as classical surfactants, except that the peptide regions can form secondary structures that bring in a wider range of interactions among headgroups and can also more readily stabilize kinetically trapped states. PAMs may adopt either spherical or cylindrical nanostructures analogous to a protein's globular or fibrous state. PAMs can display a variety of cell epitopes on the nanoparticle periphery, making them both multifunctional, multivalent, and in certain cases, responsive to external stimuli such as pH, temperature, light, or even molecular recognition. A schematic of spherical and cylindrical PAMs is shown in Figure 6.1. These properties, combined with the ease of tailoring this platform to various biomedical applications, illustrate why PAMs have great potential as targeted diagnostics, therapeutics, and tissue-engineering scaffolds. In addition to discussing key physicochemical characteristics of PAMs, this chapter will focus on recent advances in designing PAMs as spherical or cylindrical nanostructures to suit specific applications, or in other words, designing structure to impart function.

Self-Assembly: From Surfactants to Nanoparticles, First Edition. Edited by Ramanathan Nagarajan.

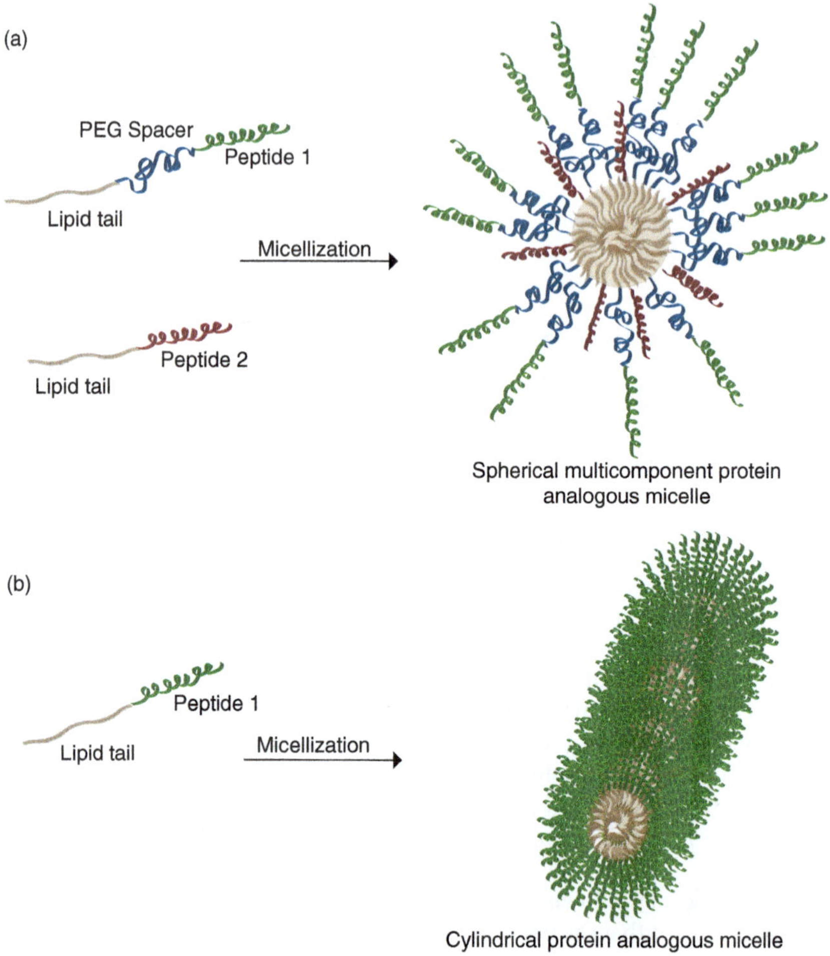

Figure 6.1 (a) A schematic of a spherical PAM formed by mixing two different PA molecules. One of the molecules contains a polyethylene glycol (PEG) spacer between the peptide and lipid tail. (b) A cylindrical PAM formed using a single PA molecule. In both cases, the hydrophobic lipids pack into the core whilst the peptides are displayed in the corona of the micelle.

PAs are ideal candidates for the formation of biomimetic assemblies [3, 4] because they combine the ability to self-assemble, imparted by the lipid and molecular recognition capabilities from the protein fragment. In addition, lipidated peptide and proteins exist in nature via post- and co-translational modifications that are related to cell signaling [1, 5], illustrating their innate biofunctionality and compatibility. Their use in forming PAMs has been explored in the last two decades, originating with the development of facile solid-phase chemical synthesis [2, 6]. One of the unique properties of PAMs is an enhancement of secondary structure compared to the unconjugated peptide. For example, model collagen peptide sequences that do not form triple helices as free peptides form stable triple helices when conjugated to hydrocarbon tails [7–10]. Short amyloid mimicking peptides that form β-sheet fibrils (cylindrical micelles) that are disrupted upon heating to 80 °C can be stabilized at the same temperature by conjugating a C_{12} hydrocarbon chain to the peptide [11, 12]. The crowding within the peptide corona of a PAM drives this enhancement of secondary structure [13–16]. Often, the folded state is required for biological activity [17]; thus, the enhanced stability of this state in a PAM can be beneficial in the use of peptides as targeting or therapeutic agents.

6.2 Physicochemical Properties of Peptide Amphiphiles

At first glance, the thermodynamic behavior of PAs follows the same principles of self-assembly that allow weak intermolecular interactions, such as the hydrophobic effect, to determine the shape and size of aggregates. For example, a classical surfactant containing a lipid tail and a polar headgroup will self-assemble into different morphologies, depending on the molecular geometry of the assembling molecule itself. The largest thermodynamic driving force for the hydrophobic self-assembly of amphiphilic molecules originates from the increased entropic contribution of associated water molecules to the overall Gibbs free energy. When an amphiphile is placed in water, the existing water structure is disrupted because water molecules are restructured around the hydrophobic portion of the molecule to maximize hydrogen bond formation [18]. Therefore, the self-assembly of amphiphilic molecules into structures, such as micelles, that shield the hydrophobic portion from water are entropically favorable. The extent to which the hydrophobic effect is favorable originates from the number of reoriented water molecules, and thus the surface area of the hydrophobic portion of the molecule [18]. This also affects the critical micelle concentration (CMC), that is, the concentration at which any additional monomer entered into the system will automatically become part of a micelle. The CMC is defined as:

$$CMC \approx \exp\left[-\frac{\mu_1^o - \mu_N^o}{kT}\right],$$

where μ_1^o and μ_N^o are the chemical potential of the monomer and aggregate, respectively. It can be thought of as the measure of stability of a micelle; molecules with lower CMC values are more stable since they require fewer molecules to make assembly into micelles favorable.

The self-assembled structures formed via hydrophobic interactions can be micelles, bilayers, vesicles, etc., depending on the critical packing parameter of the molecule. This parameter is defined as: $v/a_0 l_c$ where v is the volume of the lipid chain, a_0 is the effective surface area of the headgroup, and l_c is the length of the lipid chain. From this parameter, one can deduce that the formation of spherical micelles usually involves surfactants that contain a relatively large head group area and a proportionately small lipid volume ($v/a_0 l_c < 1/3$), while that of cylindrical micelles contain a more symmetric ratio of head group area to lipid volume ($1/2 < v/a_0 l_c > 1/3$).

6.2.1 The Role of Secondary Structures in PAMs

PAs form spherical or cylindrical micelles based on the same principles of self-assembly discussed above except that additional factors can contribute to the morphology of the aggregate structure. Transmission electron microscopy (TEM) images of two different PAM morphologies can be seen in Figure 6.2. One additional factor, imparted by peptide secondary structure, comes in the form of hydrogen bonding. Peptides, like proteins, can form both intramolecular and intermolecular hydrogen bonds. Intramolecular hydrogen bonding results in the formation of α-helical secondary structure, while intermolecular hydrogen bonding results in the formation of β-sheet secondary structure. In particular, β-sheet formation can favor the formation of cylindrical micelles or nanofibers via the creation of intermolecular hydrogen bonds, as will be subsequently discussed in detail. One way to think about this is that attractive interactions among the peptide headgroups effectively reduce the headgroup area.

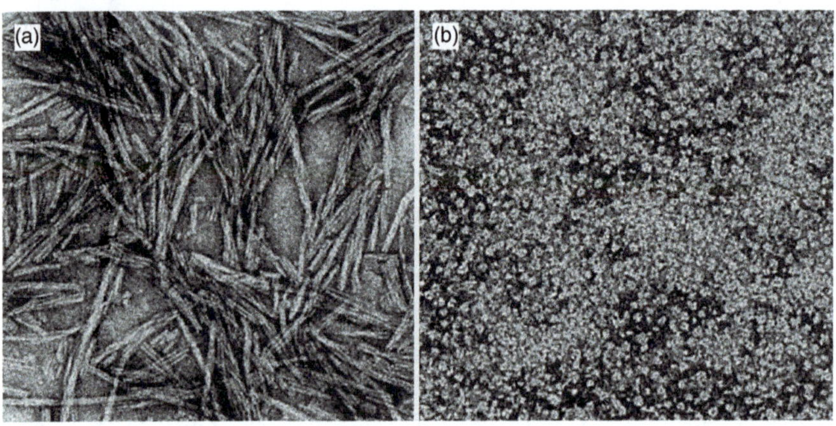

Figure 6.2 Negatively stained TEM images of (a) cylindrical and (b) spherical micelles. *Source*: Adapted from Marullo et al. [19].

In pioneering work using PAs, Hartgerink et al. designed a multifunctional molecule containing an alkyl tail to drive self-assembly, cysteine residues that allow for molecular cross-linking, a flexible linker, and a phosphoserine residue to bind calcium ions [20]. These particular molecules became the first example of PA self-assembly into nanofibers with a β-sheet structure. Additionally, the incorporation of the cysteine and phosphoserine residues added the ability to reversibly cross-link the fibers using pH and also template the growth of hydroxyapatite crystals along the direction of the long axis of the fibers [20]. Using this molecule and several similar designs that all contained a net charge, it was shown that the fibers could be physically cross-linked into gels by changing pH or introducing divalent ions [21]. The ability of the same molecule to form both a discrete singular object and bulk material widens the scope of its applicability.

A fundamental understanding of how these PA nanofibers form is necessary to be able to utilize them. The internal structure of PA nanofibers was investigated using polarization modulation infrared reflection absorption spectroscopy (PM-IRRAS). PAs were found to form β-sheets that are oriented parallel to the long axis of the fiber [22], similar to other fibrillar assemblies such as those formed by the amyloid protein Aβ [23]. The ability of these nanostructures to form β-sheets was also used to enhance the stiffness of the assemblies. By systematically varying the proximity to the hydrophobic core of amino acids with β-sheet propensity, it was found that stiffer gels had their β-sheet-forming residues closest to the hydrophobic core [24, 25]. These studies illustrate the importance of precise positioning of amino acids and also the ability to tune rheological properties.

β-sheet-forming nanofibers or cylindrical micelles are the most commonly formed PAM. This has been shown using molecular simulations, in which a phase diagram of hydrophobic strength versus hydrogen bonding strength was created for PA molecules, elucidating that nonspherical structures were the dominant morphology of PA-based assemblies [26]. Additionally, using a PA consisting of a peptide with an α-helical forming tendency, it was discovered that spherical micelles can transition to cylindrical micelles simultaneously as the secondary structure transitions from α-helix to β-sheet. This transition was accelerated with increasing temperature, indicating that the β-sheet was the more thermodynamically stable state. It is believed that the formation of a β-sheet structure reduces the effective area of the head group, therefore favoring more elongated structures [27]. Compared to the native peptide, PAMs enable the peptides to be tethered to a surface and thus in greater proximity than an average peptide in solution, allowing for an increased probability of β-sheet formation and enhanced secondary structure.

Besides nanofibers, other β-sheet morphologies have been discovered. Using a PA composed of an alternating hydrophobic/hydrophilic peptide sequence (a strong β-sheet former), a flat nanobelt morphology was created due to the fact that the hydrophilic and hydrophobic residues were on opposite sides of the β-sheet allowing the hydrophobic sides to pack together [28]. However, at lower concentrations this PA molecule self-assembled into a twisted nanoribbon structure [28]. The formation of these different PA-based β-sheet morphologies originate from the variety of conformations exhibited by β-sheets as a result of balancing hydrogen bonding, electrostatic, and hydrophobic interactions [29].

Most β-sheet-forming PA structures have nanometer-sized diameters, but often polydisperse micron-sized lengths since there is nothing restricting the formation of consecutive hydrogen bonds along the axis. To control and form monodisperse PA nanofibers, Bull et al. used a template molecule consisting of a hydrophobic core with bulky hydrophilic end caps in the shape of a dumbbell. When this molecule was mixed with a β-sheet-forming PA, the hydrophobic core of the template molecule was coated with the PA, while the bulky end caps limited growth and enabled the formation of PA fibers with controllable length [30].

β-sheet-forming PAMs tend to be more prevalent because the high density of peptides on the nanostructure surface lead to a high propensity for intermolecular hydrogen bonding. However, helical PAMs have also been developed and often impart biofunctionality through the retention of secondary structure. For example, a peptide derived from the α-helical yeast transcription factor GCN4 containing a basic leucine zipper (bZip) domain conjugated to an alkyl tail showed enhanced α-helical secondary structure, which increased upon binding of DNA. The bZip PA was able to bind DNA below the CMC in an orientation similar to that of the native protein and consequently assemble into large lamellar aggregates [31].

In another example using the model β-sheet nanofiber platform, a 14-residue cationic peptide ($(KLAKLAK)_2$ or KLAK) designed to form a membrane-disrupting amphiphilic helix was conjugated to a lipid tail containing a β-sheet-forming region. The PA formed nanofibers that had enhanced α-helical secondary structure compared to the peptide alone despite its β-sheet-forming region. Additionally, this PA system was able to disrupt cell membranes unlike the peptide alone, possibly due to the active helical configuration of the PA [32].

In a systematic study on the effect of different linkers between the peptide and lipid, the secondary structure of peptide $p53_{14\text{-}29}$ (which binds to human double minute-2 proteins in helical form) could be controlled in the PAM based on the choice of linker used in the PA. Ethylene oxide (EO) linkers of varying lengths showed a decrease in α-helicity with increasing EO lengths, whereas alanine linkers resulted in a β-sheet morphology. This ability to selectively control secondary structures is particularly important when a specific structure is needed for biomolecular recognition [17].

In work aimed at producing a synthetic, viruslike particle, a heptad repeat region from the F1 glycoprotein of respiratory syncytial virus, which has been shown to form a three-helix bundle, was coupled to a phospholipid tail. Upon assembly, enhanced secondary structure was observed for the PAM compared to the peptide alone. Additionally, the CMC of this particle was less than 20 nM, considerably lower than other PAMs, suggesting that added stability is imparted via the coiled-coil interactions of the various polypeptides in the corona of the PAM [33].

In another system, the effect of the peptide contour length was investigated by using a heptad sequence designed to form an amphipathic helix that forms coiled-coils through electrostatic interactions. By designing a series of PAs that contained a different number of repeats of this heptad sequence and attaching them to hydrocarbon tails, the influence of overall length was assessed. With only one heptad, the PAM had β-sheet secondary structure and the morphology was cylindrical. However, with increasing heptad length, the secondary structure became increasingly α-helical and the morphology became more persistently spherical. Interestingly, the two-heptad PA initially formed spheres with a mixture of both α-helical and β-sheet structure that transitioned into cylinders with 100% β-sheet structure over time, indicating that a two-heptad α-helix is less stable than a β-sheet when peptides are tethered in close proximity [19]. These results suggest that when an α-helical sequence is needed for molecular recognition in a particular application, the length and stability of the helix-forming region are important considerations.

Overall secondary structure can enhance the stability of PAMs and also provide bioactivity via the presentation of a folded structure on the micelle periphery. Specifically, a β-sheet secondary structure has been incorporated into many PA designs in order to facilitate the formation of elongated nanofibers with adjustable stiffness. On the other hand, an α-helical structure tends to impart biofunctionality, even in cases that also contain a β-sheet-forming region used to facilitate the formation of nanofibers [32, 34, 35].

6.2.2 The Role of Different Tails in PAMs

As the critical packing parameter dictates, additional design parameters in the self-assembly of PAs are the size and shape of the hydrophobic tail. One consideration is the overall length of the tail. Using PAs with model-collagen headgroups, the triple helical secondary structure was shown to be systematically enhanced with increasing tail length from C_6 to C_{12} [9]. However, if the tail length was further increased to C_{18} or C_{20}, the triple helical configuration was disrupted due to crystalline packing of the alkyl tails. This conclusion was supported by the restoration of the triple helical structure at elevated temperatures that prevent crystalline packing [10]. Using this same system, a comparison was also made using single- and double-tailed amphiphiles, concluding that all single-tailed amphiphiles had spherical morphology while double-tailed amphiphiles longer than C_{16} resulted in disclike morphologies that stacked to form extended structures [10].

A PA system based on the tumor suppressor $p53_{14-23}$ was also used to differentiate double- versus single-tailed amphiphiles. Monomer desorption rates were measured using fluorescently labeled PAs and albumin assays. The results of the study showed that double-tailed PAs were more kinetically stable. Furthermore, double-tailed PAs were shown to internalize into cells at greater concentrations compared to single-tailed PAs because of their increased ability to graft to the plasma membrane [36].

Another investigation into different types of tails involved dendritically increasing the number of C_{18} lipids attached to the N-terminus of a transcriptional activator (TAT) cell-penetrating peptide, derived from the HIV-1 trans-activating TAT protein. Increasing the number of tails from two to three decreased the CMC by an order of magnitude, and altered the morphology from spherical structures to cylindrical structures of uniform length, respectively. Other uniform length cylindrical structures were formed using four-tailed PAs and eight-tailed PAs with shorter lipid chains, alluding to the unique packing exhibited by these dendritic structures. Additionally, an analog of the four-tailed structure was created that had the ability to hydrogen bond the tails at the opposite end of the peptide, which self-assembled into long nanofibers, indicating that complex tail geometry can be used to tune morphology [37].

Hydrophobic tails can impact both the thermodynamic driving force and geometric shape of the PA molecule. These considerations lead to PAMs of various morphologies and stabilities.

6.2.3 The Role of Multiple Headgroups in PAMs

Asymmetric peptide bolamphiphiles, containing two distinct hydrophilic head groups separated by a hydrophobic section containing a β-sheet-forming region, were able to self-assemble into nanofibers in which one of the headgroups was buried in the core of the nanofibers. Therefore, the nanofiber contained both a hydrophilic core and surface. Claussen et al. attributed the driving force of this unique self-assembled structure to the β-sheet-forming region. This example shows a method of controlling the composition of the PAM core [38].

Branched amphiphiles are another way to incorporate multiple headgroups. Guler et al. have developed branched amphiphiles that allow the inclusion of different arginine-glycine-aspartic acid (RGD) epitopes on the same nanofiber structure. By comparing the fluorescence anisotropy of tryptophan residues on branched and unbranched PAs, they determined that branched molecules retain more mobility compared to their linear analogs and also provide a mechanism to incorporate multiple peptide epitopes without the necessity of making mixed micelles [39].

In another effort to impart bifunctionality, Morelli et al. designed a Y-shaped branched molecule that contains a hydrophobic tail, bioactive peptide designed to target cholecystokinin receptors (CCK8), and a gadolinium chelating agent. The molecule self-assembled into ellipsoidal micelles both in the presence and absence of Gd, indicating that the addition of the magnetic resonance imaging (MRI) contrast agent did not alter the micellar structure [40]. In an extension of this work, Gemini surfactants produced by linking two PAs using a cysteine residue were made using the same CCK8 peptide and gadolinium chelating agent. These Gemini PAs were able to self-assemble into cylindrical structures that became ellipsoids at temperatures above 50 °C [41].

Overall, using multiple headgroups imparts multifunctionality into one PA and also varies the self-assembly by changing the packing parameter of the molecule.

6.2.4 Stabilizing Spherical Structures

A stabilizing spherical structure may be necessary when considering applications for targeting specific cells and delivering cargo inside the human body using small soluble particles. In particular, spheres and short cylindrical structures have been shown to be taken up more readily by cells because longer structures are extended along streamlines in blood flow conditions which minimizes interactions with cell surfaces [42]. For this reason, methods for stabilizing the spherical structure of PAMs using PAs was a key concept discussed at length in a previous review [43]. However, we will highlight a few of the key concepts that illustrate clever ways to design with a particular application in mind. The study involving different heptads illustrated that peptide length can stabilize a α-helical secondary structure and consequently eliminate the β-sheet morphology that facilitates the formation of elongated structures [19]. Another technique that promotes the formation of spherical structures is the addition of linkers that increase the headgroup area– for example, the addition of EO between the peptide and hydrophobic tail, as discussed previously [17]. However, the use of linkers to stabilize spherical structure comes with the loss of peptide secondary structure enhancement [17]. In order to stabilize spherical structure without compromising secondary structure, hydrophobically modified dendrimers can be used as a template. This was illustrated using the bZip PA, and another PA comprised of a peptide containing a nuclear localization sequence (NLS) that self-assemble into elongated micelles and vesicles, respectively. However, using fifth-degree hydrophobically modified dendrimers as a template facilitated the formation of spherical structures in both cases while still maintaining their desired secondary structure. In addition, the CMCs of these structures, termed protein resembling dendrimer templated nanospheres

(PRTNs), were an order of magnitude lower. Furthermore, using a bovine serum albumin assay, the PRTN PAMs showed a decrease in dissociation, illustrating that PRTN PAMs are more stable than their untemplated counterparts [44]. This strategy provides a clear mechanism for adequately balancing the necessity for spherical structures while maintaining bioactive secondary structures, as was shown with the ability of the bZip PRTN to bind DNA [44].

Spherical particles have been shown to target cells more efficiently; however, elongated particles can have the benefit of increased circulation times [42]. For this reason, soluble elongated structures may be more favorable in applications that warrant a complete body response, as is the case for soliciting an immunological reaction. Similarly, if a therapeutic PA is to be injected intradermally rather than intravascularly, an elongated micelle may be a more effective choice. Long insoluble nanofibers have a tendency to form gels that can be used as scaffolds for tissue engineering purposes. Techniques aimed at creating elongated structures hinge primarily on the insertion of β-sheet-forming regions that simultaneously impact stability and shape.

6.2.5 Electrostatic Interactions

Electrostatic interactions always play a role in the self-assembly of PAs into PAMs because like-charged groups often increase the effective head group area, which then dictates micelle morphology through the critical packing parameter. Additionally, electrostatic interactions may need to be screened to overcome charge-repulsion between monomer units. In other cases, electrostatic interactions can facilitate self-assembly. For example, in a mixed micellar system, PAs with opposite charges assemble into nanofibers at neutral pH because of electrostatic attraction and charge neutralization between the different PAs. This result is in contrast to attempts to form PAMs from the individual components, which needed acidic or basic conditions to neutralize the charge on the PAs and enable nanofiber assembly. This mechanism enabled the incorporation of two different peptide epitopes into a single nanofiber structure [45].

6.2.6 Mixed Micelles

We have already discussed how incorporating different peptides into a PA can result in multifunctionality, as in the case of branched PAs incorporating a bioactive peptide and gadolinium-chelating agent in the same molecule [40]. PAMs can also provide another avenue that make them both multifunctional and modular by mixing different PA molecules into a single structure.

For example, Accardo et al. combined a Gemini surfactant containing the CCK8 binding sequence and another Gemini surfactant containing a gadolinium-chelating sequence using a disulfide bond. The linked molecules formed ellipsoidal and cylindrical nanostructures depending on the temperature. They were subsequently used for *in vivo* visualization of tumor cells that overexpress cholecystokinin receptors [41].

In another example, the cytotoxic KLAK PA was mixed with another PA containing a large polyethylene glycol (PEG) block in order to protect the bioactive KLAK from enzymatic degradation. The PAs, that contain hydrogen bonding regions to facilitate assembly, were coassembled by first treating them with hexafluoroisopropanol to disrupt hydrogen bonding between their own monomers and then lyophilizing and resuspending them in water. The KLAK and PEG PAs were confirmed to have self-assembled into the same structure through the use of diffusion-ordered spectroscopy, which demonstrated that both PAs had the same diffusion constant. This protection strategy was tested via exposure to the protease trypsin, which showed that increasing amounts of PEG PA in the coassembled structures led to increased amounts of intact KLAK PA via liquid chromatography – mass spectrometry (LC/MS). *In vitro* and *in vivo* studies showed that the PEG PA/KLAK PA mixed micelles provided the same tumorcidal activity as the KLAK PA alone. This example illustrates the shielding ability of PEG, which may also help in making this cytotoxic nanostructure inert until arrival to malignant tissue [34].

A similar shielding strategy using PEG-lipids mixed with lipids containing the cell penetrating peptide, TAT, has been employed to form micelles and liposomes with various multifunctional qualities. One PEG-lipid molecule was also modified to include a *p*-nitrophenylcarbonyl group in order to attach a monoclonal antimyosin antibody to the particle. Another separate PEG molecule with a longer PEG chain was then modified to have a pH sensitive hydrazone bond between the PEG and the lipid. Mixing of these components allowed the formation of particles that had pH-cleavable PEG shielding effects, antibody conjugation, and the cell penetrating function of TAT. *In vitro* results of these multifunctional particles with an additional fluorescently labeled monomer showed specific binding and increasing cellular uptake at lower pH once the shielding PEG was cleaved, illustrating the incorporation of multiple features into a single nanocarrier [1].

Another reason to make mixed micelles is to vary the peptide epitope density on the nanoparticle surface. Using the β-sheet nanofiber platform, Webber et al. showed that PAMs composed of mixtures of both active and inactive PAs were more effective than PAMs containing only active PAs in the ability to adhere bone marrow stem and progenitor cells using the RGD peptide epitope [3].

The ability to make multifunctional micelles by simply mixing different PA monomers is one of the key advantages of the PAM platform. This allows the creation of nanoscale objects that can incorporate imaging, targeting, therapies, and protection with ease.

6.2.7 Stimuli-Responsive PAMs

The designability of PAMs to respond to different stimuli makes them great candidates for many biomedical applications. One of the mechanisms to make these structures stimuli-responsive is through the use of pH, as was illustrated in the previous section as part of a mixed micelle containing a cleavable hydrazone bond that resulted in the exposure of the cell penetrating peptide TAT [1].

In addition to cleaving linkers, pH can also play an important role in controlling the morphology of a PAM. For example, by designing a set of model RGD-containing PAs, Jin et al. showed that pH influenced morphology via the ionization or deionization of different amino acid groups that could either enlarge the headgroup area or, in other cases, decrease charge-repulsion that prevented hydrogen bonding. By balancing these interactions, and using different hydrophobic tails, spherical micelles, cylinders, and vesicles were observed [6]. Similarly, using amphiphiles composed of multiple serine and histidine residues on different branches, PAs can transition from a liquid to a hydrogel at physiological pH when the histidine residues are no longer protonated and can participate in β-sheet hydrogen bonding. This system illustrates the potential of designing injectable hydrogel scaffolds [7].

Furthermore, a PA composed of residues 11–17 of amyloid β-peptide attached to a C_{12} tail also exhibits pH dependent behavior. At an acidic pH, the PA forms cylindrical nanofibrils, which transitioned to tapes at higher concentrations. However, at basic pH, the PA forms twisted fibrils that transitioned into twisted tapes at higher concentrations. These changes in morphology with pH were accompanied with changes in secondary structure from a random coil at low pH to a β-sheet at high pH, illustrating the ability of charged groups to inhibit β-sheet hydrogen bonding causing morphological changes [11].

In addition to pH changes, charge screening can also provide a mechanism for stimuli-responsive PAMs. By designing a PA with a net negative charge, gelation was demonstrated when a PA solution was subjected to biologically relevant ionic strength conditions, indicating another mechanism to create injectable scaffolds [13, 15]. In certain cases, the valence of the ions added to the solution can also play a role in the self-assembly of the PAMs. For example, a PA designed with a −3 net negative charge at pH 7 was found to form nanofibers that physically cross-linked into gels with the addition of Ca^{2+} but not K^{+} ions [21].

Temperature provides another avenue to tune the supramolecular structure of PAs. We have already discussed how temperature can accelerate a change in secondary structure and morphology to more thermodynamically favorable states [27] and disrupt crystalline packing of hydrophobic tails [10]. Recently, a PA consisting of a C_{16} tail coupled to a short amino acid sequence KTTKS was shown to transition from insoluble fibrillar aggregates (containing β-sheet structure) to micelles by increasing the temperature. The higher temperature disrupts hydrogen bonding and allows a reversible transition to occur at 30 °C (Kraft temperature), which had not previously been observed in PA-based systems [46].

Another interesting example involves a coiled-coil motif in which one of the peptides was conjugated to a polystyrene block and the other to a PEG block. Upon assembly of the coiled-coil motif, a noncovalent triblock copolymer was formed that self assembled into cylindrical micelles. By disrupting the coiled-coil motif at elevated temperatures, the morphology of the assemblies transitioned to spheres, the morphology of the individual components [47].

Enzymatic activity can also influence morphological changes. A PA molecule was designed using a peptide containing an enzymatically cleavable sequence and a β-sheet-forming region that had previously been shown to form nanofibers that physically crosslinked into gels. This enzymatically cleavable sequence contained a proline residue, which disrupts β-sheet formation and facilitated the self-assembly of this PA into spheres. However, upon addition of matrix metalloproteinase-7, the portion of the PA including the proline residue was cleaved. Elimination of the proline enabled the formation of β-sheets and resulted in the formation of hydrogels, illustrating potential to drive structural transitions *in vivo* [48].

Another example of enzyme-triggered PA hydrogel formation was achieved by making proteases, which normally hydrolyze peptides, work in reverse to form a peptide bond via stabilization of the reaction product relative to its precursors. Toledano et al. designed a system consisting of the enzyme, thermolysin (which has a preference for aromatic amino acids), the dipeptide diphenylalanine (FF), and different Fmoc-protected amino acids. The addition of hydrophobic Fmoc amino acids to the dipeptide via the enzyme promotes self-assembly into higher ordered structures, therefore shifting the reaction equilibrium into the production of more peptide synthesis. This consequently causes an increased concentration of the amphiphiles eventually leading to gels consisting of self-assembled nanofibers. This system illustrates a novel mechanism of forming *in situ* nanofiber hydrogel scaffolds [49].

Another mechanism to change both morphology and secondary structure is using shear flow. We previously discussed the ability of a PA containing a C_{16} tail and the W3K peptide to transition from α-helical and spherical to β-sheet and fibrous with increasing temperature [27]. The same structural and shape transition was observed beyond a critical sheer rate, providing a parallel mechanism for driving such a morphological change. Shear flow is a particularly useful mechanism for controlling morphology in the case of injectable materials such as tissue engineering matrices [50].

In other instances, light can trigger morphological transitions using mechanisms similar to acid-labile linkers. As an example, a photocleavable nitrobenzyl group was added to a PA that self-assembles into spherical micelles. Upon exposure to light, the bulky linker was cleaved allowing the

PA monomers to hydrogen bond facilitating the transition to nanofiber structures [51]. Additionally, using the same photocleavable nitrobenzyl group, the cell adhesion motif RGD could be selectively removed from a nanofiber scaffold upon exposure to light. This illustrates a mechanism for controlling the spatial regulation of the presented signals in an effort to control the cell proliferation of scaffolds that mimic the extracellular matrix [52].

The interaction of PAMs with other molecules can often trigger structural changes. For example, the bZip PA discussed previously assembles into helical ribbons and tubules in the absence of DNA. However, binding to DNA causes a morphological transition to lamella with increased relative α-helical structure [53].

Overall, to design PAMs effectively, one must consider carefully all the factors contributing to the self-assembly of the PA molecule: the peptide segment, the hydrophobic tail, and in certain cases the spacer between the two. This encompasses the need to balance hydrophobic, electrostatic, and hydrogen bonding interactions that can become dynamic in response to external stimuli.

6.3 PAMs in Biomedical Applications

Understanding how PAs self-assemble and respond to external stimuli is key to designing PAMs for specific applications. Long, insoluble fibers that physically crosslink into hydrogels with tunable rheological properties serve as excellent templates for tissue engineering where one can position bioactive groups periodically along a scaffold. PAMs can also be used in drug delivery applications to deliver insoluble drugs to a specific cell type or as diagnostic agents, where imaging capabilities are integrated into the PAM. In these applications, arriving at the site of interest though the circulation system and binding of nanoparticles to specific cells or proteins is particularly important, and spherical particles that do not get stretched out along streamlines under flow conditions offer an advantage at the expense of decreased circulation times [42]. Soluble cylindrical PAMs may present a unique balance between the ability to be taken up by cells and the circulation time, and could thus be used for applications in which the increased circulation time could be necessary to stimulate the adaptive immune system. In this section, we will discuss how different types of PAMs have been used in a variety of biomedical applications.

6.3.1 Tissue Engineering and Regenerative Medicine

PAMs have been used as scaffolds for tissue engineering where the PAM-based hydrogel acts as an extracellular matrix in a variety of different contexts. Several strategies have relied upon incorporation of the RGD cell adhesion moiety that facilitates integrin recognition. For example, RGD-based nanofibers used as scaffolds for bone marrow stem and progenitor cells both *in vitro* and *in vivo* have shown enhanced cell viability, proliferation, and adhesion [3]. Using a similar system but incorporating polymerizable acetylene groups into the hydrophobic tail allowed the formation of PA gels that were fixed into micropatterned molds, creating networks of topographical patterns with either randomly or flow-aligned PAs. Using human mesenchymal stem cells, it was shown that the stem cells aligned in the direction of the PA fibers even when the micropattern was changed from stripes to micron-sized holes. An image of randomly and flow aligned PAs and stem cell growth along the PA fibers is shown in Figure 6.3. This illustrates the influence of the PAM and also highlights different ways to manipulate cell adhesion [54].

In another example, Kokkoli and coworkers developed a fibronectin-mimetic PA containing a peptide that is a specific ligand for α5β1 integrin. The PA self-assembles into nanofibers that form hydrogels that were used as a scaffold for human umbilical vascular endothelial cells. Cells grown on this scaffold showed increased cell density, fibronectin production, and active cell spreading with more prominent actin-stress fibers in comparison to controls [55].

In another example of RGD-based PAs acting as cell scaffolds, Hartgerink and coworkers combined a RGD sequence with a sequence enzymatically cleavable by matrix metalloproteinase2 (MMP-2) in a series of PAs with different peptide lengths and bioactivities, and mixed them together with varying ratios. They discovered that using 75% of the shorter PA containing only the enzyme degradable sequence and 25% of the PA containing both RGD and the enzyme cleavable sequence resulted in 60-fold increase in mechanical properties and the same bioactivity. In addition, the presence of RGD enhanced the expression of MMP-2 enzymes that, in turn, increased the degradation of the gels, creating a feedback loop in which the cells initially bind to the hydrogel via the RGD

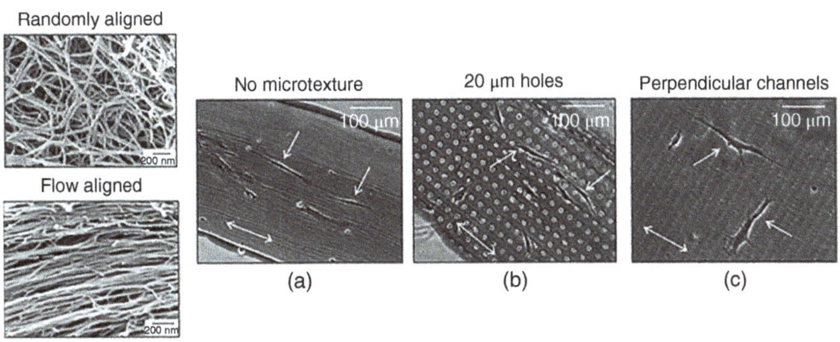

Figure 6.3 Left: SEM images of randomly aligned and flow aligned nanofibers. Right: Stem cells growing in the direction of nanofiber orientation (indicated by white arrows) using no micropattern (a) 20 μm holes (b) and perpendicular channels (c). *Source*: Adapted from Mata et al. [54].

sequence but subsequently degrade the gel in order to migrate and proliferate more effectively [56].

Creating biofunctional scaffolds using PAMs allows them to also be used in regenerative medicine such as neural or cartilage regeneration. In such an effort, PAs were designed to include the neurite-promoting laminin epitope, IKVAV, and self-assembled into a 3D network of nanofibers. Neural progenitor cells were encapsulated into the 3D nanofiber network and were shown to preferentially differentiate into neurons rather than astrocytes [13]. In follow-up *in vivo* work, mice treated with the PAMs following a spinal cord injury showed reduced astrogliosis and cell death, in addition to an increased number of oligodendroglia at the site of injury. These improvements led to regeneration of both descending motor fibers and ascending sensory fibers, as well as significant behavioral improvement, indicating that 3D nanofiber networks can promote functional recovery [15].

Using the same laminin epitope, the overall stiffness of the PA network was modified by varying hydrogen-bonding interactions near the hydrophobic tail. These studies illustrated that softer substrates containing weaker intermolecular hydrogen bonds led to weaker adhesion of neurites promoting increased cell motility and ultimately axon differentiation [57]. Additionally, laminin epitope density and mechanical properties were tuned by creating nanofibrous scaffolds containing various levels of the laminin-containing PA and collagen. Results indicated that the granule cell and Purkinje cell (two major neuronal subtypes of the cerebellar cortex that are tightly regulated during development) density had maxima at different PA concentrations. Moreover, Purkinje cells developed into either surface or penetrating dendrites, depending on the scaffold composition, and showed maximal optimal development at different PA concentrations than axons, illustrating the necessity to carefully tune scaffolds for specific subregions of the brain [58]. Subsequent, *in vivo* work further illustrated the synergistic effect of combining the signaling capacity of laminin PA and collagen to create a functional extracellular matrix [59].

Using the same versatile PA nanofiber hydrogel platform with a binding epitope to transforming growth factor β1 (TGFβ-1), Shah et al. showed slower release of TGFβ-1 and ultimately chondrogenic differentiation of human mesenchymal stem cells. Additionally, articular cartilage was regenerated in a rabbit model [60].

Using specially designed scaffolds composed of nanofibrous PAMs can also aid in biomineralization. Hartgerink et al. showed hydroxyapatite crystals grown along the length of PA-based nanofibers containing acidic and phosphorylated residues [20]. Building on this work, a 3D gel matrix designed to emulate collagen was constructed and investigated in the presence of the natural enzyme alkaline phosphatase. Incorporation of the enzyme prevented uncontrolled nonspecific mineralization by regulating the concentration of free phosphate. Creating this synergistic effect between the 3D scaffold and the enzyme led to spatially selective biomimetic mineralization [61]. Recently, the role of the nanoscale architecture using fibrous PAMs was investigated, concluding that curved cylindrical surfaces, as opposed to flat surfaces, were necessary to nucleate oriented hydroxyapatite over multiple length scales [62]. Additionally, this nanofiber-based template has been successfully compared to other bone regeneration scaffolds in balancing various factors related to osteoinductivity [63].

In the previous examples, PAM nanofiber matrices were designed to resemble collagenous networks. PAMs can also be used to increase the production of collagen. Using a PA that incorporates the sequence KTTKS, which stimulates the production of collagen and fibronectin, a dense network of extended nanotapes and fibers are formed that serve as a scaffold for collagen deposition. This PA is used in anti-aging skin creams under the trade name Matrixyl and is the first commercially available, PA-based product [64].

PA scaffolds can also aid in angiogenesis. For example, a PA system using a sequence that binds heparin, which, in turn, binds angiogenic growth factors, has been developed to promote the formation of blood vessels and was shown to successfully develop neovascularization *in vivo* using a rat corneal model [65].

Other approaches using PAMs focus on engineering blood vessels instead of stimulating their growth. For example, using PA nanofibers that align themselves in the direction of sheer flow and gel upon addition of Ca^{2+}, a macroscopic PA-based tube was assembled using a specially designed shear chamber. This PA-based tube was modified to incorporate smooth muscle cells during the fabrication process leading to the growth of the cells oriented in the direction of the nanofibers after 2 days and subsequent doubling of cell density after 12 days. The cell alignment matched the circumferential alignment of arterial cells demonstrating the potential use of these materials as replacement for damaged blood vessels [66].

6.3.2 Diagnostic and Therapeutic PAMs

Controlled angiogenesis may lead to a good pathway for the treatment of ischemic tissue, which is often developed due to atherosclerosis. Other methods of treatment for atherosclerosis do not rely on the regeneration of tissue but rather the delivery of medication to the tissue. For these applications, spherical assemblies have been generally used. In one interesting example, a short peptide sequence, cysteine–arginine–glutamic acid–lysine–alanine (CREKA) that was identified via phage display to bind to fibrin [67] was conjugated to a lipid tail and self-assembled into spherical micelles. The micelles bound to clotted plasma proteins. Additionally, mixed micelles containing CREKA and hirulog, a peptide designed to be a thrombin inhibitor, were delivered to the plaque, illustrating the potential of treating late-stage heart disease [68].

The ability of CREKA to bind to fibrin makes it a useful targeting strategy for tumors because of their characteristic fibrin deposits. Recently, our group has shown using a glioblastoma mouse model that spherical micelles containing the CREKA peptide and a Cy7 fluorophore actively targeted tumors within one hour of administration. Analysis of the biodistribution of the micelles, found accumulation in the kidneys and liver leading to secretion via renal clearance and the reticuloendothelial system with no signs of cytotoxicity or tissue damage. These findings illustrate potential for a targeted glioma therapy [69].

In the previous example, the peptide acted as a targeting mechanism, but in other cases, the peptide can be the treatment itself. For example, the KLAK peptide we previously mentioned, can form PA nanofibers and be used to induce cell apoptosis in cancer cells [32, 34]. Additionally, KLAK PAs have also been mixed with hyaluronic acid at different PA concentrations creating robust membranes containing either spherical or cylindrical PAMs. The membrane containing spherical PAMs could potentially act as a reservoir for sustained release of the therapeutic. In contrast, the membrane containing cylindrical PAMs may act as a surface with bound cytotoxicity [35].

In a different approach, cancer cell proliferation was inhibited in several cancer cell lines using spherical PAMs. The PAMs were designed to interfere with the interaction between HOX genes (a subset of homeobox genes that control the body formation and segmentation during embryogenesis and regulate hematopoiesis) and Pre-B cell leukemia transcription (PBX) factors. This interaction changes the selection of DNA binding sites and may alter transcription regulation. In addition to the peptide sequence designed to disrupt HOX/PBX/DNA complex formation and therefore DNA transcription, this PAM also incorporated a cell penetrating peptide for internalization which was shown to be necessary for the success of the PAM in the inhibition of pancreatic, leukemia, and melanoma cancer cell lines [70].

PAMs formed using PA nanofiber gels have also been investigated for controlled drug release. Several PA molecules containing a ε-lysine-amine derivatized hydrazide, which can form hydrazone linkages with ketone containing model drugs, were designed with the hydrazide moiety located at different positions within the PA molecule. Addition of the model drug Prodan to the PA showed an overall decrease in the β-sheet character compared to that of the unbound PA due to the disruption of hydrogen bonds. Additionally, the different locations of the bound Prodan led to different packing abilities of the molecules and consequently different release rates via hydrazone hydrolysis of the drug. From a design perspective, greater packing density was shown to result in a slower release rate [71]. Similarly, dexamethasone, a ketone based anti-inflammatory drug, has also been conjugated to a PA-based nanofiber system using a hydrazone linkage and evaluated for its ability to decrease inflammation after implantation both *in vitro* and *in vivo* [72].

Another nanofiber-based PA system has also been evaluated for its ability to release CO, which is known to play a protective role in tissue via the inhibition of reactive oxygen species, as well as additional anti-inflammatory, anti-apoptotic, and anti-proliferative functions. This was achieved using a peptide sequence containing a β-aspartic acid residue that reacted with $[Ru(CO_3)Cl_2]_2$ to form a CO-releasing PA similar to metal carbonyl complexes. The effectiveness of this PAM in releasing CO *in vitro* was evaluated as both soluble nanofibers and as physically crosslinked gels through the addition of Ca^{2+} ions that screened the negatively charged PAs. The PAM system performed similarly to other CO releasing molecules as a sol, but prolonged the release significantly as a gel, illustrating a potential for using this system as a localized therapy to a tissue of interest [73].

Cell penetration and internalization can be necessary for the success of PAMs as targeted deliverable therapeutics. Various studies have aimed at elucidating mechanisms for internalization, leading to design strategies. One of these strategies takes advantage of the ability of the TAT cell-penetrating peptide to translocate efficiently in the cell nucleus and cytoplasm. By incorporating TAT into a short cylindrical PA-based system loaded with Nile red, internalization of the cargo was shown in the cytoplasm and nucleus [37]. Other cell penetrating peptide motifs targeting mammalian membrane constituents that mediate uptake of their own ligands have been developed and used in a PA-based micellar system showing internalization to endothelial cells [74].

Other peptides have also been explored for their use in cellular internalization. An example of these is GALA, a fusogenic peptide that self-assembles into an amphipathic helix in mildly acidic conditions and forms transmembrane pores in lipid membranes. At physiological pH, the peptide conjugated to a C_{12} tail ruptured cell membranes using a surfactant-like mechanism. However, in mildly acidic conditions, the PA retained the peptides ability to form transmembrane pores and ultimately led to greater cell internalization without cytotoxicity compared to the peptide alone [75].

Using the $p53_{14\text{-}29}$ PA system that self-assembles into small rod-like micelles, a detailed investigation into internalization was conducted. The findings showed the increased ability of PAs to enter cells compared to the peptide alone. Interestingly, fluorescent labeling demonstrated that internalization involved PA monomers, rather than micelles, illustrating the dynamic nature of these assemblies via insertion of the PA into proteins and cell membranes causing their dissolution [76]. Although PAs were internalized and not PAMs, a different experiment using PAs composed of the CREKA peptide and a peptide that binds to the protein p32 that is overexpressed in cancer cells, showed that the PAMs did arrive to a tumor intact. This was illustrated by making a variety of mixed micelles in which the targeting monomer was not the fluorescently labeled component [77]. Therefore, PAMs still possess the ability to deliver encapsulated therapeutics to the site of interest and utilize the dynamic nature of the assembly and amphiphilic nature of the molecule for membrane insertion. Using the KLAK PA, membrane insertion leads to cell death via disruption of the cell membrane [32, 78]. However, cell viability was achieved by strengthening hydrogen bonding interactions between PA monomers illustrating the necessity to design PAMs with a specific application in mind [78].

PAMs can also be employed in immunotherapeutic applications where they can be used to multivalently present antigens on the particle. For example, the trimeric coiled-coil heptad region 1 from F1 glycoprotein of respiratory syncytial virus has been coupled to a phospholipid tail with the goal of forming synthetic virus-like micelles that induce an antigen specific response. Two synthetic antigens, Lucifer yellow dye and a disulfide bridged peptide from the V3 region of glycoprotein 120 of HIV-1, were coupled by introducing a terminal cysteine to the original PA. The immunological properties of these assemblies were tested using New Zealand white rabbits. High titers of antigen-specific antibodies containing high avidity were found, without the need for additional adjuvant [33].

In another example, the cytotoxic T-cell epitope from the model tumor antigen ovalbumin was conjugated to a double C_{16} tail that self-assembled into short cylindrical micelles. These micelles exhibited an enhanced α-helical structure compared to the native peptide and were tested both *in vitro* and *in vivo*. *In vitro* results showed that the PAM does not activate toll-like receptor 2 (TLR2) on dendritic cells. Additionally, other pattern recognition elements of dendritic cells and the upregulation of CD-40 were screened to illustrate the absence of a toll-like receptor mechanism of action. These results are in contrast to those obtained using a PA designed to be a group A streptococcal vaccine that relied on the lipid portion of the PA to activate TLR2 pathways [79]. *In vivo*, mice that received treatment with this PAM increased specific T_c-cells and reduced tumor growth without the need for additional adjuvant. Therefore, the mechanism of action may rely on the ability of PAMs to display many antigens on the micelle surface or interact directly with dendritic cells via insertion into the cell membrane. This study using short cylindrical morphologies illustrates the potential of PAMs as vaccines [80]. Overall, these examples illustrate the advantages of PAMs over other vaccine platforms that often require additional adjuvants to stimulate immune responses. Improvements in the design of PAMs could lead to modular and multifunctional vaccines.

Depending on the application, PAMs can be designed to have different morphologies and internal structures that stem from the design of the PA molecule and consideration of the environment of its use. For tissue engineering applications, long cylindrical PAMs that can be crosslinked into gels are often used as mimics for the extracellular matrix. The reasons for this stem from the ability to tune stiffness of the gel to the tissue of choice by manipulating hydrogen-bonding interactions combined with the ability to present multiple epitopes at varying densities. Furthermore, hydrogels can be designed with the ability to release therapeutics or degrade over time. Spherical or short cylindrical morphologies are often preferred for the delivery of intravenous PAMs because of their ability to navigate the circulation system, display different antigens on their surface, and be internalized into cells as monomers, allowing these PAMs to be used in both therapeutic and/or diagnostic applications.

6.4 Conclusions

In order to design a PAM properly for a particular application, one must consider all aspects that contribute to the self-assembly and stability of PAMs. Similar to surfactant molecules, these factors involve the length and number of hydrophobic tails and the shape and size of the headgroup. Additionally, secondary structure can influence PAM morphology and the stability of PAM assemblies while imparting biofunctionality. Understanding these molecular interactions enable the development of PAMs with controlled morphology and dynamic responses to light, pH, temperature, enzymes, and other stimuli.

The modularity of the PAM platform is clearly illustrated with the ability to use PAMs in applications as versatile as cartilage regeneration, drug delivery, and vaccines. Long nanofibrous PAMs that form hydrogels create bioactive interfaces that can stimulate cell or mineral growth for tissue engineering and regenerative medicine applications. Short cylindrical and spherical PAMs form discrete nanoscale objects that can circulate and localize in the body for diagnostic and therapeutic applications. Both of these approaches take advantage of the multivalent presentation of bioactive sequences that can be patterned with molecular precision and the modular benefits of self-assembly. In addition, they both benefit from the disintegration of PAMs into amino acids and lipids thus offering a biocompatible advantage over self-assembled polymeric micelles. Over the past decade, the development of PAMs has begun transitioning from material development to *in vivo* testing leading to at least one commercial product. Continued efforts in the design and implementation of PAMs could lead to novel and dynamic materials for a variety of biomedical applications.

Acknowledgments

Our work in the field of self-assembled materials has been supported in part by the US Department of Energy, Office of Science, Basic Energy Sciences, Materials Sciences, and Engineering Division.

References

1 Sawant, R.M., Hurley, J.P., Salmaso, S. et al. (2006). "SMART" drug delivery systems: double-targeted pH-responsive pharmaceutical nanocarriers. *Bioconjugate Chem.* 17: 943–949.
2 Berndt, P., Fields, G.B., and Tirrell, M. (1995). Synthetic lipidation of peptides and amino acids: monolayer structure and properties. *J. Am. Chem. Soc.* 117: 9515–9522.
3 Webber, M.J., Tongers, J., Renault, M.-A. et al. (2010). Development of bioactive peptide amphiphiles for therapeutic cell delivery. *Acta Biomater.* 6: 3–11.
4 Tu, R. (2004). Bottom-up design of biomimetic assemblies. *Adv. Drug Delivery Rev.* 56: 1537–1563.
5 Casey, P.J. (1995). Protein lipidation in cell signaling. *Science* 268: 221–225.
6 Jin, Y., Xu, X.D., Chen, C.S. et al. (2008). Bioactive amphiphilic peptide derivatives with pH triggered morphology and structure. *Macromol. Rapid Commun.* 29: 1726–1731.
7 Lin, B.F., Megley, K.A., Viswanathan, N. et al. (2012). pH-responsive branched peptide amphiphile hydrogel designed for applications in regenerative medicine with potential as injectable tissue scaffolds. *J. Mater. Chem.* 22: 19447–19454.
8 Yu, Y.-C., Berndt, P., Tirrell, M., and Fields, G.B. (1996). Self-assembling amphiphiles for construction of protein molecular architecture. *J. Am. Chem. Soc.* 118: 12515–12520.
9 Yu, Y.-C., Tirrell, M., and Fields, G.B. (1998). Minimal Lipidation stabilizes protein-like molecular architecture. *J. Am. Chem. Soc.* 120: 9979–9987.
10 Gore, T., Dori, Y., Talmon, Y. et al. (2001). Self-assembly of model collagen peptide amphiphiles. *Langmuir* 17: 5352–5360.
11 Deng, M., Yu, D., Hou, Y. et al. (2009). Self-assembly of peptide–amphiphile C 12–Aβ(11–17) into nanofibrils. *J. Phys. Chem. B* 113: 8539–8544.
12 Meijer, J.T., Roeters, M., Viola, V. et al. (2007). Stabilization of peptide fibrils by hydrophobic interaction. *Langmuir* 23: 2058–2063.
13 Silva, G.A., Czeisler, C., Niece, K.L. et al. (2004). Selective differentiation of neural progenitor cells by high-epitope density nanofibers. *Science* 303: 1352–1355.
14 Kastantin, M. and Tirrell, M. (2011). Helix formation in the polymer brush. *Macromolecules* 44: 4977–4987.
15 Tysseling-Mattiace, V.M., Sahni, V., Niece, K.L. et al. (2008). Self-assembling nanofibers inhibit glial scar formation and promote axon elongation after spinal cord injury. *J. Neurosci.* 28: 3814–3823.
16 Buhot, A. and Halperin, A. (2000). On the helix-coil transition in grafted chains. *EPL (Europhys. Lett.)* 50: 756.

17 Missirlis, D., Farine, M., Kastantin, M. et al. (2010). Linker chemistry determines secondary structure of p53 14–29in peptide Amphiphile micelles. *Bioconjugate Chem.* 21: 465–475.

18 Israelachvili, J.N. (2011). *Intermolecular and Surface Forces*, 3e. Burlington, MA: Academic Press.

19 Marullo, R., Kastantin, M., Drews, L.B. et al. (2013). Peptide contour length determines equilibrium secondary structure in protein-analogous micelles. *Biopolymers* 99: 573–581.

20 Hartgerink, J., Beniash, E., and Stupp, S. (2001). Self-assembly and mineralization of peptide-amphiphile nanofibers. *Science* 294: 1684.

21 Hartgerink, J.D., Beniash, E., and Stupp, S.I. (2002). Peptide-amphiphile nanofibers: a versatile scaffold for the preparation of self-assembling materials. *Proc. Natl. Acad. Sci. U.S.A.* 99: 5133–5138.

22 Jiang, H., Guler, M.O., and Stupp, S.I. (2007). The internal structure of self-assembled peptide amphiphiles nanofibers. *Soft Matter* 3: 454.

23 Schmit, J.D., Ghosh, K., and Dill, K. (2011). What drives amyloid molecules to assemble into oligomers and fibrils? *Biophys. J.* 100: 450–458.

24 Paramonov, S.E., Jun, H.-W., and Hartgerink, J.D. (2006). Self-assembly of peptide-amphiphile nanofibers: the roles of hydrogen bonding and amphiphilic packing. *J. Am. Chem. Soc.* 128: 7291–7298.

25 Pashuck, E.T., Cui, H., and Stupp, S.I. (2010). Tuning supramolecular rigidity of peptide fibers through molecular structure. *J. Am. Chem. Soc.* 132: 6041–6046.

26 Velichko, Y.S., Stupp, S.I., and de la Cruz, M.O. (2008). Molecular simulation study of peptide amphiphile self-assembly. *J. Phys. Chem. B* 112: 2326–2334.

27 Shimada, T., Lee, S., Bates, F.S. et al. (2009). Wormlike micelle formation in peptide-lipid conjugates driven by secondary structure transformation of the headgroups. *J. Phys. Chem. B* 113: 13711–13714.

28 Cui, H., Muraoka, T., Cheetham, A.G. et al. (2009). Self-assembly of giant peptide nanobelts. *Nano Lett.* 9: 945–951.

29 Deechongkit, S., Powers, E.T., You, S.-L. et al. (2005). Controlling the morphology of cross β-sheet assemblies by rational design. *J. Am. Chem. Soc.* 127: 8562–8570.

30 Bull, S.R., Palmer, L.C., Fry, N.J. et al. (2008). A Templating approach for monodisperse self-assembled organic nanostructures. *J. Am. Chem. Soc.* 130: 2742–2743.

31 Tu, R.S., Marullo, R., Pynn, R. et al. (2010). Cooperative DNA binding and assembly by a bZip peptide-amphiphile. *Soft Matter* 6: 1035.

32 Standley, S., Toft, D., Cheng, H. et al. (2010). Induction of cancer cell death by self-assembling nanostructures incorporating a cytotoxic peptide. *Cancer Res.* 70: 3020–3026.

33 Boato, F., Thomas, R.M., Ghasparian, A. et al. (2007). Synthetic virus-like particles from self-assembling coiled-coil lipopeptides and their use in antigen display to the immune system. *Angew. Chem. Int. Ed.* 119: 9173–9176.
34 Toft, D.J., Moyer, T.J., Standley, S.M. et al. (2012). Coassembled cytotoxic and pegylated peptide amphiphiles form filamentous nanostructures with potent antitumor activity in models of breast cancer. *ACS Nano* 6: 7956–7965.
35 Zha, R.H., Sur, S., and Stupp, S.I. (2013). Self-assembly of cytotoxic peptide amphiphiles into supramolecular membranes for cancer therapy. *Adv. Healthcare Mater.* 2: 126–133.
36 Missirlis, D., Krogstad, D.V., and Tirrell, M. (2010). Internalization of p53 14–29 peptide amphiphiles and subsequent endosomal disruption results in SJSA-1 cell death. *Mol. Pharm.* 7: 2173–2184.
37 Lim, Y.B., Lee, E., and Lee, M. (2007). Controlled bioactive nanostructures from self-assembly of peptide building blocks. *Angew. Chem. Int. Ed.* 46: 9011–9014.
38 Claussen, R.C., Rabatic, B.M., and Stupp, S.I. (2003). Aqueous self-assembly of unsymmetric peptide bolaamphiphiles into nanofibers with hydrophilic cores and surfaces. *J. Am. Chem. Soc.* 125: 12680–12681.
39 Guler, M.O., Hsu, L., Soukasene, S. et al. (2006). Presentation of RGDS epitopes on self-assembled nanofibers of branched peptide amphiphiles. *Biomacromolecules* 7: 1855–1863.
40 Mangiapia, M.V.G., Paduano, L. et al. (2007). Structural and relaxometric characterization of peptide aggregates containing gadolinium complexes as potential selective contrast agents in MRI. *ChemPhysChem* 8: 2526–2538.
41 Accardo, A., Tesauro, D., Morisco, A. et al. (2009). Micelles obtained by aggregation of gemini surfactants containing the CCK8 peptide and a gadolinium complex. *J. Biol. Inorg. Chem.* 14: 587–599.
42 Geng, Y., Dalhaimer, P., Cai, S. et al. (2007). Shape effects of filaments versus spherical particles in flow and drug delivery. *Nat. Nanotech.* 2: 249–255.
43 Trent, A., Marullo, R., Lin, B. et al. (2011). Structural properties of soluble peptide amphiphile micelles. *Soft Matter* 7: 9572.
44 Lin, B.F., Marullo, R.S., Robb, M.J. et al. (2011). De novo design of bioactive protein-resembling nanospheres via dendrimer-templated peptide amphiphile assembly. *Nano Lett.* 11: 3946–3950.
45 Niece, K.L., Hartgerink, J.D., Donners, J.J.J.M. et al. (2003). Self-assembly combining two bioactive peptide-amphiphile molecules into nanofibers by electrostatic attraction. *J. Am. Chem. Soc.* 125: 7146–7147.
46 Miravet, J.F., Escuder, B., Segarra-Maset, M.D. et al. (2013). Self-assembly of a peptide amphiphile: transition from nanotape fibrils to micelles. *Soft Matter* 9: 3558–3564.

47 Marsden, H.R., Korobko, A.V., van Leeuwen, E.N.M. et al. (2008). Noncovalent triblock copolymers based on a coiled-coil peptide motif. *J. Am. Chem. Soc.* 130: 9386–9393.

48 Koda, D., Maruyama, T., Minakuchi, N. et al. (2010). Proteinase-mediated drastic morphological change of peptide–amphiphile to induce supramolecular hydrogelation. *Chem. commun.* 46: 979–981.

49 Toledano, S., Williams, R.J., Jayawarna, V. et al. (2006). Enzyme-triggered self-assembly of peptide hydrogels via reversed hydrolysis. *J. Am. Chem. Soc.* 128: 1070–1071.

50 Shimada, T., Megley, K., Tirrell, M. et al. (2011). Fluid mechanical shear induces structural transitions in assembly of a peptide–lipid conjugate. *Soft Matter* 7: 8856.

51 Muraoka, T., Koh, C.-Y., Cui, H. et al. (2009). Light-triggered bioactivity in three dimensions. *Angew. Chem. Int. Ed.* 121: 6060–6063.

52 Sur, S., Matson, J.B., Webber, M.J. et al. (2012). Photodynamic control of bioactivity in a nanofiber matrix. *ACS Nano* 6: 10776–10785.

53 Bitton, R., Schmidt, J., Biesalski, M. et al. (2005). Self-assembly of model DNA-binding peptide amphiphiles. *Langmuir* 21: 11888–11895.

54 Mata, A., Hsu, L., Capito, R. et al. (2009). Micropatterning of bioactive self-assembling gels. *Soft Matter* 5: 1228–1236.

55 Shroff, K., Rexeisen, E.L., Arunagirinathan, M.A. et al. (2010). Fibronectin-mimetic peptide-amphiphile nanofiber gels support increased cell adhesion and promote ECM production. *Soft Matter* 6: 5064–5072.

56 Jun, H.-W., Paramonov, S.E., Dong, H. et al. (2008). Tuning the mechanical and bioresponsive properties of peptide-amphiphile nanofiber networks. *J. Biomat. Sci. Polym. Ed.* 19: 665–676.

57 Sur, S., Newcomb, C.J., Webber, M.J. et al. (2013). Tuning supramolecular mechanics to guide neuron development. *Biomaterials* 34: 4749–4757.

58 Sur, S., Pashuck, E.T., Guler, M.O. et al. (2012). A hybrid nanofiber matrix to control the survival and maturation of brain neurons. *Biomaterials* 33: 545–555.

59 Sur, S., Guler, M.O., Webber, M.J. et al. (2014). Synergistic regulation of cerebellar Purkinje neuron development by laminin epitopes and collagen on an artificial hybrid matrix construct. *Biomaterials* doi: 10.1039/C3BM60228A.

60 Shah, R.N., Shah, N.A., Del Rosario Lim, M.M. et al. (2010). Supramolecular design of self-assembling nanofibers for cartilage regeneration. *Proc. Natl. Acad. Sci. U.S.A.* 107: 3293–3298.

61 Spoerke, E.D., Anthony, S.G., and Stupp, S.I. (2009). Enzyme directed templating of artificial bone mineral. *Adv. Mater.* 21: 425–430.

62 Newcomb, C.J., Bitton, R., Velichko, Y.S. et al. (2012). The role of nanoscale architecture in supramolecular templating of biomimetic hydroxyapatite mineralization. *Small* 8: 2195–2202.

63 Hsu, E.L., Ghodasra, J.H., Ashtekar, A. et al. (2013). A comparative evaluation of factors influencing osteoinductivity among scaffolds designed for bone regeneration. *Tissue Eng. Part A* 19: 1764–1772.

64 Castelletto, V., Hamley, I.W., Perez, J. et al. (2010). Fibrillar superstructure from extended nanotapes formed by a collagen -stimulating peptide. *Chem. Commun.* 46: 9185–9187.

65 Rajangam, K., Behanna, H.A., Hui, M.J. et al. (2006). Heparin binding nanostructures to promote growth of blood vessels. *Nano Lett.* 6: 2086–2090.

66 McClendon, M.T. and Stupp, S.I. (2012). Tubular hydrogels of circumferentially aligned nanofibers to encapsulate and orient vascular cells. *Biomaterials* 33: 5713–5722.

67 Simberg, D., Duza, T., Park, J.H. et al. (2007). Biomimetic amplification of nanoparticle homing to tumors. *Proc. Natl. Acad. Sci. U.S. A.* 104: 932–936.

68 Peters, D., Kastantin, M., Kotamraju, V.R. et al. (2009). Targeting atherosclerosis by using modular, multifunctional micelles. *Proc. Natl. Acad. Sci. U.S.A.* 106: 9815–9819.

69 Chung, E.J., Cheng, Y., Morshed, R. et al. (2014). Fibrin-binding, peptide amphiphile micelles for targeting glioblastoma. *Biomaterials* 35: 1249–1256.

70 Aulisa, L., Forraz, N., McGuckin, C. et al. (2009). Inhibition of cancer cell proliferation by designed peptide amphiphiles. *Acta Biomater.* 5: 842–853.

71 Matson, J.B., Newcomb, C.J., Bitton, R. et al. (2012). Nanostructure-templated control of drug release from peptide amphiphile nanofiber gels. *Soft Matter* 8: 3586–3595.

72 Webber, M.J., Matson, J.B., Tamboli, V.K. et al. (2012). Controlled release of dexamethasone from peptide nanofiber gels to modulate inflammatory response. *Biomaterials* 33: 6823–6832.

73 Matson, J.B., Webber, M.J., Tamboli, V.K. et al. (2012). A peptide-based material for therapeutic carbon monoxide delivery. *Soft Matter* 8: 6689–6692.

74 Keller, S., Sauer, I., Strauss, H. et al. (2005). Membrane-mimetic nanocarriers formed by a dipalmitoylated cell-penetrating peptide. *Angew. Chem. Int. Ed.* 44: 5252–5255.

75 Lin, B.F., Missirlis, D., Krogstad, D.V. et al. (2012). Structural effects and lipid membrane interactions of the pH-responsive GALA peptide with fatty acid acylation. *Biochemistry* 51: 4658–4668.

76 Missirlis, D., Khant, H., and Tirrell, M. (2009). Mechanisms of peptide amphiphile internalization by SJSA-1 cells in vitro. *Biochemistry* 48: 3304–3314.

77 Karmali, P.P., Kotamraju, V.R., Kastantin, M. et al. (2009). Targeting of albumin-embedded paclitaxel nanoparticles to tumors. *Nanomed. Nanotech. Biol. Med.* 5: 73–82.

78 Newcomb, C.J., Sur, S., Ortony, J.H. et al. (2014). Cell death versus cell survival instructed by supramolecular cohesion of nanostructures. *Nat. Commun.* 5: doi: 10.1038/ncomms4321.
79 Abdel-Aal, A.-B.M., Batzloff, M.R., Fujita, Y. et al. (2008). Structure–activity relationship of a series of synthetic lipopeptide self-adjuvanting group a streptococcal vaccine candidates. *J. Med. Chem.* 51: 167–172.
80 Black, M., Trent, A., Kostenko, Y. et al. (2012). Self-assembled peptide Amphiphile micelles containing a cytotoxic T-cell epitope promote a protective immune response in vivo. *Adv. Mater.* 24: 3845–3849.

7

Self-Assembly of Protein–Polymer Conjugates

Xuehui Dong, Aaron Huang, Allie Obermeyer, and Bradley D. Olsen

Massachusetts Institute of Technology, Department of Chemical Engineering, 25 Ames Street, Cambridge, MA 02139, USA

7.1 Introduction

Block copolymers are one of the most widely investigated molecular technologies for self-assembly. Originally synthesized in the 1950s using relatively poorly controlled polymerization techniques [1, 2], these molecules have represented one of the most explosive growth areas in the soft matter community over the second half of the twentieth century and continuing today. Developments in controlled polymerization techniques [3–8] have facilitated the development of materials from a wide variety of different chemistries and with precisely tuned polymer architecture and dispersity. Industrially, block copolymers have found widespread applications as surfactant nanostructures and nanostructured elastomers, represented by the large commercial product lines Pluronics™ (PEO-PPO) and Kraton™ (PS-PI, PS-PB, PS-PEB, and PS-PEP) copolymers.

A sustained research effort has led to large advances in our physical understanding of these materials [9, 10], including experimental [11] and theoretical [12] knowledge of a phase diagram, an understanding of interfacial and geometric confinement effects in thin films [13, 14], and work on controlling alignment and defect structures using shear [15], graphoepitaxy [16], chemoepitaxy [17], and magnetic [18], or electric fields [19]. A wide variety of phases have been discovered both as the number of blocks and chemical complexity of the polymers is increased [20, 21] and more recently through a more thorough understanding of processing and phase transition kinetics [22]. This work has led to many potential emerging applications as membranes [23–25], battery separators [26], and templates for semiconductor nanopatterning [27, 28].

Currently, there is a great deal of interest in applying block copolymer self-assembly concepts to functional polymer systems that may not follow

Self-Assembly: From Surfactants to Nanoparticles, First Edition. Edited by Ramanathan Nagarajan.

Gaussian chain statistics. Many semiconducting polymers have rigid or semiflexible chain configurations [29–31], which introduces liquid crystalline interactions and can significantly impact the phase behavior of the block copolymers. Block copolymers with rigid blocks have also attracted interest in the area of engineering plastics, incorporating polymers such as aramids and aromatic polyesters [32, 33]. There has also been great interest in self-assembling "macro atoms" such as fullerenes or polyoctahedral silsesquioxanes (POSS) to yield block copolymers where one half is a rigid nanoscale object [34–36], and these polymer-nanoparticle conjugates have also attracted significant interest [37]. Finally, the large interest in and application of polymers for biotechnology has motivated a desire to incorporate DNA [38], polysaccharides [38], and proteins [39] into block copolymers.

Among the biomolecules incorporated as blocks in block copolymers, proteins are clearly the most diverse due to the advanced state of protein synthesis and bioconjugate chemistry [40–43] and the diversity of applications for proteins in functional materials [44–46]. Although proteins are linear polymers, they are very different than their synthetic polymer analogues. Proteins are copolymers of 20 different amino acids, and because they are grown from DNA templates [45, 46] or by step-by-step additive polymerizations, they have perfectly defined sequences and zero or very minimal dispersity. These properties enable proteins to adopt a wide variety of chain folds that go significantly beyond the spectrum of Gaussian chain to wormlike chain to rigid polymer that is accessible with less well-defined linear polymers. It is these folded chain sequences, from the simple secondary structural elements to more complex tertiary folds and quaternary assemblies of protein subunits, that lead to the diversity of different properties and functions that proteins have evolved.

Because these folded-chain shapes are inseparable from the function of the protein, the folded-chain shape must be preserved in any sort of biofunctional self-assembly. Therefore, the challenge of self-assembling protein–polymer block copolymers amounts to a complex problem of packing materials of different shapes into nanostructures. However, the problem is more complex than simply geometric shape. While models of proteins as spherical colloids capture some aspects of their thermodynamics, proteins are widely recognized to have patchy surfaces [47, 48] that can introduce specific, directional interactions; these interactions are often manifest in the relatively complex unit cell packing of proteins in crystals. This anisotropy in interactions also results from the protein fold and specific placement of amino acid residues; therefore, the sequence-specific nature of proteins is fundamental to complexities in their thermodynamics.

This chapter roughly divides protein and polypeptide segments into five shape categories: helices, β-sheets, globular proteins, disks, and coil-like proteins (Figure 7.1). The self-assembly of block copolymers containing each different shape category is treated in a different section, providing a clear

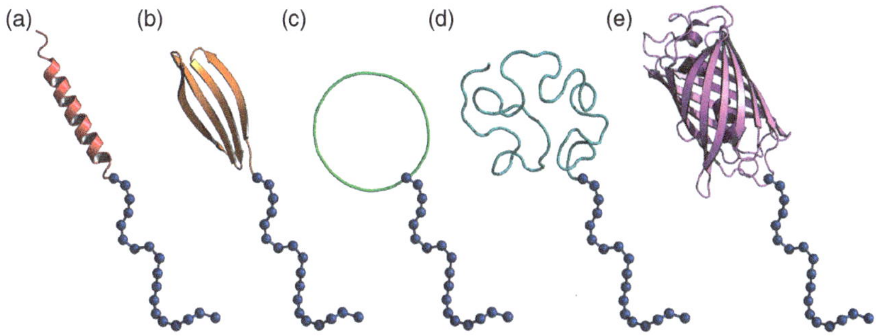

Figure 7.1 Shapes of protein–polymer conjugates (a) α-helical rod–coil conjugate, (b) β-sheet conjugate, (c) cyclic protein conjugate, (d) coil-like protein conjugate, and (e) globular protein conjugate.

comparison as to how the molecular shape and intermolecular interactions in different protein block copolymers influence their self-assembly. While the thermodynamics and nanoscience of the different bioconjugate structures is stressed, the requisite synthetic approaches and applications of different structures are eluded to frequently, motivating the use of these complex molecules to address important engineering challenges.

7.2 Helical Protein Copolymers

Perhaps the earliest interest in protein–polymer conjugates was in the self-assembly of rodlike conjugates, facilitated by the facile synthesis of α-helical polypeptides using *N*-carboxy anhydride (NCA) polymerization. Gallot et al. produced early demonstrations that such materials could self-assemble into block copolymer nanostructures, showing the ability to form lamellar nanostructures from several different polypeptide-synthetic polymer conjugates [49–51]. Since this pioneering work, there has been a great deal of interest in protein–polymer conjugates containing an α-helical protein block. Homopolypeptide helical sequences are readily accessible by NCA polymerization [52], and using blocks of L and blocks of D/L isomers provides a convenient method to produce helix-coil systems all from chemically similar monomers [53]. In addition, a diversity of helical sequences may also be accessed through protein expression [54, 55] and step-by-step synthesis techniques. In addition to individual α-helices, many helices associate into bundles commonly known as coiled-coils [56–59], providing a relatively well-understood and easily engineered aggregation motif that can be incorporated into bioconjugates and block copolymers (see Figures 7.2 and 7.3).

The simplest case of a helix-containing block copolymer is a single, non-associating α-helix conjugated to a single flexible polymer. These

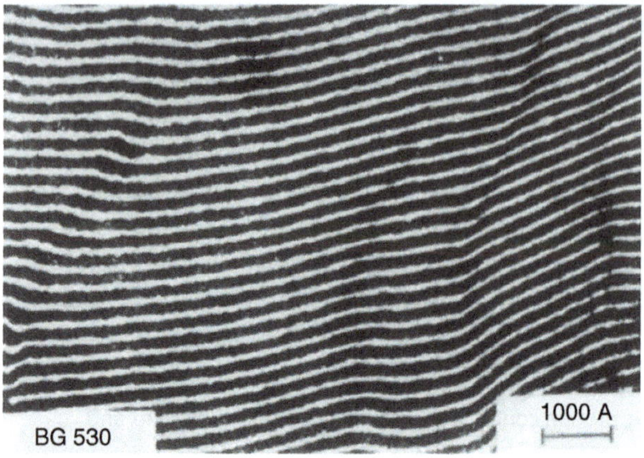

Figure 7.2 Transmission electron micrograph of poly(benzyl-L-glutamate)-coil diblock copolymer. *Source*: Reproduced with permission from Ref. [51].

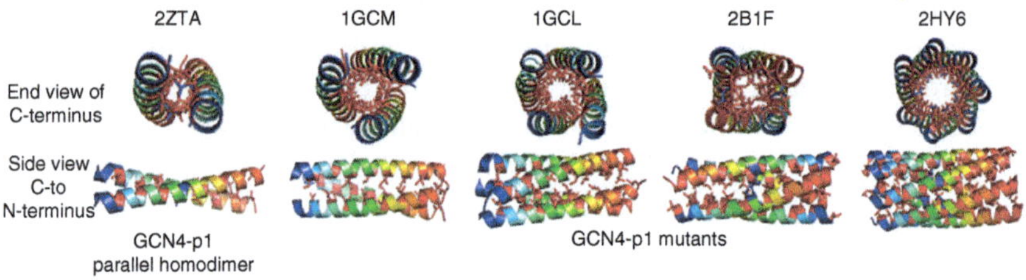

Figure 7.3 Examples of engineered coiled-coil domains to control valency and orientation of the protein aggregate. *Source*: Reproduced with permission from Ref. [60].

helix-coil conjugates can be considered as rod-coil block copolymers [29], a well-studied category of block copolymers with examples found in semi-conducting polymers, engineering plastics, and bioconjugates. Because the shape of the polypeptide is well-defined, theoretical efforts have met with success in modeling these rod-coil systems [61–64], and several thorough experimental efforts have produced detailed phase diagrams for self-assembly of rod-coil systems in the neat state [65, 66]. This work clearly demonstrates that there are important effects of liquid crystalline interactions between rod blocks that contribute to the type of self-assembled structures. The phase diagrams for rod–coil polymers are highly asymmetric, and nanostructures with low interfacial curvature tend to be preferred, such as bulk lamellar phases [50, 51, 67]. Nonlamellar phases are observed only at high-volume fraction of the coil block. The traditional hexagonal cylinder phase or body-centered cubic sphere phase are replaced with hexagonal arrays of strips and a hockey-puck phase [68, 69]. The rod block aggregates in these phases have a rectangular cross-section due to alignment of the rodlike blocks within the core of the domain. In addition to liquid crystalline interactions between rod blocks affecting the types of nanostructures formed, it is possible to observe nematic phases in the block copolymers due to nematic ordering between the rod blocks even in the disordered state [67]. Such α-helical protein–polymer conjugates also show interesting new reptation dynamics due to the presence of the rodlike block [70, 71].

Within the typical lamellar nanostructures formed by self-assembly of the neat block copolymers, the cylindrical helices will usually pack in a hexagonal geometry [67, 72, 73], forming a hierarchical cylindrical within lamellar arrangement of the polypeptide blocks. Depending on the identity of the rod block, the block may also be folded within the lamellae: poly(carbobenzoxy-L-lysine) (PZLLys) has been demonstrated to form larger lamellar domains than poly(benzyl-L-glutamate) (PBLG) due to folding of the PBLG within the lamellar domains [73]. In contrast, the helices of PZLLys are likely to be interdigitated within the lamellae [74], and increasing polydispersity can lead to a disordered zigzag structure in the lamellae as longer rods cluster together within the nanostructure [75]. Therefore, the specific properties of the helix are critical to the exact type of structure formed and the relationship between domain spacing and helix length. In some cases the structure formed depends strongly on the processing of the material [53], and care must always be taken to understand the relative contributions of kinetic and thermodynamic effects on the final structure.

While in the folded state an α-helix can behave as a very high persistence length rod and show strong liquid crystalline interactions, helices are somewhat unique among the family of rodlike polymers because they are capable of making structural transitions due to changes in the protein fold. These transitions are most commonly observed in short sequences, where a variety of different nanostructured phases have been observed due to secondary structure changes in solid state materials. These helix to sheet transitions are driven by the length of the protein block, with helicity increasing as the block length increases [76]. For short peptides, the helical secondary structure is stabilized by conjugation to the end of the peptide chain [77], but even still, peptides shorter than 10–20 amino acids tend to prefer a sheet configuration [78]. Therefore, increases in chain length typically trigger a β-sheet to α-helix secondary structural transition within the block copolymer materials. In many common helical polypeptides such as PZLLys or PBLG, the polypeptide adopts an α-helical form when it is in a hydrophobic state, such as uncharged or having a large hydrophobic group grafted as a side chain. The type of hydrophobic group has a large impact on secondary structure, and long alkyl chains can drive a helix to β-sheet transition within the protein blocks of self-assembled protein–polymer diblock copolymers [79] (see Figure 7.4).

Helical protein blocks have also been incorporated into more complex chain structures. Rod–rod polymers have been prepared using two polypeptide

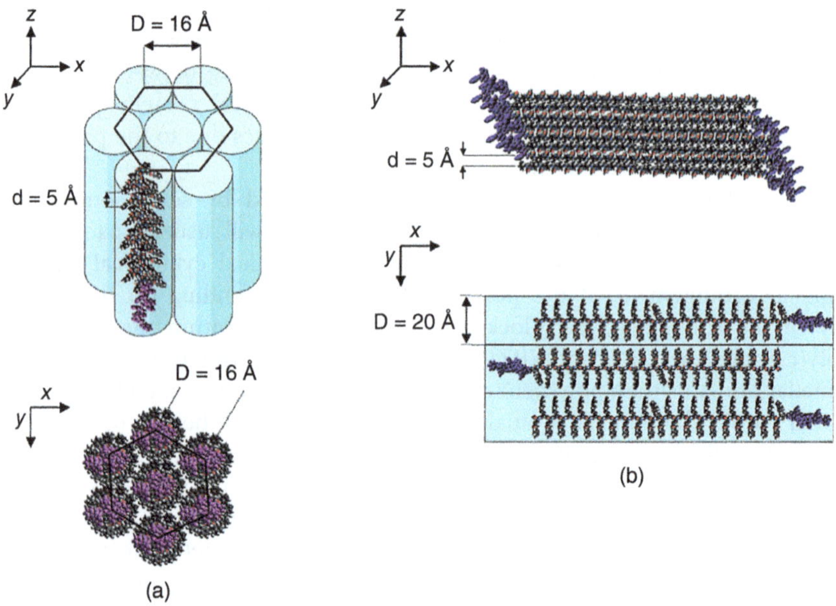

Figure 7.4 Schematic illustration of the (a) hexagonal packing of helices and (b) layered packing of β-sheets in PBLG-oligostyrene conjugates. *Source*: Reproduced with permission of Ref. [78].

blocks or a polypeptide and a synthetic rod block; lamellar phases have been observed in these systems [80]. In an alternative creative design, α-helical and β-sheet segments were incorporated into polyureas or polyurethanes [81]. These sequences have a large impact on the mechanical properties of the polymers in part due to self-assembly of the peptide groups.

Block copolymers containing one helical block have also attracted a great deal of attention for the self-assembly of solution nanostructures, motivated in large part by applications in drug delivery [82, 83]. As with traditional block copolymer micelles, the block fraction and solvent quality for each block have a large influence on the type of nanostructure formed [84]. Similar to self-assembly in neat systems, aligning interactions between helices play a strong role in determining the type of solution nanostructure that is formed. When the solvent is good for the coil block but poor for the rod block, the rods tend to segregate to the core of vesicles, as they have a relatively large radius of curvature that minimizes distortion of the rod director orientation. Increasing rod fraction above about 65% leads to dominantly vesicles [85–88]. Spherical micelles [89, 90] and pancake micelles are observed at lower rod fractions (higher coil fractions). The high aspect ratio of the rodlike blocks can lead to strong aligning interactions when the rods pack in the core of the micelles [91]. Making the solvent rod-selective promotes the formation of vesicles at low rod fraction where the rods are now in the corona of the vesicle, and spherical micelles with a rod corona form at high rod fraction [29]. Beyond these general trends, a variety of different nanostructures may be formed, depending on the molar mass, hydrophobicity, charge, and chemical nature of the blocks, including tapes, ribbons, suprahelices, or fibrils [92–95]. When fibrils or other highly asymmetric structures are present at a sufficiently high concentration, they can form fibrillar gels [93]. These gels have been extensively investigated in diblock copolypeptide systems with one hydrophobic (leucine, valine) and one hydrophilic (lysine, glutamic acid) block, demonstrating shear-thinning and rapid recovery of mechanical properties [96, 97]. Copolymers containing helical and sheet-forming blocks in the same molecule have a particular tendency to fibrillize due to aggregation of the β-sheet domains [98].

The ability of the proteins to change secondary structure also leads to responsive transitions in solution self-assembly due to changes in chain configuration. Homopolypeptide α-helices tend to form under conditions where the amino acid residues are uncharged; with all hydrogen bonding satisfied internally, these helices tend to have low water solubility [54]. Many of the responsive blocks such as PZLLys and PBLG undergo a transition with pH triggered by charging, where the uncharged state is helical and the charged state is a random coil [99, 100]. These can result in changes in the size of a vesicle or in the type of solution nanostructure formed. When both blocks are responsive, structures can be fully inverted and two separate transitions are achieved [101]. Therefore, pH-driven responsive transitions between helix and coil can be used to trigger responsive micellar self-assembly in these systems [102, 103], and tuning the length of the different blocks can lead to the formation of micelles, vesicles, and other solution nanostructures. Because charge modulates both secondary structure formation and protein-protein interactions, ionic strength is also effective in some systems as a trigger for responsive changes in self-assembly [92], but high ionic strength can also decrease the magnitude of pH-induced changes [85, 99, 104]. Calcium ions [105], trifluoroethanol [106], and trifluoroacetic acid [107] have also been shown to be effective triggers for a helix-coil transition.

Helix-to-sheet transitions can be achieved using similar environmental triggers, leading to changes in the type of micellar aggregate formed and the aggregation number [94, 108, 109]. In a short helix-alkyl chain diblock system, it was shown that initially the conjugates self-assemble into helical bundles that slowly rearrange over time into sheet secondary structures, triggering a change from spherical to wormlike micelles [110]. Changes in temperature often trigger the change from helix to sheet, with sheets observed at higher temperatures. Block length has a large impact on these conformational transitions, with longer protein or peptide blocks promoting α-helices and shorter blocks promoting β-strand or random coil configurations [111].

When the helices bundle into coiled-coils during self-assembly, these interactions are able to direct the self-assembly process. The protein sequence determines the aggregation number of helices in the coiled-coil, and the aspect ratio of the protein aggregate becomes somewhat smaller, with anywhere from two to potentially seven helices per aggregate [60]. The outer surface of a coiled-coil is typically coated with charged and hydrophilic residues, imparting strong solubility to the aggregate. This leads to repulsive interactions between the aggregates that are thought to be partly responsible for micellar stabilization, but it also changes significantly the role of liquid crystalline interactions in inducing molecular alignment. In solution, the coiled-coils drive the aggregation of bioconjugates into small nanoaggregates where the size of the aggregate is governed by the association number of the coiled coil, programmed into the protein sequence [112–114]. The aggregates can be further structured based on collapse of a thermoresponsive polymer block [115, 116], hydrophobic interactions, or polyelectrolyte coacervation [117]. Such transitions in the coil aggregation state can potentially have a stabilizing effect on the protein fold [116]. When incorporated into telechelic block copolymers, the isolated aggregates formed in solution act as crosslink points, resulting in a very robust mechanism for producing a gel with controlled junction functionality and geometry [59, 118–121].

Conjugation of multiple synthetic groups to a coiled-coil can lead to the hierarchical self-assembly of the protein clusters into larger micellar objects in solution. Xu et al. have designed a unique system where poly(ethylene glycol) PEG is conjugated to the side of a coiled-coil that has a lipid attached at one of the ends. The coiled-coils aggregate into three or four-helix bundles, and the bundles further self-assemble into micelles [122–124]. These micelles have been shown to be exceptionally stable in solution, remaining intact even under challenging biological conditions. As a result, they are showing great promise for applications in drug delivery [125, 126]. The location of polymer conjugation is critically important, with meridinal conjugation inferior to terminal conjugation. Simulations indicate that the PEG forms a shroud around the coiled-coil, enhancing its stability due to specific interactions with lysine residues [127] and shielding of hydrophobic residues [128].

This extensive work on single α-helix and coiled-coil containing conjugates demonstrates several key engineering design principles for the use of helices as blocks in block copolymers. First, aligning interactions between helices have a strong impact on the type of nanostructure formed, promoting assemblies with a high degree of helical alignment. Therefore, the use of helices as an engineering tool can be used to target different phases or typical phases under different conditions. Second, changes from helix to coil or helix to sheet chain structures often trigger changes in self-assembly, particularly in solution. These mechanisms provide a powerful basis for engineering responsive soft matter based on changes in chain configuration. Finally, association of helices into coiled-coils places additional constraints on self-assembly that can stabilize specific nanostructures and enhance micellar stability, providing a method to hierarchically engineer bioconjugate assembly.

7.3 β-Sheet Protein Copolymers

Unlike α-helical secondary structures which satisfy all of their hydrogen bonding internally, chains in β-strands arrange in an all-transconfiguration that promotes inter-chain hydrogen bonding and crystallization. Therefore, β-sheet-forming proteins and peptides have a tendency to self-organize into fibrils and fibers. Typical examples of β-sheet forming proteins are silks and amyloid proteins [129], and the incorporation of these semi-crystalline proteins or derivative peptides into hybrid block copolymers can enhance control over molecular architecture and impart materials with novel properties and functions. They have a potential to combine properties of the peptide block, such as assembly, recognition, or bioactivity, with the specific chemical and physical properties, such as the processability, solubility, and bulk mechanical properties, of conventional polymers.

The first example of a β-sheet-forming peptide attached to a synthetic PEG block was reported by Lynn and meredith [130]. A central amino acid sequence derived from amyloid peptide was conjugated with a PEG block at its *C*-terminus using standard Fmoc protocols, enhancing the water solubility of the conjugates [130–132]. By adjusting concentration, ionic strength, and pH, fibrils were observed in solution. While the peptide sequence alone formed bundles of fibrils irreversibly, fibril aggregation didn't occur in the copolymer or copolymer/peptide mixtures [130, 131]. It was suggested that the PEG block coated the surface of the fibrils, thereby preventing or retarding lateral aggregation [130–132]. The propensity of β-strand proteins or peptides to form fibrils in solution has been observed across a large variety of different studies. One drawback of the β-sheet-forming peptides is uncontrollable aggregation into disordered tangles during self-assembly in almost all cases. Attachment of water-soluble blocks helps to control peptide fibril aggregation in solution. The fibrils can take many forms, including fibrils, tapes, and twisted helices that may subsequently aggregate into larger hierarchical structures [133]. PEG conjugation to a peptide can also stabilize the peptide secondary structure against pH variation [133]. Conjugates consisting of a β-sheet-forming peptide with a short sequence of alternating glycine and alanine residues (GAGA) and a metallopolymer block, polyferrocenylsilane (PFS) [134] are able to form fibrillary structures in organic solvents.

Lateral aggregation between fibrils is only retarded for sufficiently long polymer chains in the conjugates. A comparative study of the solution self-assembly of a series of peptide-PEG conjugates with different PEG chain

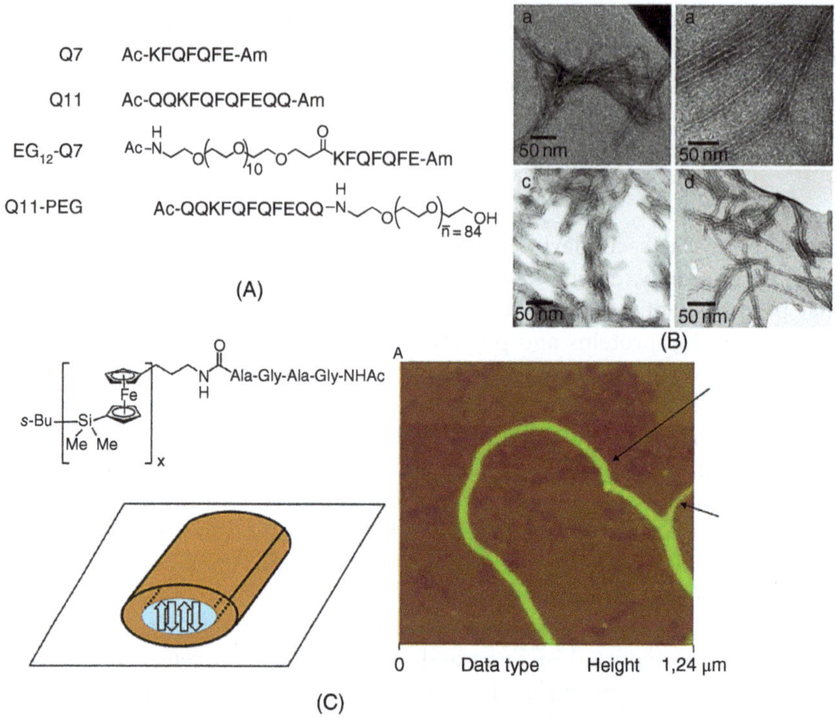

Figure 7.5 (A) The peptide sequences and chemical structure of peptides and peptide–polymer block copolymer derived from amyloid protein. (B) TEM images with negative staining of Q11 fibrils (a), Q11-PEG fibrils (b), Q7 ribbons (c), and EG_{12}-Q7 ribbons (d). (C) Fibrous structure formed in solution by PFS-GAGA block copolymer. *Source*: Reproduced with permission of Refs. [134–135].

lengths (Figure 7.5A) [135] showed that conjugates with a PEG block of 3600 Da formed individual straight fibrils with a regular helical twist and uniform width (Figure 7.5B). In contrast, considerable aggregation was observed in the oligo(ethylene glycol)-peptide conjugates with short polymer blocks. Therefore, conjugation of soluble polymer chains is an effective approach to modulate the assembly of β-sheet-forming peptides, particularly their size, uniformity, and lateral aggregation. Similar results were observed in a series of ABA triblock copolymer consisting of a central β-sheet-forming peptide block with tandem repeats of $(AG)_3EG$ sequence and two PEG end blocks with molar mass ranging from 750 to 5000 Da [136, 137]. A fibrillar nanostructure was observed for all the conjugates with heights of ~2 nm, indicating that PEG chains prevent side-to-side aggregation without interfering strongly with the desired β-sheet interactions, but at higher molar mass the fiber length was shorter presumably due to steric effects of the longer PEG chains in the self-assembled structure.

Multiblock copolymers consisting of alternating β-sheet-forming peptides and synthetic polymer blocks have also attracted a great deal of attention to produce silk-mimetic hybrid nanostructures. When the amorphous peptide domain of a spider silk *Nephila clavipes* was replaced with a PEG block, resulting in poly(alanine)-PEG multiblock block copolymers [138], the polymers retained the antiparallel β-sheet structure. Atomic force microscopy (AFM) revealed a microphase-separated morphology with irregularly shaped 100–200 nm poly(alanine) nanodomains interspersed within the PEG matrix. Similar multiblock copolymers with GAGA sequences of *Bombyx mori* were also retained the β-sheet structures and had a similar microphase-separated morphology [139]. It was thus concluded that replacing the amorphous peptide domain with PEG could be used to make hybrid silk-like materials. Multiblocks inspired by silk protein have also been prepared with hydrophobic polymers; condensing poly(alanine) (A5) blocks with polyisoprene (molecular weight of 2200 and 5000 Da) [140] yielded polymers with significantly improved solubility but similar secondary structure to that of the natural spidroin. Micelle-like aggregates were observed in the block copolymers with shorter isoprene blocks (2200 Da), showing that molar mass is a critical design variable in multiblock β-strand polymers as well.

The use of genetic engineering to construct silk-mimetic block copolymers provides an alternative route to access protein block copolymer sequences with β-sheet forming blocks. Silks are natural block copolymers consisting of characteristic crystalline β-sheet secondary structures with amino acid sequences of multiple repeats of mainly alanine (A), glycine-alanine (GA), or glycine-alanine-serine (GAS), and hydrophilic, noncrystalline regions commonly composed of β-spirals (GPGXX, where X is mostly glutamine) or helical structures (GGX). Martin et al. reported the first example of silk based triblock copolymer with the crystalline segment (GAGAGS) attached to the both ends of the cell attachment domain of human fibronectin in 1994 [141]. Kaplan studied the self-assembly of a series of genetically engineered spider silk block copolymers with a hydrophobic polyalanine-rich block and a hydrophilic glycine-rich block [142]. The β-sheet content increased with an increase of the length of the polyalanine blocks, and a large variety of self-assembled nanostructures, including spheres, rods, bowl-shaped micelles, and giant compound micelles, were observed by transmission electron microscopy (TEM) and AFM as a function of the size of the hydrophobic block, the presence of the histidine tag, and the solvent properties.

A more widely utilized approach than the above designs is to use silk domains as crystalline blocks and elastin domains as flexible blocks to form silk-elastin-like protein polymers (SELPs), which combine the mechanical and biological properties of both silk and elastin [143, 144]. SELP-based materials have been developed for drug delivery, tissue engineering, and plasmonics [145–147]. Silk blocks GAGAGS mimicking the natural silkworm heavy chain tends to form insoluble β-sheets, while elastin blocks VPGXG are soluble in water below the thermoresponsive transition temperature [148]. Driven by silk block aggregation, SELPs could self-assemble into micellar-like particles with a silk block core and an elastin block corona. This first step of self-assembly is spontaneous in solution. Heating the solution above the transition temperature of the elastin blocks triggered a second assembly step, which was driven by the hydrophobic interaction between elastin blocks (Figure 7.6d). Various nanostructures including nanoparticles, hydrogels, or nanofibers were generated through tuning of the ratio of silk-to-elastin blocks (Figure 7.6b,c). The addition of hydrophobic compounds can also trigger the self-assembly of these SELPs into uniform micellar-like nanoparticles [149]. The silk-to-elastin

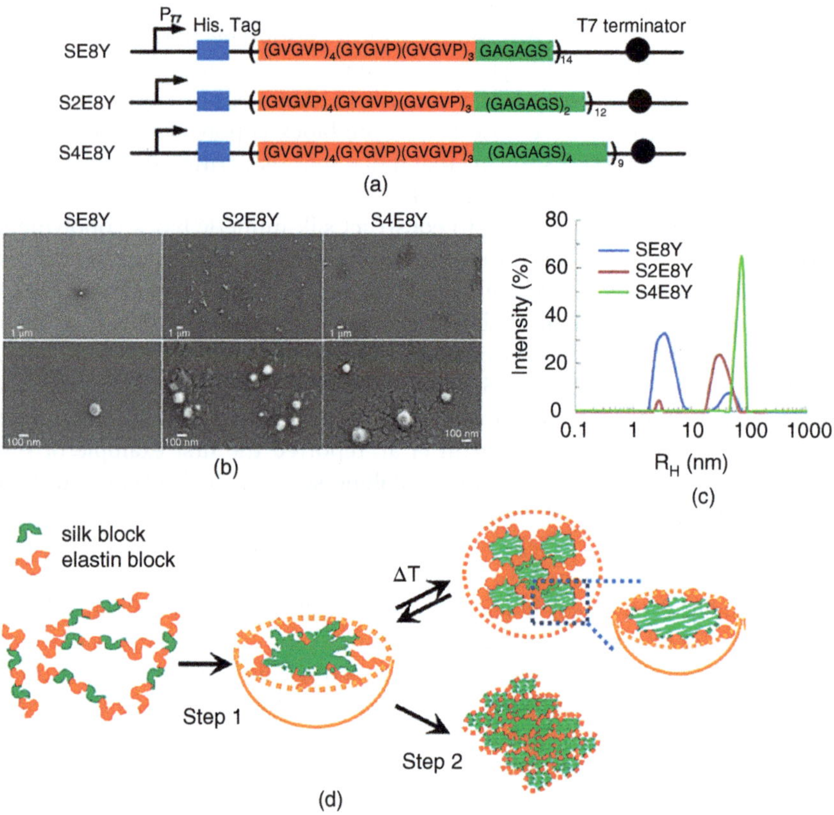

Figure 7.6 (a) SELPs sequences with varying rations of the silk-to-elastin blocks. (b) Representative SEM images of the micellar-like particles in solution. (c) DLS size distributions of the micellar-like particles. (d) Proposed two-step self-assembly process. *Source*: Reproduced with permission of Ref. [148].

block ratio was found to be critical on the assembly (Figure 7.6a) [147–149]. Similar to the role of synthetic polymers, the soluble elastin blocks prevent uncontrolled lateral aggregation of the silk block in a temperature-sensitive manner (aggregation is prevented only below the thermoresponsive transition temperature). Therefore, in SELPs the silk block assembly can be induced by a trigger such as pH to form fibrils, and the fibrils can be driven to form bundles or aggregates by an increase in temperature [150].

Finally, silk-like blocks have been incorporated into copolymers with collagen-inspired random coil blocks to yield de novo designed recombinant triblock polypeptides [151, 152]. Histidine residues in the silk-like middle block imparted pH-sensitivity, with fiber formation only above pH 6 when the histidine is uncharged. At high concentrations, the fibers become physically entangled and formed self-healing physical gels [151] analogous to those formed by low-molar-mass peptide fibril self-assembly [96, 153, 154]. A series of triblock copolymers with different middle-block length $(GAGAGAGH)_n$ (n = 8, 16, 24, 48) were also synthesized to understand the effect of β-strand length on fibril formation [152]. All the block copolymers showed pH-responsive self-assembly behavior, but the triblock copolymers with short middle block (n = 8 and 16) formed spherical micelles (Figure 7.7c,d) while filaments were observed for those with long middle block ($n = 24$ and 48). The CD spectra indicated that the short middle block adopted a random coil configuration as compared to a β-roll structure in the case of $n = 24$ and 48 (Figure 7.7a,b) [152].

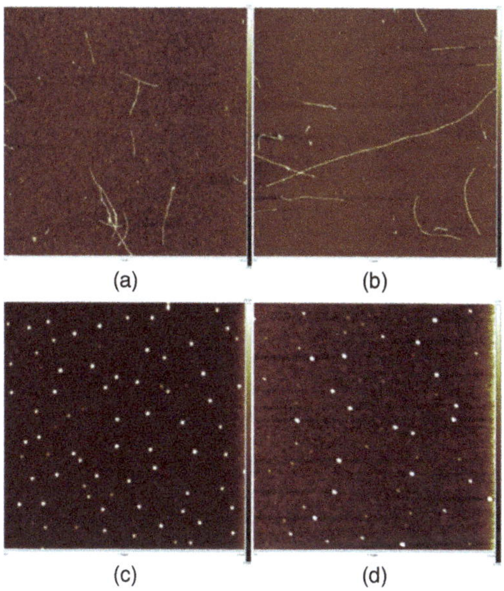

Figure 7.7 AFM images of self-assembled recombinant triblock polypeptide at pH 8 with different middle block length: $n = 48$ (a), 24 (b), 16 (c), and 8 (d). Images are 5 × 5 μm (a, b) or 2 × 2 μm (c, d). *Source*: Reproduced with permission of Ref. [152].

7.4 Cyclic Protein Copolymers

Constraining linear peptides in a cyclized form restricts the conformational freedom. This restriction can enhance the preference for β-sheet formation, which when the peptide is cyclized occurs through the self-assembly into one-dimensional nanotubes templated by intermolecular hydrogen bonding between the backbone amides along the tube growth axis (Figure 7.8a,b) [155].

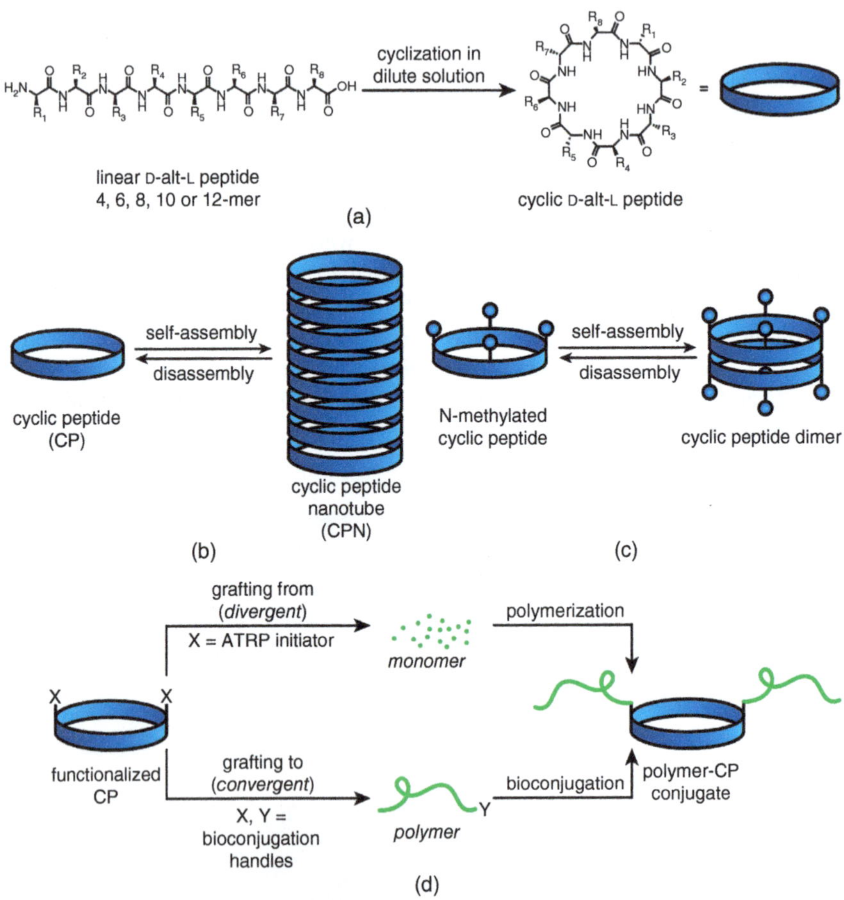

Figure 7.8 Cyclic peptides (CPs) from D-alt-L linear peptides. (a) Cyclization of linear peptides with alternating stereochemistry (D-alt-L) generates peptides that readily form β-sheets and assemble to form nanotubes with a controlled inner diameter (b). (c) Lateral self-assembly can be inhibited by methylating alternate amide nitrogen atoms to block H-bonding interactions on one face of the peptide. (d) Strategies for the synthesis of conjugate materials include grafting polymers from ("grafting from") CPs and grafting polymers to ("grafting to") CP scaffolds. *Source*: Reproduced with permission of Ref. [159].

Several peptide scaffolds promote this well-defined assembly, with 8-mer D-alt-L cyclic peptides serving as the canonical example [156]. Cyclic peptides (CP) formed from 6 to 12 α-amino acids with alternating chirality (D-alt-L) assemble in solution to form high aspect ratio nanotubes with a well-defined pore size dictated by the number of amino acids in the polypeptide [157, 158]. The amino acid side chains are projected on the exterior of the CP nanotube enabling surface functionalization. Control over the self-assembly process and improvements to the solubility of the CP nanotubes can be achieved through the synthetic modification of the amide backbone or amino acid side chains. Methylation of alternating amide nitrogen atoms prevents H-bonding on one face of the cyclic peptide and results in dimer formation (Figure 7.8c) [159]. Tunable control over the assembly of the peptide nanotubes can also be accomplished through the conjugation of polymers to the exterior [160, 161].

Because cyclic peptides are synthesized using solid-phase synthesis [162], sites for polymer attachment on the cyclic peptide scaffold can be defined precisely. Polymers can be conjugated to the functionalized cyclic peptide using either a divergent ("grafting from") [163, 164] or convergent ("grafting to") [160, 165] strategy (Figure 7.8d). For both of these strategies, the efforts have been primarily focused on the conjugation of polymers to 8-mer D-alt-L CPs. In the divergent approach, atom transfer radical polymerization (ATRP) initiators are attached to amino acid side chains, typically through the modification of the ε amine of lysine [166]. Polymerization can be initiated from one to four sites on the cyclic peptide, but the characterization of the resulting CP–polymer conjugate is difficult. Due to the fact the peptide is cyclized before polymerization, the chain growth occurs from the assembled CP nanotube and not from a soluble monomeric cyclic peptide. Alternately, end-functionalized polymers can be grafted to CPs in a convergent fashion. This synthetic strategy enables the complete characterization of the individual components prior to conjugation and in many cases bioconjugation can be performed on monomeric peptides [160, 165, 167]. The bioconjugation strategy relies on one to four lysine or functionalized lysine residues incorporated in the CP, but the yield for bioconjugation of four polymers to a single cyclic peptide is low due to steric crowding, which causes the current practical limit for this strategy to be three polymers/peptide [168].

Using both of these strategies a wide range of polymers – with varying chemical functionality and degrees of polymerization – have been attached to cyclic peptides. The attached polymer chains can be used to improve the solubility of the nanotubes in water, impart responsive properties, or provide crosslinkable functionality to stabilize the supramolecular assembly [169–171]. In order to provide control over the assembly of the CP–polymer conjugates, the bioconjugates first need to be dissolved in strongly H-bond forming solvents (competing solvents), such as *N,N*-dimethylformamide (DMF) or trifluoracetic acid (TFA), and then subsequently diluted in to poor solvents, such as tetrahydrofuran (THF) or methanol. Importantly, the conjugated polymer improves the solubility of the cyclic peptide and thus permits controlled self-assembly and monitoring of the self-assembly process [161]. The assembly of the cyclic peptides into nanotubes is modulated by the properties of the attached polymer, including its length, polydispersity, and the number of polymers per cyclic peptide (Figure 7.9a) [172]. Increasing the polymer molar mass leads to a decrease in the length of the self-assembled nanotube due to steric hinderance of the self-assembly process. In solvents with intermediate hydrogen bonding competitiveness, CP nanotube length is governed by polymer degree of polymerization, but maximum nanotube lengths are achieved with intermediate polymer lengths [161]. This is due to a balance between shielding the peptide nanotube from the competing solvent (favored by longer polymers) and minimizing steric disruption of the nanotube assembly (favored by shorter polymers). Conjugation of polymers to preformed cyclic peptide nanotubes also results in a decrease in nanotube length, indicating that the polymer is capable of destabilizing the intermolecular hydrogen bonds of a pre-assembled nanotube [173]. Maintenance of the anti-parallel β-sheet structure in the CP nanotube can be monitored by attenuated total reflectance Fourier transform infrared spectroscopy (ATR-FTIR) [172, 174].

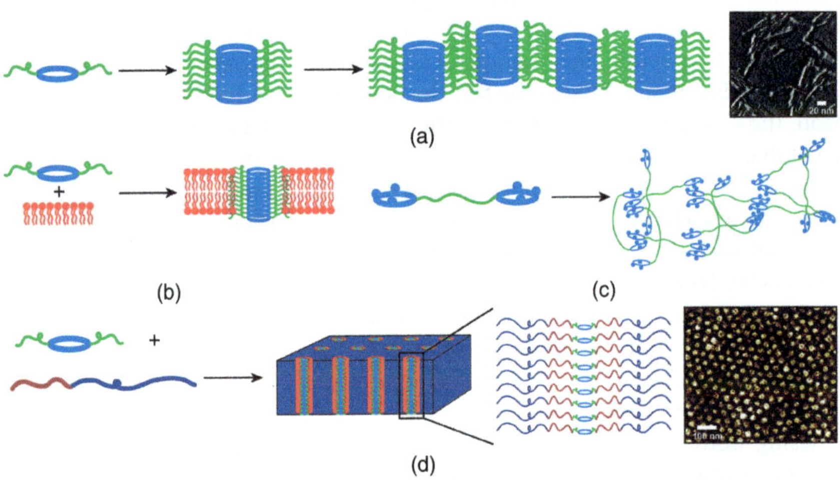

Figure 7.9 Assembly of CP–polymer conjugates. (a) Cyclic peptide conjugates assemble into nanotubes driven by hydrogen bonding interactions between the peptide backbone. These nanotubes can undergo further assembly governed by interactions between the polymer coating. (b) Depending on the hydrophobicity of the CP–polymer conjugate the self-assembled nanotubes can partition into lipid bilayers (membranes). (c) Polymer conjugates of N-methylated cyclic peptides can self-assemble into larger order aggregates. (d) CP–polymer conjugates can be co-assembled with coiled-coil block copolymers. *Source*: Reproduced with permission of Ref. [172].

The amide N-H stretch (~3280 cm^{-1}) and C=O stretch (~1600–1700 cm^{-1}) have characteristic frequencies when in β-sheet structure and can be used to confirm the mechanism of self-assembly.

Depending on the hydrophobicity of the amino acid side chains in the cyclic peptide, the CP nanotubes can partition in to lipid bilayers and serve as a synthetic size-selective pore [175]. Attachment of hydrophobic polymers, such as poly(n-butyl acrylate), to CP scaffolds can promote insertion in to lipid bilayers, while conjugation of hydrophilic polymers, such as poly(2-hydroxyethyl acrylate), inhibits bilayer partitioning [176]. Additionally, when the grafted polymer conjugates are thermoresponsive, partitioning into the membrane can be thermally gated to provide responsive membrane puncturing. The polymer chemistry can also be used to control whether the CP–polymer conjugates form unimeric channels or aggregate to form a larger channel (e.g. barrel staves). Most conjugates form unimeric channels, but Perrier et al. were able to form larger channels using Janus CP–polymer nanotubes [167] where they sequentially modified the cyclic peptide with two immiscible polymers using orthogonal bioconjugation reactions.

N-methylation of the cyclic peptides controls their aggregation by eliminating hydrogen bond donation from the nitrogen of the amide group, leading to the rapid formation of dimers. When incorporated into protein–polymer conjugates with a cyclic peptide at both ends of a polymer midblock, they can promote supramolecular polymerization via hydrogen bonding. The cyclic peptides can further aggregate to form physical branching points, leading to the formation of large networked molecules (Figure 7.9c) [177].

Control over both the nanotube length and orientation can also be accomplished through the co-assembly of CP–polymer conjugates and coiled-coil block copolymers [178]. The conjugated polymers can be used to mediate the cyclic peptide interactions to selectively partition into one domain of the block copolymer (BCP). By balancing the interactions between the two components, the block copolymer can undergo microphase separation and the CP–polymer conjugate can assemble within a single domain of microphase separated BCP (Figure 7.9d). This co-assembly has been demonstrated in thin films and has generated well-ordered films with ~ 0.7–0.8 nm pores governed by the cyclic peptide nanotubes [178].

7.5 Coil-Like Protein Copolymers

Unlike the previously discussed protein and peptide blocks that have specific secondary structures that directly affect the self-assembly of the protein, another class of proteins is structurally disordered in solution, best approximated as Gaussian coils. By far the most common protein in this category is the elastin-like polypeptides (ELP), derived from natural elastins. Elastin is a class of naturally occurring structural proteins that provides elastic properties to many organs and tissues [179–181]. Among them, tropoelastin is one of the most studied elastin with a penta-peptide repeat sequence, Val-Pro-Gly-Val-Gly [182]. EPLs with a similar penta-peptide repeat sequence, (Val/Ile)-Pro-Xaa-Yaa-Gly, are widely studied biomaterials. The third residue is either Ala or Gly. Replacing the Gly residual with Ala leads to a change in the mechanical response of the material from elastomeric to plastic. The fourth residual of the sequence (Yaa) could be any amino acid other than proline.

ELP has been widely incorporated as a stimuli responsive block into block copolymers for therapeutics delivery and tissue engineering because of its high elasticity and thermal responsive behavior [39, 179, 183]. At a specific temperature, ELPs undergo a sharp solubility transition. This behavior is similar to the well-known thermo-responsive polymers such as poly(N-isopropylacrylamide) (PNIPAM). The temperature of the transition, also known as lower critical solution temperature (LCST), varies with ELP sequence, molecular weight, and concentration [184–187]. By carefully tuning these parameters, ELPs exhibiting desired transition temperatures could be designed and synthesized for particular applications. For example, adjusting the fourth residue (Yaa) tunes the thermoresponsive transition temperature of the ELP in aqueous solution [187].

Because the ELP transition temperature is easily tuned with small changes in protein sequence, it represents one of the most flexible thermoresponsive polymers [188] among the broader field of both synthetic and natural systems for the design of temperature-responsive self-assemblies and nanostructures. The transition temperature of ELP can be readily tuned by adjusting the fourth residue (Yaa), molecular weight, and ELP concentration [187, 189, 190]. ELP block copolymers consisting of two ELP blocks with different transition temperatures can self-assemble into micelles with varied morphologies in the temperature window between the LCST of two blocks [39, 183]. The relatively hydrophobic ELP block collapses to form the micellar core, which is stabilized by the relatively hydrophilic ELP corona. The use of genetic engineering and biological protein expression to design and produce the polymers affords precise control of composition, sequence, and size of each peptide blocks [191]. For example, a system with a hydrophobic alanine-rich block and a hydrophilic valine-rich block formed both spherical and cylindrical micelles in solution above the transition temperature of the more hydrophobic block (Figure 7.10) [191]. The cylinders formed though coalescence of the spheres (Figure 7.10b). A systematic study of a series of 10 different ELP block copolymers with different hydrophilic-to-hydrophobic block ratios and overall molar masses by Chilkoti et al. [192] provides a set of design rules for the self-assembly of ELP–ELP diblock copolymers in solution. This study showed that two-phase transitions could be observed in the block copolymers with a hydrophobic-to-hydrophilic ratio between 1 : 2 and 2 : 1 upon increasing the solution temperature. A unimer to spherical micelle transition was observed

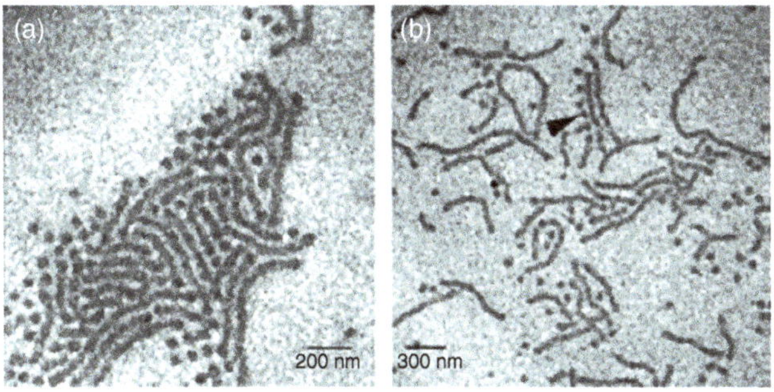

Figure 7.10 TEM image of the spherical and cylindrical micelles of the ELP–ELP block copolymer self-assembled in aqueous solution (a, b). The arrow in (b) indicates the cylinders were evolved from coalescence of spheres [191].

at an intermediate temperature, and a micelle to bulk aggregate transition at a higher temperature as the temperature is increased above the transition temperature of the second block such that both blocks become insoluble [192]. The critical micelle temperature could be tuned by the length of the hydrophobic ELP block, while the size of the micelle is controlled by both the overall molecular weight and hydrophilic-to-hydrophobic block ratio.

Micelles formed through hydrophobic interactions are typically unstable in a complex environment containing naturally occurring amphiphiles, such as plasma proteins and glycolipids, which limits the application of the micelles. To enhance the stability, the micelle may be chemically crosslinked [193, 194]. Chaikof et al. demonstrated the first example of thermally responsive and crosslink-stabilized ELP block copolymer micelles using cysteine residues located at the junction of two ELP blocks (Figure 7.11a,b) [195]. After self-assembly, disulfide bonds crosslinked the shell and stabilized the resultant micelles against environmental disturbances (Figure 7.11c). Shell-crosslinked micelles did not show a significant change in size or polydispersity over a broad temperature range between 5 and 37 °C. Adding reducing agent destabilizes the micelles by reduction of the disulfide bonds [195]. These highly stable micelles could serve as a robust vehicle for controlled drug delivery [196]. Corona crosslinked micelles have also been developed based on the strain promoted azide–alkyne cycloaddition (SPAAC) [197]. The ELP–ELP diblock copolymer was azidated at the end of hydrophilic block using a diazotransfer reaction. Subsequent self-assembly and crosslinking using a bis-cyclooctyne reagent afforded stable nanoparticles in solution, which had potential to be nanocarriers for drug delivery, as showed by hydrophobic dye encapsulation.

ELP micelles have been explored for several different applications. Incorporation of silica precipitants in the ELP block copolymers enables the preparation of hybrid organic–inorganic materials, which could potentially promote the utility and stability of self-assembled nanostructures. Biomimetic silaffin

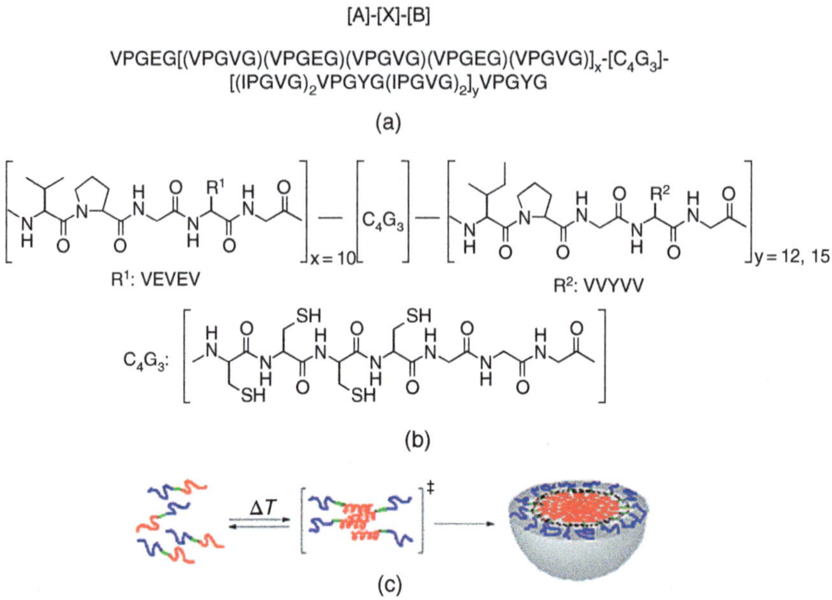

Figure 7.11 Amino acid sequence of the ELP–cysteine–ELP block copolymer (a, b), and the schematic illustration of micelle formation and crosslinking (c). *Source*: Reproduced with permission of Ref. [195].

R5, a peptide sequence with silica affinity, was attached onto the hydrophilic terminus of amphiphilic ELP–ELP diblock copolymers (Figure 7.12a) [198]. These block copolymers underwent temperature-triggered self-assembly into well-defined spherical micelles, leading to presentation of silaffin R5 peptide on the corona of the micelles (Figure 7.12b). Localized condensation of silica and subsequent silicification led to the formation of uniform, monodisperse hybrid ELP-silica nanoparticles with sub-100 nm diameters (Figure 7.12c). Various cell recognition tags have been incorporated into the end of the hydrophilic block using genetic engineering [192, 199–203]. Subsequent formation of micelles leads to the enrichment of the recognition tags on the surface (corona) of the nanoparticle, which increase avidity due to simultaneous interaction of multiple ligands with multiple receptors, enabling efficient internalization of a wide variety of cargo in diverse cell types for applications such as tumor targeting [199, 200]. A greater than eightfold increase in cellular uptake was observed after self-assembly of micelles relative to the unimers due to the high density of targeting residues on the micelle surface [200].

In addition to self-assembly in solution, ELP block copolymers have also been investigated as elastomeric materials. ELP block copolymers with a linear BAB triblock architecture, where A was hydrophilic block and B was hydrophobic block, underwent a reversible sol-gel transition over a narrow

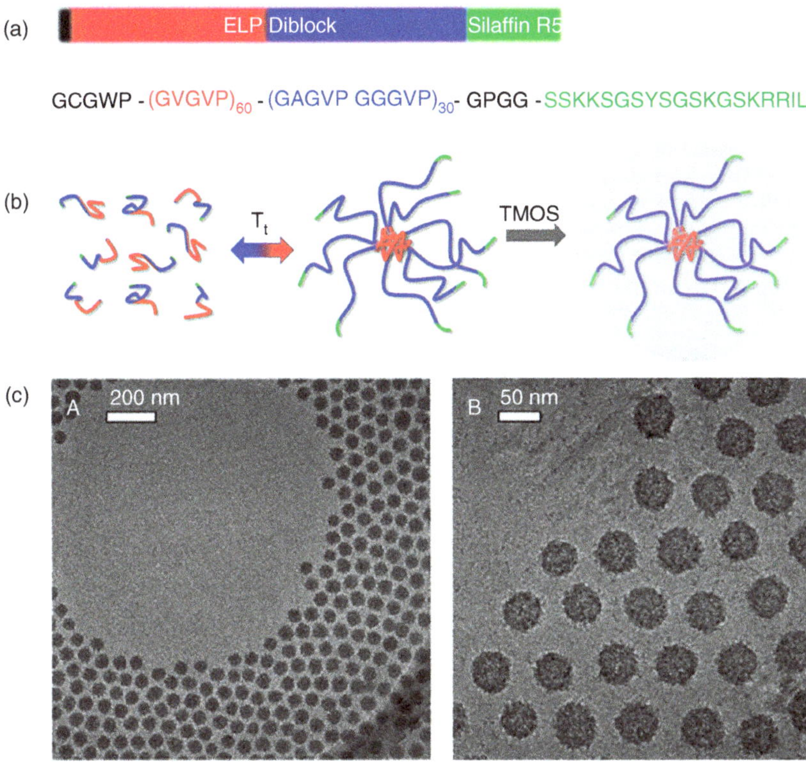

Figure 7.12 Schematic representation (a) and amino acid sequence (b) of ELP–ELP diblock copolymer with R5 peptide on the C-terminal: the hydrophobic ELP block is in red and the hydrophilic ELP block in blue, and cryo-TEM image of micelle-template silica nanoparticles after silicification (c). *Source*: Reproduced with permission of Ref. [198].

temperature range [204, 205]. A well-defined elastomeric network formed above the lower critical solution temperature of the hydrophobic block B. Selective microscopic phase separation of the hydrophobic domains from hydrophilic matrix led to the formation of physical crosslinks interconnected via hydrophilic midblock bridges, generating protein-based analogues of conventional thermoplastic elastomers [204].

Fusion proteins that combine a functional or structural protein with ELP block synthesized through genetic engineering are also commonly synthesized; they resemble the geometric features of small molecule surfactants with larger sizes and diversified biological functionalities. They can self-associate into biofunctional nanostructures based on the similar self-assembly principle of block copolymers and small molecule surfactants. For example, ELPs may be fused to a viral capsid protein (CP) to give a block copolymer ELP–CP. This fusion combined both the ability of CP to form well defined virus-like

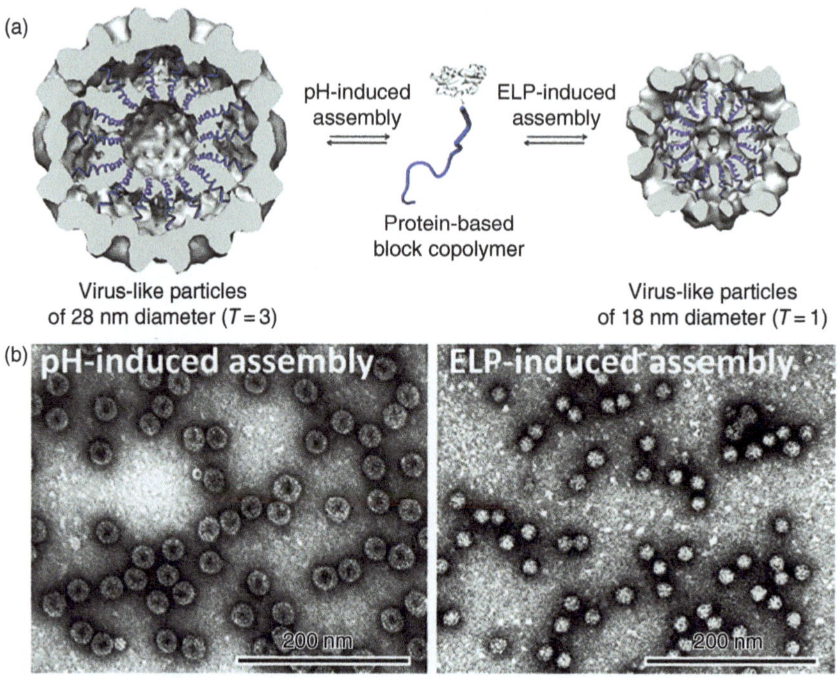

Figure 7.13 Representation of the ELP–CP and its self-assembly (a), and corresponding TEM images (b) under different conditions. *Source*: Reproduced with permission of Ref. [206].

particles and the stimuli-responsive character of the ELP block [206]. Two types of uniform self-assembled nanostructures, therefore, formed via different approaches induced by either CP or ELP aggregation (Figure 7.13). Virus-like nanoparticles with a T = 3 icosahedral architecture and a diameter of 28 nm were observed by pH-induced assembly of CP at room temperature and pH = 5.0, under which conditions ELP block would be soluble. On the other hand, thermally responsive ELP led to the efficient production of T = 1 icosahedra with a diameter of 18 nm upon adding NaCl, which lowered the transition temperature of the ELP block to below room temperature [206]. Additional examples involving globular proteins are discussed in the following section.

ELP blocks have also been conjugated with synthetic polymers, which combines thermos-responsive features and biofunctional properties of ELP with the processability and bulk mechanical properties of conventional polymers. In one approach, a series of ELP-PEG block copolymers was synthesized via grafting to pendant functionality in the ELP chain [207]. Azido functional groups were introduced by a pH-controlled diazotransfer reaction on the primary amines present in ELP (N-terminus and lysine side chains), and azide-alkyne chemistry was used to attach PEG side chains. By varying pH during the diazo transfer reaction, ELP-blocks with one or two azido functional groups were obtained, enabling ELP-PEG diblock copolymers and miktoarm star polymers to be synthesized. Self-assembly was triggered by the addition of NaCl leading to spherical micelles with a diameter of 40–70 nm [207]. Alternately, elastin-mimetic hybrid copolymers with an alternating VPGVG block and poly(*tert*-butyl acrylate) (P*t*BA) block were synthesized by condensation of telechelic P*t*BA and short, alkyne-terminated peptides with a sequence $(VPGVG)_2$ (VG2) [208]. The resultant multiblock copolymer, [P*t*BA–VG2]n contained up to six repeats of the ELP domain, and could be converted to PAA-VG2 copolymers through removal of the *tert*-butyl protecting group. PAA-VG2 diblock copolymers and VG2-PAA-VG2 triblock copolymers can be similarly prepared [209]. These polymers show pH and temperature-dependent aggregation in aqueous solution due to the responsive behaviors of both blocks.

7.6 Globular Protein Copolymers

Perhaps the greatest challenge in protein self-assembly is the assembly of fully folded and functional globular proteins. These proteins, which are typically found as soluble domains or membrane-anchored domains, contain complex tertiary or quaternary structures that are responsible for catalysis, light harvesting, and many sensing and response functions within cells. They range from smaller single domain proteins such as lysozyme or green fluorescent protein to complex multidomain structures such as cytochrome P450 and antibodies. This group of proteins is of great industrial interest for macromolecular therapeutics [210, 211], where such biologics represent a growing fraction of market share. They are also of high interest for biocatalysts, where high-volume commercial processes such as glucose fructose isomerization [212], acrylamide synthesis [213], and the synthesis of penicillin [214] rely on their catalytic activity to produce chemical transformations. Furthermore, readily accessible medical assays such as home glucose tests [215] and pregnancy tests [216] rely on globular protein biosensors. Therefore, advances in globular protein assembly that can improve the performance of these proteins within materials can potentially have a large impact.

Advances in bioconjugation chemistry and molecular biology offer a variety of methods to form globular proteins into block copolymers while preserving the folded protein structure and activity, a major challenge in the self-assembly of these molecules [40, 217]. Reactive residues on the protein surface such as lysine and cysteine offer attractive targets for bioconjugation [218, 219]. However, while the protein sequence can often be engineered to have a single instance of these residues accessible for bioconjugation, natural amino acid residues usually occur many times in a single protein making selective modification of amino acid residues difficult [43]. Alternative bioconjugation strategies such as incorporation of bioorthogonal functionalities using non-canonical amino acids and oxidation or transamination of the N-terminus to yield reactive aldehyde moieties have been developed [220, 221], offering greater control over site-specific modification of proteins. In addition to modification of the protein post-expression, the block copolymer molecular structure can be introduced biosynthetically. Fusions of globular proteins with unstructured protein domains such as ELP have been demonstrated to be useful in protein purification [222] but have also been observed to exhibit globular protein block copolymer self-assembly [223]. The numerous developments in bioconjugation chemistry, molecular biology, and protein purification have made globular protein block copolymers increasingly accessible, leading to a proliferation of studies exploring their physics and applications.

The largest interest has focused on the self-assembly of globular protein block copolymers into solution nanostructures. Termed "giant amphiphiles" by Nolte et al. who performed thorough studies of their behavior [224–227], the self-assembly of these molecules in water is generally driven by a hydrophobic-hydrophilic contrast between the two blocks of the molecule, with polymer hydrophobicity driving its aggregation into the micellar core and protein hydrophilicity locating it within the corona of the nanostructure [224]. Depending on the molecular design and kinetic process used to induce self-assembly, a wide variety of nanostructures can be formed. Bioconjugates of lipase with polystyrene have been shown to spontaneously assemble into micellar rods in water [224], while triblocks formed by bioconjugation of horseradish peroxidase with polystyrene-block-poly(ethylene glycol) (PS-b-PEG) have been demonstrated to form vesicular aggregates, Y-junctions, toroids, micellar rods, and micellar spheres [226]. Myoglobin(Mb)-b-PS-b-PEG triblocks spontaneously assemble into the same types of structures as well as into figure eights, octopi, and aggregates containing lamellae (Figure 7.14) [226].

In addition to using hydrophobic polymers, conjugates to thermoresponsive polymers or fusions to thermoresponsive ELP enable thermoresponsive assembly of such nanostructures in solution. Bioconjugates of bovine serum albumin to poly(*N*-isopropyl acrylamide) (PNIPAM) and of enhanced green fluorescent protein (EGFP) to random copolymers of methoxy ethoxy ethyl acrylate and ethylene glycol methyl ether have been observed to micellize above the lower critical solution temperatures of the polymers [226, 230]. Similarly, temperature responsive micellization has been observed in ELP fusions to thioredoxin, a fibronectin type III domain [201], and an anti-CD20 single-chain antibody (scFv) (Figure 7.14a,b) [228, 231] while maintaining the bioactivity of the proteins. These structures have attracted significant interest for targeted delivery of therapeutics due to the specific binding capability of the proteins for biomolecules that may be located in a tissue of interest or for

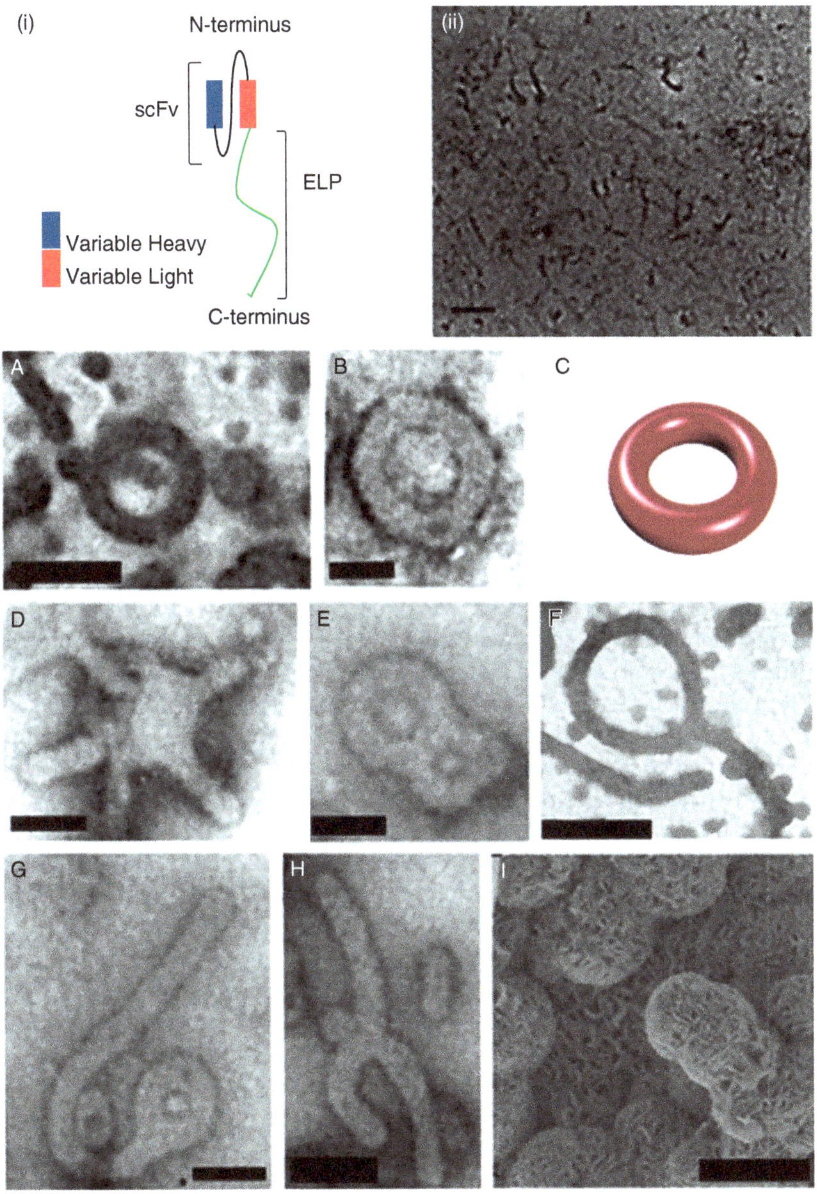

Figure 7.14 (i) An anti-CD20 scFv was fused to the N-terminus of an ELP, and (ii) the self-assembled wormlike nanoparticles in solution. Ref. [228]. Aggregates of Mb-b-PS-b-PEG forming (A, B) toroids, (C) schematic of a toroid, (D) octopi, (E) figure eights, (F, G, H) junctions, (I) spherical aggregates with lamellae. *Source*: Ref. [229]. Reproduced by permission of The Royal Society of Chemistry.

the therapeutic value of the proteins themselves. For instance, bovine serum albumin (BSA)-b-poly(methyl methacrylate) micelles have been shown to be capable of delivering chemotherapeutic agents *in vitro* [229], and ELP fusions have been proposed as promising systems for delivering pharmaceutical proteins [232]. In addition, the CD20 fusions show the ability to bind B-cell lymphomas and effectively slow tumor growth [228].

A major challenge in the design and self-assembly of such solution nanostructures is predicting the type of structure that will form as a function of the molecular design of the conjugate. Examples from the literature clearly illustrate a central challenge: it is currently difficult to predict whether a given conjugate or fusion will macrophase separate or micellize [230, 233–237]. The same thermoresponsive conjugates that can be used to form protein-functional micelles were demonstrated decades ago by Hoffman et al. [234] as a means to precipitate proteins, and similar reports have been given by other authors [201, 233]. Even relatively subtle changes such as changing from EGFP to mCherry as the protein corona can result in a change from stable to unstable micelles [236]. Theory and simulation of these complex molecules is desperately needed to provide a more fundamental understanding of the factors affecting self-assembly and to provide design rules for achieving a stable conjugate.

More recently, globular protein block copolymers have attracted interest for the self-assembly of solid-state materials, where they offer the potential to form new heterogeneous catalysts and biosensors with a very high density of protein by packing the protein in three dimensions [238], forming materials analogous to nanoporous heterogeneous catalysts used in traditional chemical engineering applications. The ability to self-assemble these materials under conditions where both the protein and polymer block are soluble in water has been demonstrated, leading to ordered nanostructures that resemble traditional coil-coil block copolymers [9, 20] such as lamellae, hexagonally packed cylinders, perforated lamellae, and gyroid morphologies. The molecules can be self-assembled in solution to form gels, and removal of water resulted in protein block copolymer plastics [239]. The phase diagram of the fluorescent protein mCherry conjugated to PNIPAM has been extensively studied (Figure 7.15a) [237, 239–241]. In contrast to the symmetry in phase behavior of traditional coil-coil block copolymers [20], globular protein block copolymers exhibit a high degree of compositional asymmetry consistent with the asymmetry of the molecule. In concentrated solution, these globular protein block copolymers predominantly exhibit lamellar phases above a critical order-disorder concentration at low temperature, and at high temperature a micellar phase is observed due to the collapse of the PNIPAM blocks above their LCST. As the concentration is further increased, the phases tend to shift to hexagonally packed cylindrical morphologies, a surprising re-entrant disordering transition at very high concentrations [241].

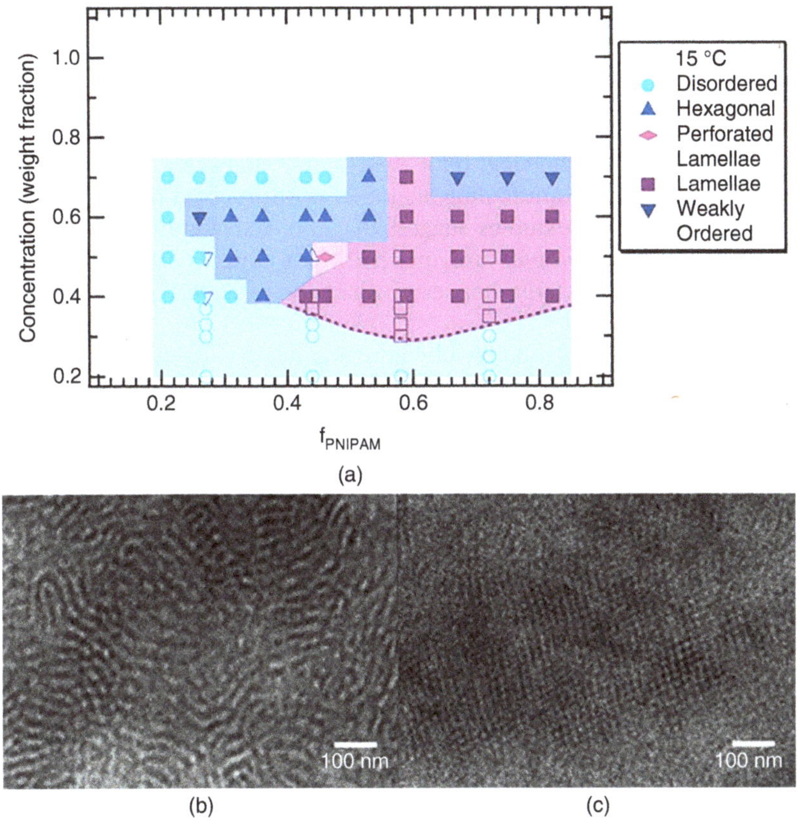

Figure 7.15 (a) Phase diagram for the self-assembly of mCherry-b-PNIPAM block copolymers at 15 °C. TEM micrographs of (b) lamellar and (c) cylindrical morphologies. *Source*: Ref. [240] – Reproduced by permission of The Royal Society of Chemistry.

A number of efforts have aimed at elucidating the specific design considerations that govern the type of nanostructure formed by these globular protein–polymer block copolymers. Small amino acid mutations modifying point charges on the surface of the protein have a minimal effect on self-assembly, and studies performed on the structurally homologous but chemically dissimilar EGFP-b-PNIPAM bioconjugate likewise results in similar phase behavior to mCherry-b-PNIPAM systems, implying that specific surface charge interactions have little effect on the self-assembly behavior of these molecules [237]. However, differences in phase behavior for proteins having a different 3D folded shape, as observed in Mb-b-PNIPAM systems, where phases remain largely disordered despite similar compositions as well-ordering mCherry-b-PNIPAM bioconjugates [238]. Similarly, bioconjugates synthesized with polymer blocks containing different hydrogen bonding motifs reveal considerably modified phase behavior compared with PNIPAM bioconjugates, assembling into gyroid morphologies. Additionally, recent studies of mCherry bioconjugates with the antifouling, zwitterionic polymer poly(3-[N-(2-methacroyloyethyl)-N,N-dimethylammonio]propane sulfonate) (PDMAPS) showing poor segregation between mCherry and PDMAPS imply that the antifouling character is outweighed by electrostatic interactions [242]. These studies suggest that the coarse-grained properties of the protein and that protein–polymer hydrogen-bonding and electrostatic interactions have a very large impact on the type of nanostructure formed [238, 243].

Recently, the same self-assembly phenomenon has also been demonstrated with mCherry-ELP fusion proteins. Unlike micellization of ELP fusions previously discussed, these systems show the ability to achieve protein-protein microphase separation under conditions where all proteins are soluble by increasing the concentration to above 20 wt% [223]. Two different fusion topologies have been explored: the "bola" fusion where ELP is at both termini of the protein and the "diblock" fusion where it is only at one terminus (Figure 7.16a). Despite the chemical similarity of the two protein blocks in the fusion, they self-assembled into nanostructures such as hexagonal and lamellar phases in concentrated solutions (Figure 7.16b,c). The topology of the molecule has a large impact on the nanostructure formed, with the bola fusion forming ordered nanostructures at a lower concentration and forming phases in a different order than the diblock fusion [223]. In addition to topology, length and hydrophobicity of the ELP have been studied. Similarly to synthetic polymer bioconjugates to mCherry, longer coil domains correspond to better ordering. Greater hydrophobicity of the ELP domain also results in higher quality of ordering due to the increased segregation strength between the blocks [244].

The engineering of globular protein block copolymers also enables globular proteins to be easily coated into thin films, overcoming longstanding challenges with loss of protein activity and low viscosity that leads to uneven coatings. This enables the formation of nanostructured films with thicknesses from tens of nanometer up to microns. Initial studies of Mb-b-PNIPAM heterogeneous biocatalysts catalyzing small molecule redox reactions show promising results [238], and there is a strong need for theory and more fundamental experiments to continue to understand the complex self-assembly process.

In addition to globular proteins, another large class of proteins exhibiting folded, 3-dimensional tertiary structures is membrane proteins. As their name implies, membrane proteins exist in biological membranes either permanently embedded within a membrane as integral membrane proteins, or associated to the surface of the membrane by hydrophobic and electrostatic interactions as peripheral membrane proteins [245]. Because they contact both

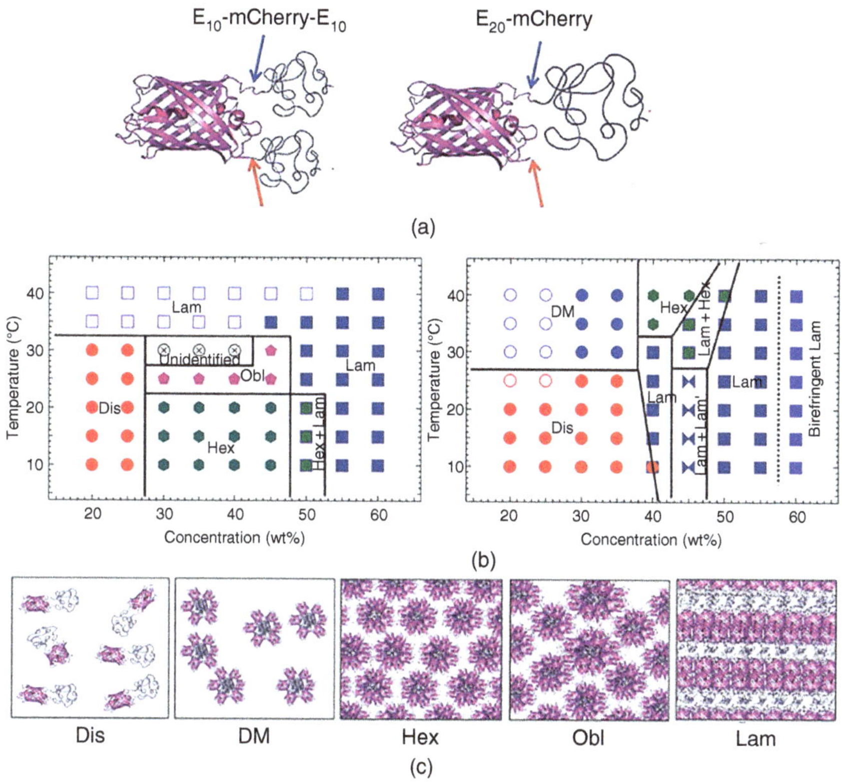

Figure 7.16 (a) Schematic of mCherry-ELP fusion block copolymers for double- and single-tailed fusions. (b) Phase diagrams of fusion proteins as a function of temperature and concentration for E_{10}-mCherry-E_{10} (left) and E_{20}-mCherry (right). (c) Schematic illustration of various phases: denoted as disordered (Dis), disordered micellar (DM), hexagonal (Hex), oblique (Obl) and lamellar (Lam). Open symbols represent regions where macrophase separation between a fusion-rich phase and a water-rich phase is observed. *Source*: Reproduced with permission of Ref. [223].

the hydrophobic interior of lipid bilayers as well as the hydrophilic exterior, membrane proteins naturally have blocky hydrophobicity. Membrane proteins often act as receptors that actuate a response to internal or external stimuli within a cell or as transporters to facilitate selective movement of molecules across a membrane [246]. As a result, there is motivation in isolating and engineering these molecules for use in responsive materials such as sensors or for incorporation into selective membranes for chemical separation. However, despite 20–30% of all cellular proteins existing as membrane proteins [247], few have been studied due to the difficulty of extracting and purifying these proteins while retaining their structure and activity [248].

Most studies on the self-assembly of membrane proteins focus on extraction of the membrane protein from the cell by co-assembly into micelles or nanodiscs. Solubilizing membrane proteins into detergent solutions is a technique that has been used for decades to isolate membrane proteins [249]; however, detergents often disrupt protein stability and interfere with protein function. Integral membrane proteins have been shown to co-assemble with phospholipids and amphipathic helical proteins referred to as membrane scaffold proteins (MSPs) into disc-like complexes. These complexes comprise a phospholipid bilayer core incorporating a single membrane protein with the MSP acting as a belt containing the structure [250, 251]. Membrane proteins have also been incorporated into block copolymer assemblies. The light harvesting membrane proteins have been templated into poly(ethylene oxide)-b-poly(propylene oxide)-b-poly(ethylene oxide) arrays and shown to be able to catalyze photodependent hydrogen production [252], and block copolymers have been used to mimic cell membranes to produce vesicles that act as bioreactors [238].

7.7 Outlook

A comparison of the many different proteins and biohybrid block copolymers discussed in this chapter clearly shows that nature has provided the materials engineer with a diverse set of shapes for the design of new materials. The comparison of the different materials clearly shows that each type has different considerations. Helical and disk proteins tend to have a strong liquid crystalline character, organizing into calamitic and discotic phases, respectively. These molecular-scale interactions translate into the formation of micellar structures in solution and bulk phases that reflect these liquid crystalline interactions, such as lamellar phases in helix-containing block copolymers and columnar phases in disk-containing block copolymers. Coil-like proteins show a large number of similarities to traditional synthetic polymers, but with complex sequence-property relationships, and β-sheet sequences provide for the ability to carefully engineer crystallization and aggregation within the materials. The self-assembly of block copolymers and bioconjugates containing globular and membrane proteins with tertiary and quaternary structure is only beginning to be understood based on colloidal protein thermodynamics, but it provides great potential for expanding biofunctionality in materials.

Even a brief glance at the natural world shows how these diverse shapes can provide inspiration for the design of new biofunctional materials that borrow their natural capabilities. The areas where this emerging science will have the greatest impact will be linked to the areas where the challenges in constructing a protein–polymer conjugate or fusion material will be compensated. Well-established areas for protein-based materials include biocatalysis and biosensors, where improvements in protein function in solid-state systems could have a major impact on commercial products and on the performance of chemical processes. Nanostructured materials may have a key role to play in this area and could potentially expand the number of processes and products that are viable through a biological route. Medical materials represent another key application, where materials for tissue culture and drug delivery are being actively formulated from protein. Tuning nanostructure using bottom-up self-assembly methods may offer new opportunities to tune biological functionality, mechanical properties, and transport rates in materials. Finally, these types of biological materials may serve as a new path to produce highly controlled nanostructures. The ability to tune each individual amino acid in a protein sequence enables careful control over molecular construction and self-assembly, enabling new types of molecular responses and nanostructures to be realized. These types of materials may be exploited as templates for the fabrication or placement of other functionalities, e.g. nanoparticles or photonic elements. In these cases, the protein groups in self-assembling block copolymers may allow access to types of structures that are not readily achievable with other molecules.

In conclusion, there have been rapid advances in our knowledge of how to self-assemble protein–polymer block copolymers since the early work of Gallot on simple helix-coil conjugates. Today scientists and engineers explore a wide variety of molecular shapes and are learning to master their design to achieve desired nanostructured materials. The future for these systems is extremely bright, with the potential to impact a number of different industries that can benefit from the unique properties of protein.

Acknowledgments

BDO acknowledges support from the Department of Energy Office of Basic Energy Sciences (DE-SC0007106). AH acknowledges support from the National Science Foundation Division of Materials Research (DMR-1253306). XD acknowledges support from the Air Force Office of Scientific Research (FA9550-12-0259). ACO acknowledges support as an Arnold O. Beckman Postdoctoral Fellow.

References

1 Woodward, A.E. and Smets, G. (1955). The preparation and characterization of block copolymers. *Journal of Polymer Science* 17: 51.
2 Dunn, A.S. and Melville, H.W. (1952). Synthesis of block copolymers. *Nature* 169: 699.

3 Moad, G., Rizzardo, E., and Thang, S.H. (2005). Living radical polymerization by the RAFT process. *Australian Journal of Chemistry* 58: 379.
4 Braunecker, W.A. and Matyjaszewski, K. (2007). Controlled/living radical polymerization: features, developments, and perspectives. *Progress in Polymer Science* 32: 93.
5 McCormack, C.L. and Lowe, A.B. (2004). Aqueous RAFT polymerization: recent developments in synthesis of functional water-soluble (Co)polymers with controlled structures. *Accounts of Chemical Research* 37: 312.
6 Hadjichristidis, N., Pitsikalis, M., and Pispas, S. (2001). Polymers with complex architecture by living anionic polymerization. *Chemical Reviews* 101: 3747.
7 Hawker, C.J., Bosman, A.W., and Harth, E. (2001). New polymer synthesis by nitroxide mediated living radical polymerizations. *Chemical Reviews* 101: 3661.
8 Young, R.N., Quirk, R.P., and Fetters, L.J. (1984). Anionic polymerizations of non-polar monomers involving lithium. *Advances in Polymer Science* 56: 1.
9 Hamley, I.W. (1998). *The Physics of Block Copolymers*. New York, NY: Oxford University Press.
10 Hamley, I.W. (2005). *Block Copolymers in Solution: Fundamentals and Applications*. New York: Wiley.
11 Khandpur, A.K., Forster, S., and Bates, F.S. (1995). Polyisoprene-polystyrene diblock copolymer phase diagram near the order-disorder transition. *Macromolecules* 28: 8796.
12 Cochran, E.W., Garcia-Cervera, C.J., and Fredrickson, G.H. (2006). Stability of the gyroid phase in diblock copolymers at strong segregation. *Macromolecules* 39: 2449.
13 Segalman, R.A. (2005). Patterning with block copolymer thin films. *Materials Science & Engineering R-Reports* 48: 191.
14 Albert, J.N.L. and Epps, T.H. (2010). Self-assembly of block copolymer thin films. *Materials Today* 13: 24.
15 Angelescu, D.E., Waller, J.H., Adamson, D.H. et al. (2004). Macroscopic orientation of block copolymer cylinders in single-layer films by shearing. *Advanced Materials* 16: 1736.
16 Segalman, R.A., Yokoyama, H., and Kramer, E.J. (2001). Graphoepitaxy of spherical domain block copolymer films. *Advanced Materials* 13: 1152.
17 Kim, S.O., Solak, H.H., Stoykovich, M.P. et al. (2003). Epitaxial self-assembly of block copolymers on lithographically defined nanopatterned substrates. *Nature* 424: 411.
18 Osuji, C., Ferreira, P.J., Mao, G.P. et al. (2004). Alignment of self-assembled hierarchical microstructure in liquid crystalline diblock copolymers using high magnetic fields. *Macromolecules* 37: 9903.

19 Morkved, T.L., Lu, M., Urbas, A.M. et al. (1996). Local control of microdomain orientation in diblock copolymer thin films with electric fields. *Science* 273: 931.
20 Bates, F.S. and Fredrickson, G.H. (1999). Block copolymers - designer soft materials. *Physics Today* 52: 32.
21 Epps, T.H., Cochran, E.W., Bailey, T.S. et al. (2004). Ordered network phases in linear poly(isoprene-b-styrene-b-ethylene oxide) triblock copolymers. *Macromolecules* 37: 8325.
22 Lee, S., Bluemle, M.J., and Bates, F.S. (2010). Discovery of a frank-Kasper sigma phase in sphere-forming block copolymer melts. *Science* 330: 349.
23 Phillip, W.A., Amendt, M., O'Neill, B. et al. (2009). Diffusion and flow across nanoporous polydicyclopentadiene-based membranes. *ACS Applied Materials & Interfaces* 1: 472.
24 Schacher, F., Ulbricht, M., and Muller, A.H.E. (2009). Self-supporting, double stimuli-responsive porous membranes from polystyrene-block-poly (*N,N*-dimethylaminoethyl methacrylate) diblock copolymers. *Advanced Functional Materials* 19: 1040.
25 Jha, A.K., Chen, L., Offeman, R.D. et al. (2011). Effect of nanoscale morphology on selective ethanol transport through block copolymer membranes. *Journal of Membrane Science* 373: 112.
26 Singh, M., Odusanya, O., Wilmes, G.M. et al. (2007). Effect of molecular weight on the mechanical and electrical properties of block copolymer electrolytes. *Macromolecules* 40: 4578.
27 Stoykovich, M.P., Muller, M., Kim, S.O. et al. (2005). Directed assembly of block copolymer blends into nonregular device-oriented structures. *Science* 308: 1442.
28 Yang, J.K.W., Jung, Y.S., Chang, J.-B. et al. (2010). Complex self-assembled patterns using sparse commensurate templates with locally varying motifs. *Nature Nanotechnology* 5: 256.
29 Olsen, B.D. and Segalman, R.A. (2008). Self-assembly of rod-coil block copolymers. *Materials Science & Engineering R-Reports* 62: 37.
30 Zhang, W., Gomez, E.D., and Milner, S.T. (2014). Predicting chain dimensions of semiflexible polymers from dihedral potentials. *Macromolecules* 47: 6453.
31 McCulloch, B., Ho, V., Hoarfrost, M. et al. (2013). Polymer chain shape of poly(3-alkylthiophenes) in solution using small-angle neutron scattering. *Macromolecules* 46: 1899.
32 Abbel, R., Schleuss, T.W., Frey, H. et al. (2005). Rod-length dependent aggregation in a series of oligo(p-benzamide)-block-poly(ethylene glycol) rod-coil copolymers. *Macromolecular Chemistry and Physics* 206: 2067.
33 de Ruijter, C., Jager, W.F., Li, L.B. et al. (2006). Lyotropic rod-coil poly(amide-block-aramid) alternating block copolymers: phase behavior and structure. *Macromolecules* 39: 4411.

34 Huang, M., Hsu, C.-H., Wang, J. et al. (2015). Selective assemblies of giant tetrahedra via precisely controlled positional interactions. *Science* 348: 424.

35 Zhang, W.-B., Li, Y., Li, X. et al. (2011). Synthesis of shape amphiphiles based on functional polyhedral oligomeric silsesquioxane end-capped poly(L-Lactide) with diverse head surface chemistry. *Macromolecules* 44: 2589.

36 Zhang, W.-B., Tu, Y., Ranjan, R. et al. (2008). Clicking fullerene with polymers: synthesis of 60 fullerene end-capped polystyrene. *Macromolecules* 41: 515.

37 Glotzer, S.C., Horsch, M.A., Iacovella, C.R. et al. (2005). Self-assembly of anisotropic tethered nanoparticle shape amphiphiles. *Current Opinion in Colloid & Interface Science* 10: 287.

38 Hermanson, G.T. (2008). *Bioconjugate Techniques*, 2e. Academic Press.

39 Rabotyagova, O.S., Cebe, P., and Kaplan, D.L. (2011). Protein-based block copolymers. *Biomacromolecules* 12: 269.

40 Kalia, J. and Raines, R.T. (2010). Advances in bioconjugation. *Current Organic Chemistry* 14: 138.

41 Krishna, O.D. and Kiick, K.L. (2010). Protein- and peptide-modified synthetic polymeric biomaterials. *Biopolymers* 94: 32.

42 Van Hest, J.C.M. (2007). Biosynthetic-synthetic polymer conjugates. *Polymer Reviews* 47: 63.

43 Obermeyer, A.C. and Olsen, B.D. (2015). Synthesis and application of protein-containing block copolymers. *ACS Macro Letters* 4: 101.

44 Olsen, B.D. (2013). Engineering materials from proteins. *AIChE Journal* 59: 3558.

45 Maskarinec, S.A. and Tirrell, D.A. (2005). Protein engineering approaches to biomaterials design. *Current Opinion in Biotechnology* 16: 422.

46 Kim, W. and Conticello, V.P. (2007). Protein engineering methods for investigation of structure-function relationships in protein-based elastomeric materials. *Polymer Reviews* 47: 93.

47 Liu, H.J., Kumar, S.K., and Douglas, J.F. (2009). Self-assembly-induced protein crystallization. *Physical Review Letters* 103: 018101.

48 Cooper, C.L., Dubin, P.L., Kayitmazer, A.B. et al. (2005). Polyelectrolyte-protein complexes. *Current Opinion in Colloid & Interface Science* 10: 52.

49 Billot, J.P., Douy, A., and Gallot, B. (1976). Synthesis and structural study of block copolymers with a hydrophobic polyvinyl block and a hydrophilic polypeptide block - copolymers polystyrene-poly(L-lysine) and Polybutadiene-poly(L-lysine). *Makromolekulare Chemie-Macromolecular Chemistry and Physics* 177: 1889.

50 Billot, J.P., Douy, A., and Gallot, B. (1977). Preparation, fractionation, and structure of block copolymers polystyrene-poly(Carbobenzoxy-L-lysine)

and Polybutadiene-poly(Carbobenzoxy-L-lysine). *Makromolekulare Chemie-Macromolecular Chemistry and Physics* 178: 1641.

51 Perly, B., Douy, A., and Gallot, B. (1976). Block copolymers Polybutadiene-poly(benzyl-L-glutamate) and polybutadiene-poly(N5-Hydroxypropylglutamine) preparation and structural study by X-ray and electron-microscopy. *Makromolekulare Chemie-Macromolecular Chemistry and Physics* 177: 2569.

52 Deming, T.J. (1997). Facile synthesis of block copolypeptides of defined architecture. *Nature* 390: 386.

53 Minich, E.A., Nowak, A.P., Deming, T.J. et al. (2004). Rod-rod and rod-coil self-assembly and phase behavior of polypeptide diblock copolymers. *Polymer* 45: 1951.

54 Farmer, R.S. and Kiick, K.L. (2005). Conformational behavior of chemically reactive alanine-rich repetitive protein polymers. *Biomacromolecules* 6: 1531.

55 Zhang, G.H., Fournier, M.J., Mason, T.L. et al. (1992). Biological synthesis of monodisperse derivatives of poly(Alpha,L-glutamic acid) - model rodlike polymers. *Macromolecules* 25: 3601.

56 Apostolovic, B., Danial, M., and Klok, H.-A. (2010). Coiled coils: attractive protein folding motifs for the fabrication of self-assembled, responsive and bioactive materials. *Chemical Society Reviews* 39: 3541.

57 Xu, T. and Shu, J. (2010). Coiled-coil helix bundle, a peptide tertiary structural motif toward hybrid functional materials. *Soft Matter* 6: 212.

58 Hodges, R.S. (1996). De novo design of alpha-helical proteins: basic research to medical applications. *Biochemistry and Cell Biology-Biochimie Et Biologie Cellulaire* 74: 133.

59 Petka, W.A., Harden, J.L., McGrath, K.P. et al. (1998). Reversible hydrogels from self-assembling artificial proteins. *Science* 281: 389.

60 Banta, S., Wheeldon, I.R., and Blenner, M. (2010). *Annual Review of Biomedical Engineering*, vol. 12 (ed. M.L. Yarmush, J.S. Duncan and M.L. Gray), 167.

61 Pryamitsyn, V. and Ganesan, V. (2004). Self-assembly of rod-coil block copolymers. *Journal of Chemical Physics* 120: 5824.

62 Holyst, R. and Schick, M. (1992). Correlations in a rigid - flexible Diblock copolymer system. *Journal of Chemical Physics* 96: 730.

63 Hammouda, B. (1993). Scattering from mixtures of flexible and stiff polymers. *Journal of Chemical Physics* 98: 3439.

64 Matsen, M.W. and Barrett, C. (1998). Liquid-crystalline behavior of rod-coil diblock copolymers. *Journal of Chemical Physics* 109: 4108.

65 Olsen, B.D., Shah, M., Ganesan, V. et al. (2008). Universalization of the phase diagram for a model rod-coil Diblock copolymer. *Macromolecules* 41: 6809.

66 Lee, M., Cho, B.K., Kim, H. et al. (1998). Self-organization of rod-coil molecules with layered crystalline states into thermotropic liquid crystalline assemblies. *Journal of the American Chemical Society* 120: 9168.

67 Wu, J., Wang, Q., Tang, S.-F. et al. (2013). Synthesis and investigation of phase-separated lamellar morphologies of rod-coil type polystyrene-B-poly (gamma-benzyl-L -glutamate) polypeptide hybrids. *Acta Polymerica Sinica* 1151.

68 Radzilowski, L.H., Carragher, B.O., and Stupp, S.I. (1997). Three-dimensional self-assembly of rodcoil copolymer nanostructures. *Macromolecules* 30: 2110.

69 Olsen, B.D. and Segalman, R.A. (2007). Nonlamellar phases in asymmetric rod-coil block copolymers at increased segregation strengths. *Macromolecules* 40: 6922.

70 Wang, M., Timachova, K., and Olsen, B.D. (2013). Diffusion mechanisms of entangled rod-coil Diblock copolymers. *Macromolecules* 46: 5694.

71 Wang, M., Timachova, K., and Olsen, B.D. (2013). Experimental measurement of coil-rod-coil block copolymer tracer diffusion through entangled coil homopolymers. *Macromolecules* 46: 1651.

72 Graf, R., Spiess, H.W., Floudas, G. et al. (2012). Conformational transitions of poly(L-proline) in copolypeptides with poly(gamma-benzyl-L-glutamate) induced by packing. *Macromolecules* 45: 9326.

73 Losik, M., Kubowicz, S., Smarsly, B. et al. (2004). Solid-state structure of polypeptide-based rod-coil block copolymers: folding of helices. *European Physical Journal E* 15: 407.

74 Schlaad, H., Kukula, H., Smarsly, B. et al. (2002). Solid-state morphologies of linear and bottlebrush-shaped polystyrene-poly(Z-L-lysine) block copolymers. *Polymer* 43: 5321.

75 Schlaad, H., Smarsly, B., and Losik, M. (2004). The role of chain-length distribution in the formation of solid-state structures of polypeptide-based rod-coil block copolymers. *Macromolecules* 37: 2210.

76 Floudas, G., Papadopoulos, P., Klok, H.A. et al. (2003). Hierarchical self-assembly of poly(gamma-benzyl-L-glutamate)-poly(ethylene glycol)-poly(gamma-benzyl-L-glutamate) rod-coil-rod triblock copolymers. *Macromolecules* 36: 3673.

77 Klok, H.A., Langenwalter, J.F., and Lecommandoux, S. (2000). Self-assembly of peptide-based diblock oligomers. *Macromolecules* 33: 7819.

78 Lecommandoux, S., Achard, M.F., Langenwalter, J.F. et al. (2001). Self-assembly of rod-coil diblock oligomers based on alpha-helical peptides. *Macromolecules* 34: 9100.

79 Ren, K., Cheng, Y., He, C. et al. (2013). The effect of alkyl side groups on the secondary structure and crystallization of poly(ethylene glycol)-block-polypeptide copolymers. *Polymer* 54: 2466.

80 Gkikas, M., Haataja, J.S., Seitsonen, J. et al. (2014). Extended self-assembled long periodicity and Zig-Zag domains from helix-helix Diblock copolymer poly(gamma-benzyl-L-glutamate)-block-poly(O-benzyk-hydroxyproline). *Biomacromolecules* 15: 3923.
81 Johnson, J.C., Wanasekara, N.D., and Korley, L.T.J. (2014). Influence of secondary structure and hydrogenbonding arrangement on the mechanical properties of peptidic-polyurea hybrids. *Journal of Materials Chemistry B* 2: 2554.
82 Hehir, S. and Cameron, N.R. (2014). Recent advances in drug delivery systems based on polypeptides prepared from *N*-carboxyanhydrides. *Polymer International* 63: 943.
83 Shi, S.-Y., He, Y.-G., Chen, W.-W. et al. (2015). Polypeptide-b-poly(phenyl Isocyanide) hybrid rod-rod copolymers: one-pot synthesis, self-assembly, and cell imaging. *Macromolecular Rapid Communications* 36: 1511.
84 Koga, T., Koike, M., Kamiwatari, S. et al. (2011). Morphological variation of peptide-synthetic hybrid block copolymer assemblies in nonaqueous media. *Chemistry Letters* 40: 1244.
85 Checot, F., Brulet, A., Oberdisse, J. et al. (2005). Structure of polypeptide-based diblock copolymers in solution: stimuli-responsive vesicles and micelles. *Langmuir* 21: 4308.
86 Checot, F., Rodriguez-Hernandez, J., Gnanou, Y. et al. (2006). Responsive micelles and vesicles based on polypeptide diblock copolymers. *Polymers for Advanced Technologies* 17: 782.
87 Checot, F., Rodriguez-Hernandez, J., Gnanou, Y. et al. (2007). pH-responsive micelles and vesicles nanocapsules based on polypeptide diblock copolymers. *Biomolecular Engineering* 24: 81.
88 Babin, J., Rodriguez-Hernandez, J., Lecommandoux, S. et al. (2005). Self-assembled nanostructures from peptide-synthetic hybrid block copolymers: complex, stimuli-responsive rod-coil architectures. *Faraday Discussions* 128: 179.
89 Ray, J.G., Ly, J.T., and Savin, D.A. (2011). Peptide-based lipid mimetics with tunable core properties via thiol-alkyne chemistry. *Polymer Chemistry* 2: 1536.
90 Yoon, Y.-R., Lim, Y.-B., Lee, E. et al. (2008). Self-assembly of a peptide rod-coil: a polyproline rod and a cell-penetrating peptide tat coil. *Chemical Communications* 1892.
91 Mochida, Y., Cabral, H., Miura, Y. et al. (2014). Bundled assembly of helical nanostructures in polymeric micelles loaded with platinum drugs enhancing therapeutic efficiency against pancreatic tumor. *ACS Nano* 8: 6724.
92 Popescu, M.-T., Liontos, G., Avgeropoulos, A. et al. (2015). Stimuli responsive fibrous hydrogels from hierarchical self-assembly of a triblock copolypeptide. *Soft Matter* 11: 331.

93 Boerner, H.G., Suetterlin, R.I., Theato, P. et al. (2014). Topology- dependent Swichability of peptide secondary structures in bioconjugates with complex architectures. *Macromolecular Rapid Communications* 35: 180.

94 Top, A., Roberts, C.J., and Kiick, K.L. (2011). Conformational and aggregation properties of a PEGylated alanine-rich polypeptide. *Biomacromolecules* 12: 2184.

95 Cai, C., Lin, J., Chen, T. et al. (2009). Super-helices self-assembled from a binary system of amphiphilic polypeptide block copolymers and polypeptide homopolymers. *Chemical Communications* 2709.

96 Nowak, A.P., Breedveld, V., Pakstis, L. et al. (2002). Rapidly recovering hydrogel scaffolds from self-assembling diblock copolypeptide amphiphiles. *Nature* 417: 424.

97 Pakstis, L.M., Ozbas, B., Hales, K.D. et al. (2004). Effect of chemistry and morphology on the biofunctionality of self-assembling diblock copolypeptide hydrogels. *Biomacromolecules* 5: 312.

98 Gibson, M.I. and Cameron, N.R. (2008). Organogelation of sheet-helix diblock copolypeptides. *Angewandte Chemie-International Edition* 47: 5160.

99 Checot, F., Lecommandoux, S., Gnanou, Y. et al. (2002). Water-soluble stimuli-responsive vesicles from peptide-based diblock copolymers. *Angewandte Chemie-International Edition* 41: 1339.

100 Checot, F., Lecommandoux, S., Klok, H.A. et al. (2003). From supramolecular polymersomes to stimuli-responsive nano-capsules based on poly(diene-b-peptide) diblock copolymers. *European Physical Journal E* 10: 25.

101 Rodriguez-Hernandez, J. and Lecommandoux, S. (2005). Reversible inside-out micellization of pH-responsive and water-soluble vesicles based on polypeptide diblock copolymers. *Journal of the American Chemical Society* 127: 2026.

102 Gebhardt, K.E., Ahn, S., Venkatachalam, G. et al. (2007). Rod-sphere transition in polybutadiene-poly(L-lysine) block copolymer assemblies. *Langmuir* 23: 2851.

103 Rao, J., Luo, Z., Ge, Z. et al. (2007). Schizophrenic micellization associated with coil-to-helix transitions based on polypeptide hybrid double hydrophilic rod-coil diblock copolymer. *Biomacromolecules* 8: 3871.

104 Kukula, H., Schlaad, H., Antonietti, M. et al. (2002). The formation of polymer vesicles or peptosomes by polybutadiene-block-poly(L-glutamate)s in dilute aqueous solution. *Journal of the American Chemical Society* 124: 1658.

105 Kuehnle, R.I., Gebauer, D., and Boerner, H.G. (2011). Calcium ions as bioinspired triggers to reversibly control the coil-to-helix transition in peptide–polymer conjugates. *Soft Matter* 7: 9616.

106 Zhao, Z., Wang, Y., Han, J. et al. (2014). Self-assembled micelles of amphiphilic poly(L-phenylalanine)-b-poly(L-serine) polypeptides for tumor-targeted delivery. *International Journal of Nanomedicine* 9: 5849.
107 Itoh, T., Hatanaka, T., Ihara, E. et al. (2012). Helix-coil transformation of poly(gamma-benzyl-L-glutamate) with polystyrene attached to the N or C terminus in trifluoroacetic acid-chloroform mixtures. *Polymer Journal* 44: 189.
108 Triftaridou, A.I., Checot, F., and Iliopoulos, I. (2010). Poly(*N*,*N*-dimethylacrylamide)-block-poly(L-lysine) hybrid block copolymers: synthesis and aqueous solution characterization. *Macromolecular Chemistry and Physics* 211: 768.
109 Gebhardt, K.E., Ahn, S., Venkatachalam, G. et al. (2008). Role of secondary structure changes on the morphology of polypeptide-based block copolymer vesicles. *Journal of Colloid and Interface Science* 317: 70.
110 Shimada, T., Lee, S., Bates, F.S. et al. (2009). Wormlike micelle formation in peptide-lipid conjugates driven by secondary structure transformation of the Headgroups. *Journal of Physical Chemistry B* 113: 13711.
111 Choi, Y.Y., Jang, J.H., Park, M.H. et al. (2010). Block length affects secondary structure, nanoassembly and thermosensitivity of poly(ethylene glycol)-poly(L-alanine) block copolymers. *Journal of Materials Chemistry* 20: 3416.
112 Vandermeulen, G.W.M., Tziatzios, C., and Klok, H.A. (2003). Reversible self-organization of poly(ethylene glycol)-based hybrid block copolymers mediated by a De Novo four-stranded alpha-helical coiled coil motif. *Macromolecules* 36: 4107.
113 Pechar, M., Kopeckova, P., Joss, L. et al. (2002). Associative diblock copolymers of poly(ethylene glycol) and coiled-coil peptides. *Macromolecular Bioscience* 2: 199.
114 Klok, H.A., Vandermeulen, G.W.M., Nuhn, H. et al. (2005). Peptide mediated formation of hierarchically organized solution and solid state polymer nanostructures. *Faraday Discussions* 128: 29.
115 Luo, T., He, L., Theato, P. et al. (2015). Thermoresponsive self-assembly of nanostructures from a collagen-like peptide-containing diblock copolymer. *Macromolecular Bioscience* 15: 111.
116 Krishna, O.D., Wiss, K.T., Luo, T. et al. (2012). Morphological transformations in a dually thermoresponsive coil-rod-coil bioconjugate. *Soft Matter* 8: 3832.
117 Pham, T.T.H., Wang, J., Werten, M.W.T. et al. (2013). Multi-responsive physical gels formed by a biosynthetic asymmetric triblock protein polymer and a polyanion. *Soft Matter* 9: 8923.
118 Skrzeszewska, P.J., de Wolf, F.A., Werten, M.W.T. et al. (2009). Physical gels of telechelic triblock copolymers with precisely defined junction multiplicity. *Soft Matter* 5: 2057.

119 Werten, M.W.T., Teles, H., Moers, A.P.H.A. et al. (2009). Precision gels from collagen-inspired triblock copolymers. *Biomacromolecules* 10: 1106.

120 Shen, W., Zhang, K.C., Kornfield, J.A. et al. (2006). Tuning the erosion rate of artificial protein hydrogels through control of network topology. *Nature Materials* 5: 153.

121 Glassman, M.J., Chan, J., and Olsen, B.D. (2013). Reinforcement of shear thinning protein hydrogels by responsive block copolymer self-assembly. *Advanced Functional Materials* 23: 1182.

122 Lund, R., Shu, J., and Xu, T. (2013). A small-angle X-ray scattering study of alpha-helical bundle-forming peptide–polymer conjugates in solution: chain conformations. *Macromolecules* 46: 1625.

123 Shu, J.Y., Huang, Y.-J., Tan, C. et al. (2010). Amphiphilic peptide–polymer conjugates based on the coiled-coil helix bundle. *Biomacromolecules* 11: 1443.

124 Shu, J.Y., Lund, R., and Xu, T. (2012). Solution structural characterization of coiled-coil peptide–polymer side-conjugates. *Biomacromolecules* 13: 1945.

125 Dube, N., Seo, J.W., Dong, H. et al. (2014). Effect of alkyl length of peptide–polymer amphiphile on cargo encapsulation stability and pharmacokinetics of 3-helix micelles. *Biomacromolecules* 15: 2963.

126 Dube, N., Shu, J.Y., Dong, H. et al. (2013). Evaluation of doxorubicin-loaded 3-helix micelles as nanocarriers. *Biomacromolecules* 14: 3697.

127 Jain, A. and Ashbaugh, H.S. (2011). Helix stabilization of poly(ethylene glycol)-peptide conjugates. *Biomacromolecules* 12: 2729.

128 Hamed, E., Xu, T., and Keten, S. (2013). Poly(ethylene glycol) conjugation stabilizes the secondary structure of alpha-helices by reducing peptide solvent accessible surface area. *Biomacromolecules* 14: 4053.

129 Rabotyagova, O.S., Cebe, P., and Kaplan, D.L. (2011). Protein-based block copolymer. *Biomacromolecules* 12: 269.

130 Burkoth, T.S., Benzinger, T.L.S., Jones, D.N.M. et al. (1998). C-terminal PEG blocks the irreversible step in amyloid(10-35) fibrillogenesis. *Journal of the American Chemical Society* 120: 7655.

131 Burkoth, T.S., Benzinger, T.L.S., Urban, V. et al. (1999). Self-assembly of Abeta(10-35)-PEG block copolymer fibrils. *Journal of the American Chemical Society* 121: 7429.

132 Burkoth, T.S., Benzinger, T.L.S., Urban, V. et al. (2000). Structure of the beta-amyloid(10-35) fibril. *Journal of the American Chemical Society* 122: 7883.

133 Hamley, I.W., Ansari, I.A., Castelletto, V. et al. (2005). Solution self-assembly of hybrid block copolymers containing poly(ethylene glycol) and amphiphilic beta-strand peptide sequences. *Biomacromolecules* 6: 1310.

134 Vandermeulen, G.W.M., Kim, K.T., Wang, Z. et al. (2006). Metallopolymer-peptide conjugates: synthesis and self-assembly of Polyferrocenylsilane graft and block copolymers containing a beta-sheet forming Gly-Ala-Gly-Ala Tetrapeptide segment. *Biomacromolecules* 7: 1005.

135 Collier, J.H. and Messersmith, P.B. (2004). Self-assembling polymer-peptide conjugates: Nanostructural tailoring. *Advanced Materials* 16: 907.

136 Smeenk, J.M., Schon, P., Otten, M.B.J. et al. (2006). Fibril formation by triblock copolymers of Silklike beta-sheet polypeptides and poly(ethylene glycol). *Macromolecules* 39: 2989.

137 Smeenk, J.M., Otten, M.B., Thies, J. et al. (2005). Controlled assembly of macromolecular beta-sheet fibrils. *Angewandte Chemie International Edition* 44: 1968.

138 Rathore, O. and Sogah, D.Y. (2001). Self-assembly of beta-sheets into nanostructures by poly(alanine) segments incorporated in multiblock copolymers inspired by spider silk. *Journal of the American Chemical Society* 123: 5231.

139 Rathore, O. and Sogah, D.Y. (2001). Nanostructure formation through beta-sheet self-assembly in silk-based materials. *Macromolecules* 34: 1477.

140 Zhou, C., Leng, B., Yao, J. et al. (2006). Synthesis and characterization of multiblock copolymers based on spider dragline silk proteins. *Biomacromolecules* 7: 2415.

141 Anderson, J.P., Cappello, J., and Martin, D.C. (1994). Morphology and primary crystal structure of a silk-like protein polymer synthesized by genetically engineered *Escherichia coli* bacteria. *Biopolymers* 34: 1049.

142 Rabotyagova, O.S., Cebe, P., and Kaplan, D.L. (2009). Self-assembly of genetically engineered spider silk block copolymers. *Biomacromolecules* 10: 229.

143 Nagarsekar, A., Crissman, J., Crissman, M. et al. (2003). Genetic engineering of stimuli-sensitive silkelastin-like protein block copolymers. *Biomacromolecules* 4: 602.

144 Gustafson, J.A. and Ghandehari, H. (2010). Silk-Elastinlike protein polymers for matrix-mediated cancer gene therapy. *Advanced Drug Delivery Reviews* 62: 1509.

145 Anumolu, R., Gustafson, J.A., Magda, J.J. et al. (2011). Fabrication of highly uniform nanoparticles from recombinant silk-elastin-like protein polymers for therapeutic agent delivery. *ACS Nano* 5: 5374.

146 Qiu, W., Huang, Y., Teng, W. et al. (2010). Complete recombinant silk-Elastinlike protein-based tissue scaffold. *Biomacromolecules* 11: 3219.

147 Lin, Y., Xia, X., Wang, M. et al. (2014). Genetically programmable Thermoresponsive Plasmonic gold/silk-elastin protein Core/Shell nanoparticles. *Langmuir : the ACS Journal of Surfaces and Colloids* 30: 4406.

148 Xia, X.-X., Xu, Q., Hu, X. et al. (2011). Tunable self-assembly of genetically engineered silk–elastin-like protein polymers. *Biomacromolecules* 12: 3844.

149 Xia, X.-X., Wang, M., Lin, Y. et al. (2014). Hydrophobic drug-triggered self-assembly of nanoparticles from silk-elastin-like protein polymers for drug delivery. *Biomacromolecules* 15: 908.
150 Golinska, M.D., Pham, T.T.H., Werten, M.W.T. et al. (2013). Fibril formation by pH and temperature responsive silk-elastin block copolymers. *Biomacromolecules* 14: 48.
151 Golinska, M.D., Włodarczyk-Biegun, M.K., Werten, M.W.T. et al. (2014). Dilute self-healing hydrogels of silk-collagen-like block copolypeptides at neutral pH. *Biomacromolecules* 15: 699.
152 Beun, L.H., Storm, I.M., Werten, M.W.T. et al. (2014). From micelles to fibers: balancing self-assembling and random coiling domains in pH-responsive silk-collagen-like protein-based polymers. *Biomacromolecules* 15: 3349.
153 Krysmann, M.J., Castelletto, V., Kelarakis, A. et al. (2008). Self-assembly and hydrogelation of an amyloid peptide fragment. *Biochemistry* 47: 4597.
154 Schneider, J.P., Pochan, D.J., Ozbas, B. et al. (2002). Responsive hydrogels from the intramolecular folding and self-assembly of a designed peptide. *Journal of the American Chemical Society* 124: 15030.
155 Ghadiri, M.R., Granja, J.R., Milligan, R.A. et al. (1993). Self-assembling organic nanotubes based on a cylic peptide architecture. *Nature* 366: 324.
156 Chapman, R., Danial, M., Koh, M. et al. (2012). Design and properties of functional nanotubes from the self-assembly of cyclic peptide templates. *Chemical Society Reviews* 41: 6023.
157 Bong, D., Clark, T., Granja, J. et al. (2001). Self-assembling organic nanotubes. *Angewandte Chemie-International Edition* 40: 988.
158 Brea, R., Reiriz, C., and Granja, J. (2010). Towards functional bionanomaterials based on self-assembling cyclic peptide nanotubes. *Chemical Society Reviews* 39: 1448.
159 Ghadiri, M., Kobayashi, K., Granja, J. et al. (1995). The structural and thermodynamic basis for the formation of self-assembled peptide nanotubes. *Angewandte Chemie-International Edition in English* 34: 93.
160 Cate, M.G.J.T., Severin, N., and Borner, H.G. (2006). Self-assembling peptide–polymer conjugates comprising (D-alt-L)-Cyclopeptides as aggregator domains. *Macromolecules* 39: 7831.
161 Chapman, R., Koh, M., Warr, G. et al. (2013). Structure elucidation and control of cyclic peptide-derived nanotube assemblies in solution. *Chemical Science* 4: 2581.
162 Chan, W.C. and White, P.D. (2000). *Fmoc Solid Phase Peptide Synthesis: A Practical Approach*. Oxford: OUP.
163 Couet, J., Jeyaprakash, J., Samuel, S. et al. (2005). Peptide–polymer hybrid nanotubes. *Angewandte Chemie-International Edition* 44: 3297.

164 Couet, J. and Biesalski, M. (2006). Surface-initiated ATRP of N-isopropylacrylamide from initiator-modified self-assembled peptide nanotubes. *Macromolecules* 39: 7258.

165 Chapman, R., Jolliffe, K., and Perrier, S. (2010). Synthesis of self-assembling cyclic peptide–polymer conjugates using click chemistry. *Australian Journal of Chemistry* 63: 1169.

166 Loschonsky, S., Couet, J., and Biesalski, M. (2008). Synthesis of peptide/polymer conjugates by solution ATRP of butylacrylate using an initiator-modified cyclic D-alt-L-peptide. *Macromolecular Rapid Communications* 29: 309.

167 Danial, M., Tran, C., Young, P. et al. (2013). Janus cyclic peptide–polymer nanotubes. *Nature Communications* 4.

168 Poon, C., Chapman, R., Jolliffe, K. et al. (2012). Pushing the limits of copper mediated azide-alkyne cycloaddition (CuAAC) to conjugate polymeric chains to cyclic peptides. *Polymer Chemistry* 3: 1820.

169 Gokhale, R., Couet, J., and Biesalski, M. (2010). In situ cross-linking of the shell of self-assembled peptide nanotubes. *Physica Status Solidi A-Applications and Materials Science* 207: 878.

170 Chapman, R., Bouten, P., Hoogenboom, R. et al. (2013). Thermoresponsive cyclic peptide - poly(2-ethyl-2-oxazoline) conjugate nanotubes. *Chemical Communications* 49: 6522.

171 Chapman, R., Warr, G., Perrier, S. et al. (2013). Water-soluble and pH-responsive polymeric nanotubes from cyclic peptide templates. *Chemistry-A European Journal* 19: 1955.

172 Chapman, R., Jolliffe, K., and Perrier, S. (2011). Modular design for the controlled production of polymeric nanotubes from polymer/peptide conjugates. *Polymer Chemistry* 2: 1956.

173 Couet, J. and Biesalski, M. (2008). Polymer-wrapped peptide nanotubes: peptide-grafted polymer mass impacts length and diameter. *Small* 4: 1008.

174 Hourani, R., Zhang, C., van der Weegen, R. et al. (2011). Processable cyclic peptide nanotubes with tunable interiors. *Journal of the American Chemical Society* 133: 15296.

175 Sanchez-Quesada, J., Kim, H., and Ghadiri, M. (2001). A synthetic pore-mediated transmembrane transport of glutamic acid. *Angewandte Chemie-International Edition* 40: 2503.

176 Danial, M., Tran, C., Jolliffe, K. et al. (2014). Thermal gating in lipid membranes using thermoresponsive cyclic peptide–polymer conjugates. *Journal of the American Chemical Society* 136: 8018.

177 Koh, M., Jolliffe, K., and Perrier, S. (2014). Hierarchical assembly of branched supramolecular polymers from (cyclic peptide)-polymer conjugates. *Biomacromolecules* 15: 4002.

178 Xu, T., Zhao, N., Ren, F. et al. (2011). Subnanometer porous thin films by the co-assembly of nanotube subunits and block copolymers. *ACS Nano* 5: 1376.

179 Urry, D.W., Luan, C.-H., Harris, C.M. et al. (1997). *Proteins and Modified Proteins as Polymeric Materials* (ed. K. McGrath and D. Kaplan). Boston, MA: Birkhauser.

180 Manno, M., Emanuele, A., Martorana, V. et al. (2001). Interaction of processes on different length scales in a bioelastomer capable of performing energy conversion. *Biopolymers* 59: 51.

181 van Hest, J.C.M. and Tirrell, D.A. (2001). Protein-based materials, toward a new level of structural control. *Chemical Communication* 1897.

182 Gray, W.R., Sandberg, L.B., and Foster, J.A. (1973). Molecular model for elastin structure and function. *Nature* 246: 461.

183 Wright, E.R. and Conticello, V.P. (2002). S elf-assembly of block copolymers derived from elastin-mimetic polypeptide sequences. *Advanced Drug Delivery Reviews* 54: 1057.

184 Meyer, D.E. and Chilkoti, A. (2004). Quantification of the effects of chain length and concentration on the thermal behavior of elastin-like polypeptides. *Biomacromolecules* 5: 846.

185 Christensen, T., Hassouneh, W., Trabbic-Carlson, K. et al. (2013). Predicting transition temperatures of elastin-like polypeptide fusion proteins. *Biomacromolecules* 14: 1514.

186 Li, N.K., Garcia Quiroz, F., Hall, C.K. et al. (2014). Molecular description of the LCST behavior of an elastin-like polypeptide. *Biomacromolecules* 15: 3522.

187 Urry, D.W., Luan, C.-H., Parker, T.M. et al. (1991). Temperature of polypeptide inverse temperature transition depends on mean residue hydrophobicity. *Journal of the American Chemical Society* 113: 4346.

188 Roy, D., Brooks, W.L., and Sumerlin, B.S. (2013). New directions in thermoresponsive polymers. *Chemical Society Reviews* 42: 7214.

189 Meyer, D.E. and Chilkoti, A. (2002). Genetically encoded synthesis of protein-based polymers with precisely specified molecular weight and sequence by recursive directional ligation: examples from the elastin-like polypeptide system. *Biomacromolecules* 3: 357.

190 Meyer, D.E. and Chilkoti, A. (2004). Quantification of the effects of chain length and concentration on the thermal behavior of elastin-like polypeptides. *Biomacromolecules* 5: 845.

191 Lee, T.A.T., Cooper, A., Apkarian, R.P. et al. (2000). Thermo-reversible self-assembly of nanoparticles derived from elastin-mimetic polypeptides. *Advanced Materials* 12: 1105.

192 Dreher, M.R., Simnick, A.J., Fischer, K. et al. (2008). Temperature triggered self-assembly of polypeptides into multivalent spherical micelles. *Journal of the American Chemical Society* 130: 687.

193 Thurmond, K.B., Kowalewski, T., and Wooley, K.L. (1996). Water-soluble knedel-like structures: the preparation of shell-cross-linked small particles. *Journal of the American Chemical Society* 118: 7239.

194 O'Reilly, R.K., Hawker, C.J., and Wooley, K.L. (2006). Cross-linked block copolymer micelles: functional nanostructures of great potential and versatility. *Chemical Society Reviews* 35: 1068.

195 Kim, W., Thevenot, J., Ibarboure, E. et al. (2010). Self-assembly of thermally responsive amphiphilic diblock copolypeptides into spherical micellar nanoparticles. *Angewandte Chemie International Edition* 49: 4257.

196 Kim, W., Xiao, J., and Chaikof, E.L. (2011). Recombinant amphiphilic protein micelles for drug delivery. *Langmuir : the ACS journal of Surfaces and Colloids* 27: 14329.

197 Smits, F.C.M., Castelijns, W.W.A., and van Hest, J.C.M. (2015). Crosslinked ELP-based nanoparticles, using the strain promoted azide–alkyne cycloaddition. *European Polymer Journal* 62: 386.

198 Han, W., MacEwan, S.R., Chilkoti, A., and Lopez, G.P. (2015). Bio-inspired synthesis of hybrid silica nanoparticles templated from elastin-like polypeptide micelles. *Nanoscale* 7: 12038.

199 Simnick, A.J., Valencia, C.A., Liu, R. et al. (2010). Morphing low-affinity ligands into high-avidity nanoparticles by thermally triggered self-assembly of a genetically encoded polymer. *ACS Nano* 4: 2217.

200 Macewan, S.R. and Chilkoti, A. (2012). Digital switching of local arginine density in a genetically encoded self-assembled polypeptide nanoparticle controls cellular uptake. *Nano Letters* 12: 3322.

201 Hassouneh, W., Fischer, K., MacEwan, S.R. et al. (2012). Unexpected multivalent display of proteins by temperature triggered self-assembly of elastin-like polypeptide block copolymers. *Biomacromolecules* 13: 1598.

202 MacEwan, S.R. and Chilkoti, A. (2014). Controlled apoptosis by a thermally toggled nanoscale amplifier of cellular uptake. *Nano Letters* 14: 2058.

203 Soon, A.S., Smith, M.H., Herman, E.S. et al. (2013). Development of self-assembling mixed protein micelles with temperature-modulated avidities. *Advanced Healthcare Materials* 2: 1045.

204 Wright, E.R., McMillan, R.A., Cooper, A. et al. (2002). Thermoplastic elastomer hydrogels via self-assembly of an elastin-mimetic Triblock polypeptide. *Advanced Functional Materials* 12: 149.

205 Lim, D.W., Nettles, D.L., Setton, L.A. et al. (2008). In situ cross-linking of elastin-like polypeptide block copolymers for tissue repair. *Biomacromolecules* 9: 222.

206 van Eldijk, M.B., Wang, J.C., Minten, I.J. et al. (2012). Designing two self-assembly mechanisms into one viral capsid protein. *Journal of the American Chemical Society* 134: 18506.

207 van Eldijk, M.B., Smits, F.C., Vermue, N. et al. (2014). Synthesis and self-assembly of well-defined elastin-like polypeptide-poly(ethylene glycol) conjugates. *Biomacromolecules* 15: 2751.

208 Grieshaber, S.E., Paik, B.A., Bai, S. et al. (2013). Nanoparticle formation from hybrid, multiblock copolymers of poly(acrylic acid) and VPGVG peptide. *Soft Matter* 9: 1589.

209 Paik, B.A., Blanco, M.A., Jia, X. et al. (2015). Aggregation of poly(acrylic acid)-containing elastin-mimetic copolymers. *Soft Matter* 11: 1839.

210 Greish, K., Fang, J., Inutsuka, T. et al. (2012). Macromolecular therapeutics. *Clinical Pharmacokinetics* 42: 1089.

211 Matsumura, Y. and Maeda, H. (1986). A new concept for macromolecular therapeutics in cancer chemotherapy: mechanism of Tumoritropic accumulation of proteins and the antitumor agent Smancs. *Cancer Research* 46: 6387.

212 Bhosale, S.H., Rao, M.B., and Deshpande, V.V. (1996). Molecular and industrial aspects of glucose isomerase. *Microbiological Reviews* 60: 280.

213 Yamada, H. and Kobayashi, M. (1996). Nitrile hydratase and its application to industrial production of acrylamide. *Bioscience, Biotechnology, and Biochemistry* 60: 1391.

214 Luengo, J.M., Alemany, M.T., Salto, F. et al. (1986). Direct enzymatic synthesis of penicillin G using cyclases of penicillium chrysogenum and acremonium chrysogenum. *Nature Biotechnology* 4: 44.

215 Newman, J.D. and Turner, A.P.F. (2005). Home blood glucose biosensors: a commercial perspective. *Biosensors and Bioelectronics* 20: 2435.

216 Wide, L. and Gemzell, C.A. (1960). An Immunological Pregnancy Test. *Acta Endocrinologica* XXXV: 261.

217 van Hest, J. and Sumerlin, B. (2011). Polymer bioconjugates. *Polymer Chemistry* 2: 1427.

218 Chalker, J.M., Bernardes, G.J.L., Lin, Y.A. et al. (2009). Chemical modification of proteins at cysteine: opportunities in chemistry and biology. *Chemistry – An Asian Journal* 4: 630.

219 Hermanson, G.T. (2013). *Bioconjugate techniques*. Academic Press.

220 Kiick, K.L., Saxon, E., Tirrell, D.A. et al. (2002). Incorporation of azides into recombinant proteins for chemoselective modification by the Staudinger ligation. *Proceedings of the National Academy of Sciences of the United States of America* 99: 19.

221 Gilmore, J.M., Scheck, R.A., Esser-Kahn, A.P. et al. (2006). N-terminal protein modification through a biomimetic transamination reaction. *Angewandte Chemie International Edition* 45: 5307.

222 Meyer, D.E. and Chilkoti, A. (1999). Purification of recombinant proteins by fusion with thermally-responsive polypeptides. *Nature Biotechnology* 17: 1112.

223 Qin, G., Glassman, M.J., Lam, C.N. et al. (2015). Topological effects on globular protein-ELP fusion block copolymer self-assembly. *Advanced Functional Materials* 25: 729.

224 Velonia, K., Rowan, A.E., and Nolte, R.J.M. (2002). Lipase polystyrene giant amphiphiles. *Journal of the American Chemical Society* 124: 4224.

225 Boerakker, M.J., Botterhuis, N.E., Bomans, P.H.H. et al. (2006). Aggregation behavior of Giant Amphiphiles prepared by cofactor reconstitution. *Chemistry – A European Journal* 12: 6071.

226 Lavigueur, C., Garcia, J.G., Hendriks, L. et al. (2011). Thermoresponsive giant biohybrid amphiphiles. *Polymer Chemistry* 2: 333.

227 Dirks, A.T., van Berkel, S.S., Hatzakis, N.S. et al. (2005). Preparation of biohybrid amphiphiles via the copper catalysed Huisgen [3+ 2] dipolar cycloaddition reaction. *Chemical Communications* 4172.

228 Aluri, S.R., Shi, P., Gustafson, J.A. et al. (2014). A hybrid protein–polymer Nanoworm potentiates apoptosis better than a monoclonal antibody. *ACS Nano* 8.

229 Ge, J., Neofytou, E., Lei, J. et al. (2012). Protein–polymer hybrid nanoparticles for drug delivery. *Small* 8: 3573.

230 Li, M., De, P., Gondi, S.R., and Sumerlin, B.S. (2008). Responsive polymer-protein bioconjugates prepared by RAFT polymerization and copper-catalyzed azide-alkyne click chemistry. *Macromolecular Rapid Communications* 29: 1172.

231 Reynhout, I.C., Cornelissen, J.J.L.M., and Nolte, R.J.M. (2007). Self-assembled architectures from biohybrid triblock copolymers. *J. Am. Chem. Soc.* 129: 2327.

232 Hassouneh, W., MacEwan, S.R., and Chilkoti, A. (2012). Fusions of elastin-like polypeptides to pharmaceutical proteins. *Methods in Enzymology* 502: 215.

233 Li, H., Bapat, A.P., Li, M. et al. (2011). Protein conjugation of thermoresponsive amine-reactive polymers prepared by RAFT. *Polymer Chemistry* 2: 323.

234 Chen, J.P., Yang, H.J., and Huffman, A.S. (1990). Polymer-protein conjugates: I. Effect of protein conjugation on the cloud point of poly(N-isopropylacrylamide). *Biomaterials* 11: 625.

235 De, P., Li, M., Gondi, S.R. et al. (2008). Temperature-regulated activity of responsive polymer–protein conjugates prepared by grafting-from via

RAFT polymerization. *Journal of the American Chemical Society* 130: 11288.

236 Xia, Y., Tang, S., and Olsen, B.D. (2013). Site-specific conjugation of RAFT polymers to proteins via expressed protein ligation. *Chemical Communications*.

237 Lam, C.N., Kim, M., Thomas, C.S. et al. (2014). The nature of protein interactions governing globular protein–polymer block copolymer self-assembly. *Biomacromolecules* 15: 1248.

238 Huang, A., Qin, G., and Olsen, B.D. (2015). Highly active biocatalytic coatings from protein–polymer Diblock copolymers. *ACS Applied Materials & Interfaces* 7: 14660.

239 Thomas, C.S., Glassman, M.J., and Olsen, B.D. (2011). Solid-state nanostructured materials from self-assembly of a globular protein–polymer Diblock copolymer. *ACS Nano* 5: 5697.

240 Thomas, C.S., Xu, L., and Olsen, B.D. (2012). Kinetically controlled nanostructure formation in self-assembled globular protein–polymer Diblock copolymers. *Biomacromolecules* 13: 2781.

241 Thomas, C.S. and Olsen, B.D. (2014). Coil fraction-dependent phase behaviour of a model globular protein–polymer diblock copolymer. *Soft Matter* 10: 3093.

242 Chang, D. and Olsen, B.D. (2016). Self-assembly of protein-zwitterionic polymer bioconjugates into nanostructured biomaterials. *Polymer Chemistry* 7: 2410.

243 Chang, D., Lam, C.N., Tang, S. et al. (2014). Effect of polymer chemistry on globular protein–polymer block copolymer self-assembly. *Polymer Chemistry* 5: 4884.

244 Qin, G., Perez, P.M., Mills, C.E. et al. (2016). Effect of ELP sequence and fusion protein design on concentrated solution self-assembly. *Biomacromolecules* 3: 928.

245 White, S.H. and Wimley, W.C. (1999). Membrane protein folding and stability: physical principles. *Annual Review of Biophysics and Biomolecular Structure* 28: 319.

246 Sachs, J.N. and Engelman, D.M. (2006). Introduction to the membrane protein reviews: the interplay of structure, dynamics, and environment in membrane protein function. *Annual Review of Biochemistry* 75: 707.

247 Almén, M.S., Nordström, K.J.V., Fredriksson, R. et al. (2009). Mapping the human membrane proteome: a majority of the human membrane proteins can be classified according to function and evolutionary origin. *BMC Biology* 7: 50.

248 Hunte, C., Von Jagow, G., and Schagger, H. (2003). *Membrane Protein Purification and Crystallization: A Practical Guide*. Academic Press.

249 Bordier, C. (1981). Phase separation of integral membrane proteins in triton X-114 solution. *Journal of Biological Chemistry* 256: 1604.

250 Bayburt, T.H., Grinkova, Y.V., and Sligar, S.G. (2002). Self-assembly of Discoidal phospholipid bilayer nanoparticles with membrane scaffold proteins. *Nano Letters* 2: 853.

251 Bayburt, T.H. and Sligar, S.G. (2003). Self-assembly of single integral membrane proteins into soluble nanoscale phospholipid bilayers. *Protein Science : A Publication of the Protein Society* 12: 2476.

252 Cardoso, M.B., Smolensky, D., Heller, W.T. et al. (2011). Supramolecular assembly of biohybrid photoconversion systems. *Energy & Environmental Science* 4: 181.

8

Multiscale Modeling and Simulation of DNA-Programmable Nanoparticle Assembly

Ting Li, Rebecca J. McMurray, and Monica Olvera de la Cruz

Northwestern University, Department of Materials Science and Engineering, 2220 Campus Drive Evanston, IL 60208, USA

8.1 Introduction

Using DNA-coated nanoparticles to direct the self-assembly of crystals was independently developed by Mirkin et al. and Alivisatos et al. [1, 2]. Harnessing the complementary binding ability of single-stranded DNA (ssDNA), Mirkin et al. developed a system for assembling aggregates of colloidal gold nanoparticles using short, stiff double-stranded DNA (dsDNA) sequences with single-stranded "sticky ends" (Figure 8.1) [1]. Using these sticky ends to bind complementary ssDNA sequences, these functionalized DNA-nanoparticles can self-assemble in a thermally reversible manner. In parallel, Alivisatos et al. used long and soft ssDNA to build nanocrystals [2].

During the last decade, significant advances have improved this wet chemical technique [3, 5–23]. In 2008, two independent studies by Park et al. and Nykypanchuk et al. used stiff and soft ssDNA chains, respectively, to produce body-centered cubic (BCC) and face-centered cubic (FCC) structures [3, 12]. These basic crystalline structures were the first to elegantly demonstrate the viability of DNA-programmable self-assembly, and more importantly, DNA-induced nanoparticle crystallization. Moreover, the ability to customize components of the DNA-nanoparticle structure has enabled researchers to tailor this system, creating a range of lattice structures from 25 to 255 nm in length [3, 11, 12, 15, 17].

More recently, a bottom-up approach was used to create one-, two-, and three-dimensional superlattices [17]. This strategy uses nanoparticles as "atoms" with the inherent attraction between complementary DNA sequences serving as the "chemical bond." Using this approach, MacFarlane et al. constructed a series of quasi-Pauling's rules and proposed six design rules for

Self-Assembly: From Surfactants to Nanoparticles, First Edition. Edited by Ramanathan Nagarajan.

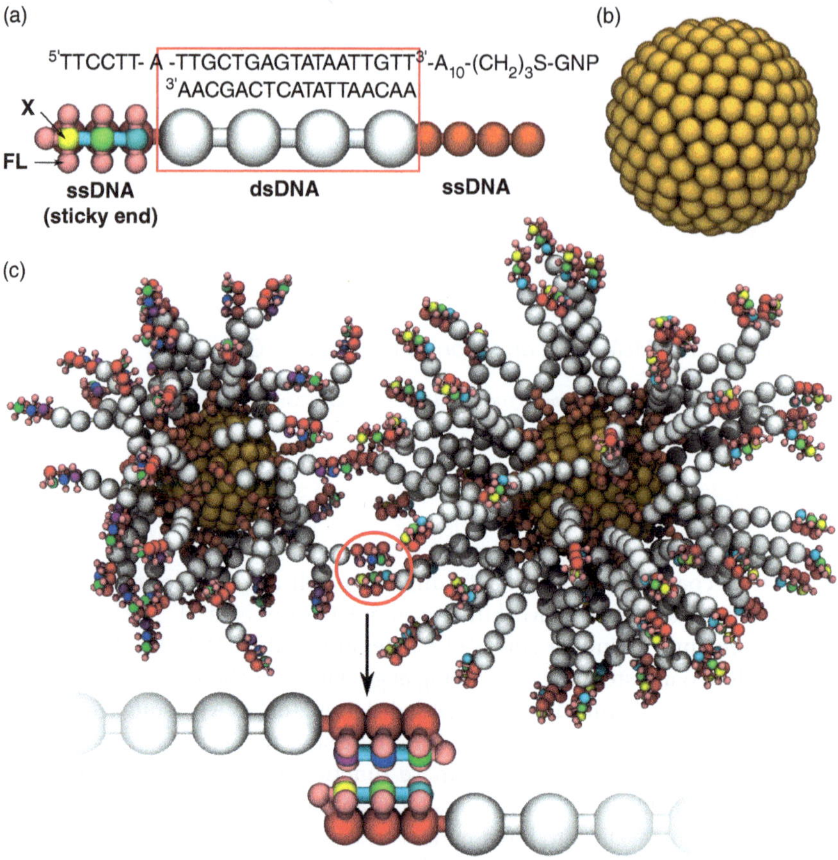

Figure 8.1 Schematic of the coarse-grained model. (a) A model DNA chain and sequence used in experiments [3]. (b) A model spherical NP core. (c) An example of two DNA-nanoparticles with an 8 nm gold core and 40 DN chains (left) and a 10 nm core and 60 DNA chains (right) [4].

the development of nine distinct colloidal crystal structures, including simple cubic, FCC, hexagonal, close-packed (HCP), BCC, CsCl, AlB_2, Cr_3Si, and Cs_6C_{60} symmetries [17].

In addition to studying nanospheres, researchers are investigating the assembly of DNA-functionalized anisotropic nanoparticles [24]. Using face-selective functionalization of particles, anisotropic interactions can be used to synthesize structures such as two-dimensional hexagonal sheets, one-dimensional lamellar, and FCC lattices.

In 2014, another major breakthrough occurred when faceted single crystals using DNA-functionalized nanoparticles were produced [18]. These single crystals form via a mechanism that mimics atomic crystallization and are confirmed to have the Wulff equilibrium crystal structure that was originally predicted using theoretical and molecular dynamics (MD) simulations.

Despite these rapid advances in the field during recent years, many aspects of these systems remain poorly understood. As such, recent advances in computation technology have given researchers unprecedented power to study the underlying mechanisms driving DNA-mediated self-assembly. Computational simulations can be used to understand the structure and microscopic interactions of molecule assemblies. In addition to conventional experimental studies, simulations can provide a level of insight that is otherwise hard to obtain. Recent studies have predicted crystal structures by computing the free energy or stability of various crystals [25, 26]. However, for more detailed models, while full atomistic models remain computationally unfeasible, it is possible to design coarse-grained models that are able to capture all the relevant physical processes whilst remaining computationally viable.

There are essentially two simulation techniques that produce the most reasonable description of the dynamics: Monte Carlo (MC) and MD [27, 28]. While the MC method uses repeated random sampling, MD simulates the evolution of the system over time, taking into account dynamic properties, such as the diffusion coefficient and time-dependent responses to perturbations. Furthermore, MD simulations require no assumptions about the thermodynamic state of the system.

In the case of DNA-programmable self-assembly, MD simulations provide two major advantages. First, in contrast to experimental studies during which nanoparticle assemblies can occur within seconds, MD simulations make it possible to easily visualize the movement of the nanoparticles within the system as well as the conformation and hybridization behaviors of the DNA chains [28]. As a result, MD simulations can provide extremely useful insight into the self-assembly process. Second, when compared to experimental systems, which can be expensive and time consuming, MD simulations provide a cost-effective, faster alternative for the design and prediction of new DNA-NP assemblies. The following sections aim to highlight recent advances in our understanding of DNA nanoparticle assembly and crystallization using MD simulations to develop (i) a scale accurate coarse-grained model with explicit DNA chains (ii) a kinetic model of the DNA hybridization process, and (iii) predicting the Wulff equilibrium crystal structure of faceted, single crystalline microcrystals.

8.2 A Molecular Dynamics Study of a Scale-Accurate Coarse-Grained Model with Explicit DNA Chains

One of the greatest advances in the field thus far is our ability to predict the superlattice structure of a crystal. In the last 20 years, research has evolved from generating amorphous products to yielding highly crystalline materials with predictable crystal arrangement and lattice parameters [1, 3, 5–10, 15]. As stated previously, MacFarlane et al. demonstrated that there are six rules that can be applied when predicting superlattice structure [17]. These rules are governed by varying the components that make up a DNA-nanoparticle, such as nanoparticle size and composition, DNA length, total number of DNA connections and sequence complementarity. The ability to fine-tune each of these parameters individually can ultimately produce a broad spectrum of crystals with distinct properties. MD simulations have provided a unique perspective on the process of guided assembly and crystallization process of DNA-nanoparticles. Using a coarse-grained model adapted from that proposed by Knorowski et al., which captures scale-accurate properties for dsDNA instead of ssDNA, Li et al. successfully demonstrated the self-assembly of DNA-nanoparticles into various types of ordered crystals, where the crystal structure is dictated by the initial design parameters [4, 29]. Using the relative size ratios and number of DNA chains per nanoparticle as described by MacFarlane et al., the same structures, including BCC, CsCl, AlB_2, and Cr_3Si were found. Results from these simulations were used to construct a detailed phase diagram for binary systems and showed excellent agreement with experimental systems [17]. Furthermore, MD simulations can provide detailed information about the assembly process by providing real-time tracking of each nanoparticle's position.

By designing a coarse-grained model, as shown in Figure 8.1, each DNA-nanoparticle is modeled as a rigid spherical core with a fixed number of beads and DNA chains attached. dsDNA chains are represented with a single bead that equates approximately to five base pairs. DNA chains are terminated by single-stranded sticky ends that have the ability to bind complementary DNA bases (i.e. A-T, C-G). Using this model, typical crystalline structures were obtained for CsCl, AlB_2, and Cr_3Si. As seen in Figure 8.2 the basic structure of CsCl contains two types of building blocks, A (with the coordinates 0, 0, 0) and B (with the coordinates $a/2$, $a/2$, $a/2$).

Regions of the phase space were constructed that illustrate the hydrodynamic size ratio of the two DNA-nanoparticle types (A and B) and the DNA coverage ratio of the NP cores for three stoichiometric ratios, 1 : 1 for CsCl, 1 : 2 for AlB_2, and 1 : 3 for the Cr_3Si lattice (Figure 8.3). When compared to the experimental results reported by MacFarlane et al. the outcomes predicted by MD simulations are in agreement [17]. Questions still remain, however, surrounding the outcome should the initial stoichiometric ratio not be compatible with any crystalline structure. Unfortunately, the computational power required to simulate such large systems is currently unfeasible.

Modeling DNA-nanoparticle crystallization using MD simulations has revealed two elements that are critical to the process and enable the most thermodynamically favorable positioning of the nanoparticles: (i) the softness

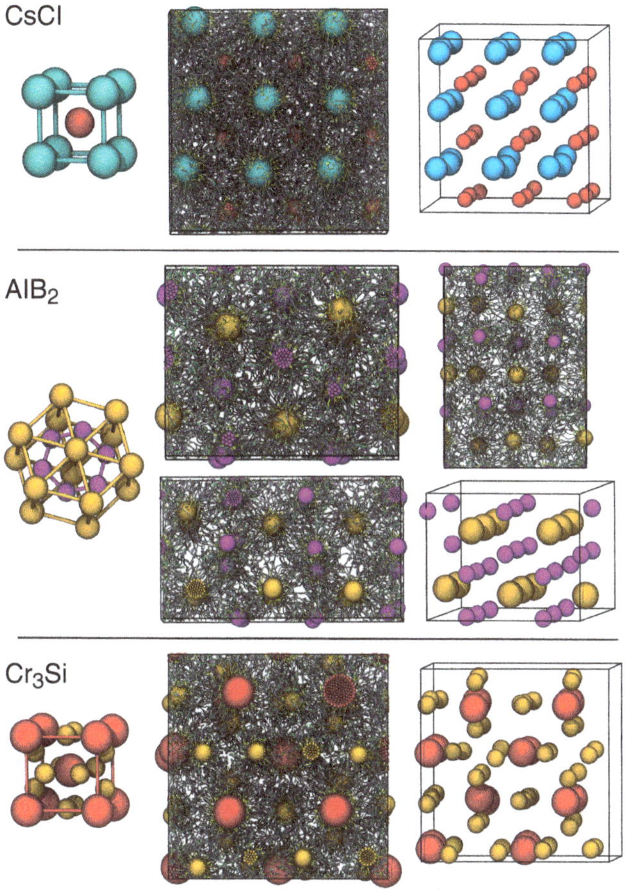

Figure 8.2 Crystal structures obtained from MD simulations. Images show the ideal crystals structure for each arrangement (left), an image of the system once it is fully crystalized (middle), and the average positions of the DNA-nanoparticles over 2×10^6 time steps (right). CsCl: A large DNA-NP is built of a 15-nm NP core and dressed with 100 DNA chains of length ≈ 19 nm; small DNA-NP is built with an 8 nm NP core and dressed with 40 DNA chains of length ≈ 17 nm. AlB_2: A large DNA-NP has a 10 nm NP core and 60 DNA chains of length ≈ 24 nm; small DNA-NP has an 8 nm NP core with 40 DNA chains of length ≈ 12 nm. Cr_3Si: Large DNA-NP contains 13 nm NP core and 80 DNA chains of length ≈ 17 nm; small DNA-NP contains an 8 nm NP core with 40 DNA chains of length ≈ 10 nm [4].

of the spherical nucleic acid (SNA) grafts and (ii) the dynamic hybridization between the sticky ends, as they allow for flexibility between particles and sampling of several positions within the lattice, respectively. Furthermore, by studying the hybridization and dehybridization events of the single-stranded DNA sticky ends, MD simulations confirmed previous experimental data

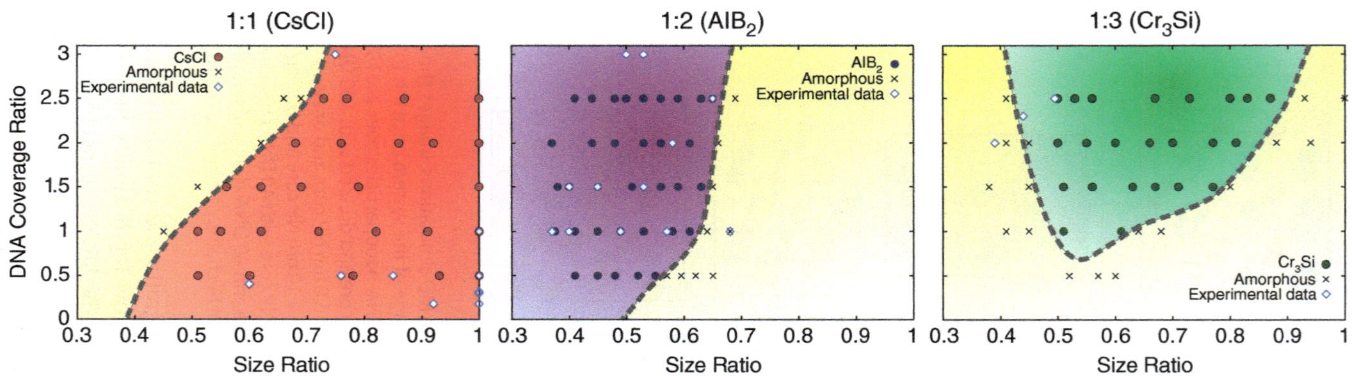

Figure 8.3 Phase diagrams as a function of the DNA-NP size ratio and the DNA coverage ratio (large to small DNA-NP) for three stoichiometric ratios, 1 : 1 (left), 1 : 2 (middle), and 1 : 3 (right). Filled symbols indicate that the corresponding crystal structure formed while the cross indicates that crystal structure did not show in the simulations; open diamonds are the experimental data points [17] with the same crystal structure as those obtained by simulations for a given stoichiometric ratio. Boundaries between different phases cannot be determined exactly from our simulations and thus are represented with dashed lines [4].

identifying that the highly dynamic hybridization process between the semi-flexible single-stranded DNA linkers enables crystallization [4].

8.3 Thermally Active Hybridization

One of the key challenges in nanoparticle self-assembly is preventing nanoparticle assemblies from becoming kinetically trapped in an amorphous state. The key to preventing this was found to be in selecting the right sequence for the single-stranded DNA "sticky end." Experimental studies found that by shortening the linker sequence from 12 bases to 6 bases for complementary linkers and to 4 bases for self-complementary sequences, a higher number of binding and unbinding events occurred [1, 3, 15, 17, 24, 30]. However, while experimental systems are not yet advanced enough to provide in depth detail about individual hybridization events, MD simulations can provide a quantitative insight into the physical properties of DNA sticky ends that are essential for successful hybridization.

Maintaining a highly dynamic level of DNA linker hybridization is essential to properly simulate the formation of nanoparticle superlattices. Active hybridization provides a crucial pathway for a successful hybridization by reducing the energy barrier between a kinetically trapped structure and a state of thermodynamic equilibrium. It is therefore essential to properly parameterize the force field for complementary sticky ends in a coarse-grained model.

In a recent study, Li et al. successfully designed a scale-accurate coarse-grained model that faithfully captures relevant DNA-pair interactions (Figure 8.4) [31]. Using this method, an investigation into the optimal linker strength for assembling distinct superlattices was carried out. Figure 8.5 shows the survival rate for a set of hybridizations as a function of the simulation time in order to directly quantify the dynamics, as well as the percentage of

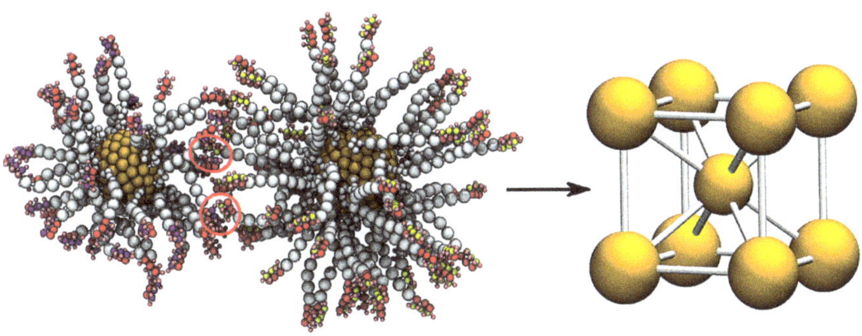

Figure 8.4 Schematic showing DNA-guided nanoparticle self-assembly. The building blocks for simulations are shown on the left with DNA hybridization events circled in red [31].

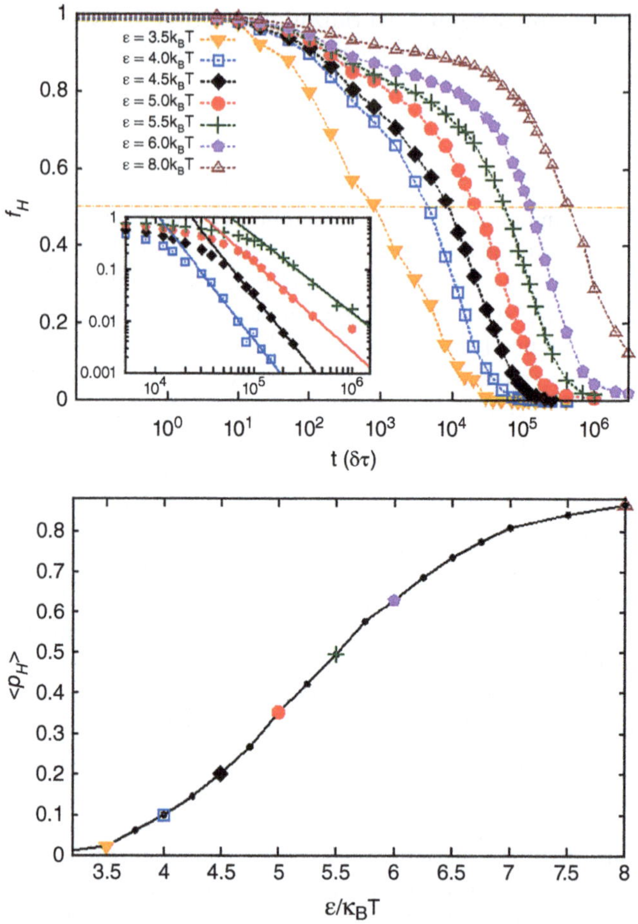

Figure 8.5 Top: Fraction of hybridizations f_H that survive up to a time t. Inset: A power-law fit of the long-time tail of $f_H \propto t^{-b}$. The exponents are between $b \approx 2.4$ (for $\varepsilon = 4k_BT$) and $b \approx 1.5$ (for $\varepsilon = 5.5k_BT$), which are in good agreement with Ref. [10]. Bottom: Percentage of hybridizations $<p_{Hi}>$ (relative to the maximum number of hybridizations in this system) vs. ε. The same type of symbols represents the same ε in both plots. ε is measured in units of k_BT. Error bars are smaller than the symbol size [31].

hybridizations relative to the maximum number of sticky end pairs in the system as a function of ε. As such, it was found that larger bond strength leads to a greater percentage of hybridizations, and ultimately proposed two primary rules for hybridization. The first rule pertains to sufficient hybridizations, with the second being thermally active hybridization. For the former, in order to overcome thermal fluctuations of the nanoparticles, which might ordinarily prevent successful assembly, a high percentage of hybridized sticky ends in equilibrium are required to allow for adequate DNA interactions. In the latter case, dynamic binding and unbinding of complementary sticky ends between nanoparticles as a result of thermal fluctuations helps prevent the development of kinetically trapped structures.

In order to expand on previously studied systems, such as those mentioned in Section 1.1, a study was conducted taking into account ~100–250 building blocks instead of ~30–60. This increase in the number of building blocks requires 5–10 times more simulation time for modeling the assembly process. However, while this was found to increase the percentage of hybridizations for larger systems, the time it takes to form a crystal once all the nanoparticles are in place appears to be independent of the size of the system (Figure 8.6). In addition, by studying the assembly mechanism in systems with up to 10^6 degrees of freedom (DoF), it was found that a slight decrease in enthalpy occurs during the crystallization process. It should be noted, however, that in a similar study conducted by Knorowski et al. a drop in enthalpy was not found [32]. It is thought that this discrepancy between the studies could be due to a difference in the overall DNA coverage of the nanoparticle cores used in the two studies. It is proposed that in such a system where the DNA coverage is significantly denser, that once the particles are in place, additional hybridizations between adjacent DNA strands can occur, thereby lowering the free energy of the system even further.

Lastly, the researchers also studied a range of optimal linker strengths and the corresponding average hybridization percentage for various binary systems such as BCC, CsCl, and AlB_2, etc., as a function of various design parameters, including DNA coverage, hydrodynamic size ratio, and stoichiometry (lattice type), as reported in previous experimental studies [4, 17]. By maintaining the structure of the small, type A building blocks for all experiments, the average hybridization percentage was calculated. As such, it was found that (i) a larger DNA coverage ratio leads to a larger average hybridization percentage due to more complementary DNA chains. However, as the coverage ratio goes beyond ~2.0, the average hybridization percentage begins to saturate. (ii) With respect to hydrodynamic size ratio, it was found that the optimal linker strength is influenced significantly instead of the average hybridization ratio. (iii) A superlattice with a higher symmetry has a broader range of optimal linker strength (Figure 8.7). Consistent with the order of superlattice symmetries, the range of linker strengths was found to decrease from BCC to CsCl to Cr_3Si to AlB_2 structures. As larger values of optimal linker strength correspond to easier crystallization, it is therefore not surprising that a design of higher symmetry and with DNA coverage close to the inverse of the stoichiometry leads to easier self-assembly and crystallization. While it is important to acknowledge the limitations of coarse-grained models to capture all the aspects of nanoparticle

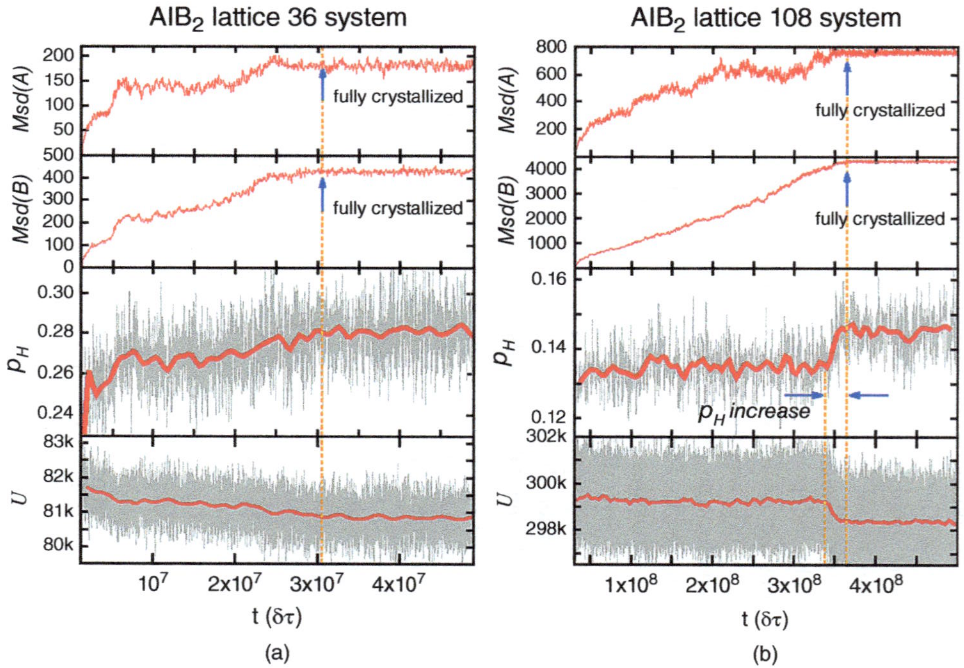

Figure 8.6 Comparison of the crystallization process between two system sizes. Two representative cases for: (a) AlB_2 lattice with 36 building blocks ($3 \times 2 \times 2$ unit cells) in the simulation box; (b) AlB_2 lattice with 108 building blocks ($4 \times 3 \times 3$ unit cells) in the simulation box. The *x*-axis plots the simulation time step. The arrows indicate that the crystallization process has been completed. For time beyond the dashed line, the system is crystalline [31].

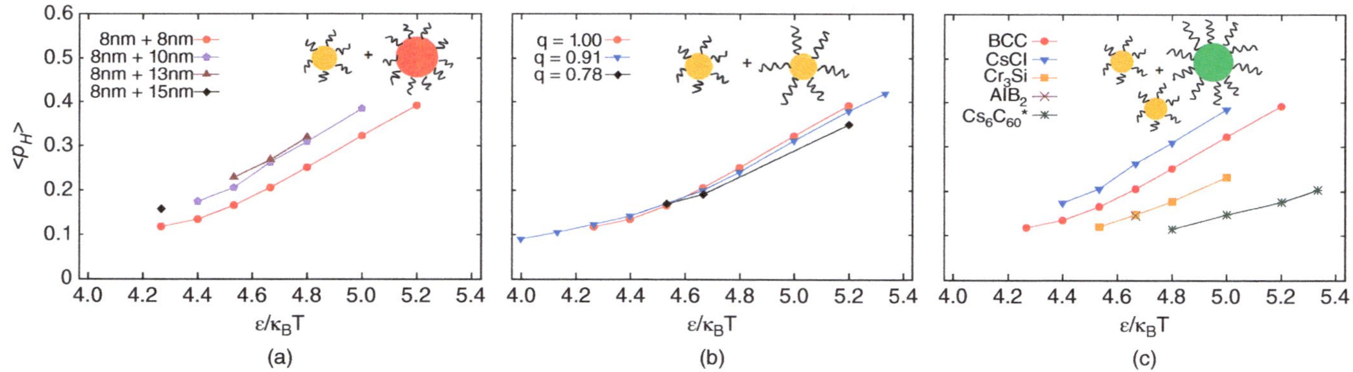

Figure 8.7 Optimal linker strength and corresponding average hybridization percentage for various binary systems. Studied parameters are: (a) DNA coverage ratio *r* (i.e. ratios of the sizes of two NP cores); (b) hydrodynamic size ratio q (DNA length); (c) lattice type. *Note that Cs_6C_{60} is the only system for which an annealing process was used. Error bars are smaller than the symbol size [31].

self-assembly, these results have thus far provided the most detailed model of the crystallization process.

8.4 DNA-Mediated Nanoparticle Crystallization in Wulff Polyhedra

Using a slow-cooling approach, researchers have for the first time produced faceted, single crystalline microcrystals. Traditional DNA-nanoparticle crystallization takes place via a two-step process: (i) initial particle assembly, during which a disordered aggregate forms, followed by (ii) reorganization of particles within the aggregate, forming into a superlattice, when thermally annealed slightly below its melting temperature (Figure 8.8). However, while an increase in annealing time results in higher-quality crystals with larger domain sizes, the overall amorphous shape of the crystal was found to remain the same. Interestingly, as traditional crystallization of atoms and molecules occurs as a result of slow cooling through its melting temperature it was hypothesized that such an approach could be used to yield single crystals of DNA-nanoparticles. Therefore, using the melting temperature of the aggregate, which is defined by the nucleic acid sequences used for assembly, DNA-functionalized gold nanoparticle solutions were heated to above the melting temperature of the DNA linkers and then slowly cooled to room temperature.

In the study conducted by Auyeung et al., MD simulations were carried out on multiscale models alongside experimental studies to study and predict crystal shapes [18]. Two scales of coarse-grained models were used to investigate properties at different levels. The first uses a coarse-grained model with explicit DNA chains, to design crystal symmetries and estimate their properties such as surface energy etc. Surface energy ratios were estimated for different surface orientations and used to predict the relevant Wulff polyhedra. In particular, the formation of rhombic dodecahedra (RD) microcrystals from nanoparticles with BCC packing in experimental systems can be explained by carrying out a theoretical analysis of the exposed facets surface energy. The standard broken bond model approximation for surface energy was used that states that exposing the closest packed plane is the most favorable as it requires breaking the least number of particle-to-particle interactions per unit area, thereby exposing the facet with the lowest surface energy. As such, the equilibrium crystal structure, a Wulff polyhedron was predicted to be an RD enclosed by (110) facets for a BCC metal and a truncated octahedron enclosed by (111) and (100) facets for a FCC metal [33] (Figure 8.9). These simulated results have been found to correlate well with experimental results. In a second, colloidal model with implicit DNA chains, each DNA-nanoparticle is represented by a single bead with a pair-wise potential to simulate the dynamics of crystal growth. Using this method, the shape of the BCC system was confirmed to be a (110) enclosed RD;

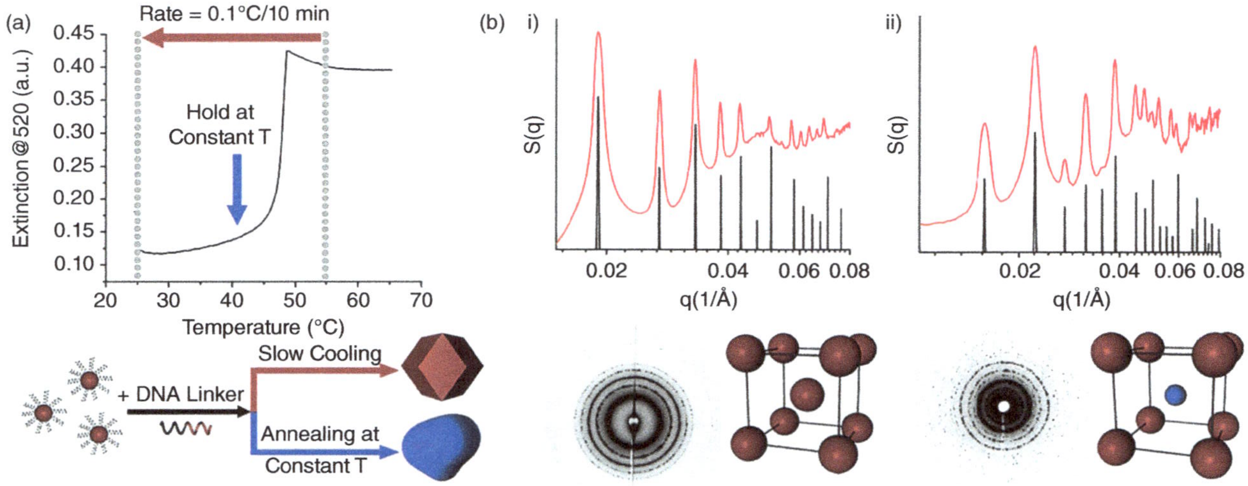

Figure 8.8 DNA-functionalized gold NP single crystals form using a slow-cooling method. (a) Two approaches for DNA-mediated nanoparticle crystallization are designated by arrows on a thermal melting curve of a DNA-linked gold nanoparticle aggregate. The traditional method (blue arrow), in which the aggregate is annealed a few degrees below the melting temperature, produces polycrystalline superlattices with no defined shape. The slow-cooling method (red arrow), in which the aggregate is heated above its melting temperature and cooled at a rate of 0.01 °C min^{-1}, produced well-defined, faceted microcrystals in each of the dozens of experiments conducted using these conditions. (b) Representative one-dimensional (top) and two-dimensional (bottom) SAXS data for BCC (i) and CsCl (ii) superlattices synthesized from the slow-cooling technique. In the one-dimensional data, the red trace is the experimentally obtained scattering pattern and the black trace is the predicted scattering pattern for a perfect lattice. SAXS data are shown as plots of superlattice structure factor (S(q), in arbitrary units) versus scattering vector (q, in units of $Å^{-1}$). (c) TEM images of shape-controlled BCC (i) and (ii) and CsCl (iii) and (iv) microcrystals. Scale bars are 1.5 μm for (i), 1.0 μm for (ii), 0.5 μm for (iii), and 0.5 μm for (iv). (d) Scanning electron microscopy image of a representative BCC microcrystal with visible faceting where constituent nanoparticles can readily be seen (20 nm gold nanoparticles shown; scale bar is 1 μm). The inset shows a high-magnification (X ~ 52 500) view of the crystal facet with labeled surface defects [18].

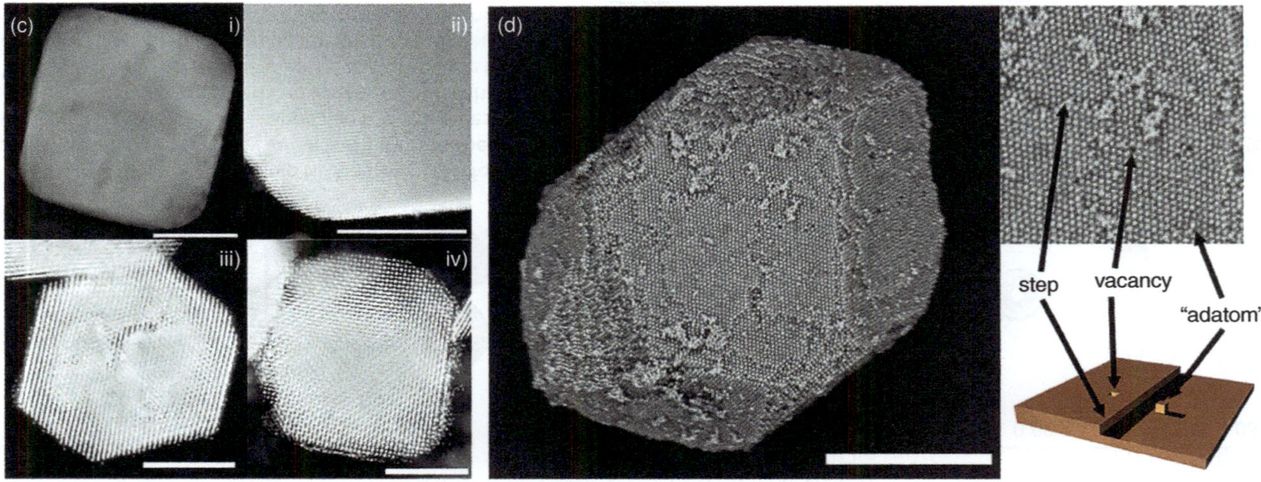

Figure 8.8 (*Continued*)

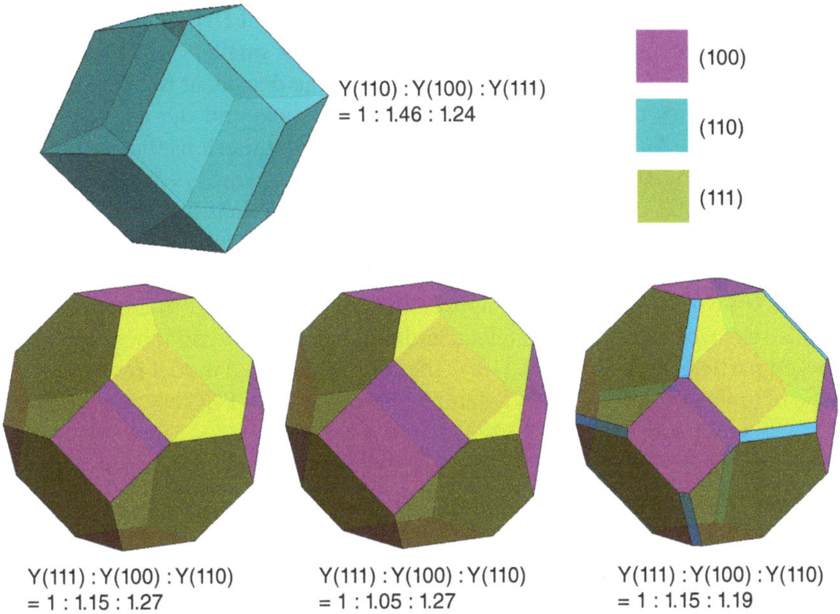

Figure 8.9 Wulff polyhedra of BCC and FCC DNA-AuNPs. Top: BCC (110) rhombic dodecahedron. Bottom: FCC (111) octahedron truncated by (100) and (110) faces [18].

however, for the FCC system no uniform shape was found except for triangular features with (111) and (100) facets that can be caused by defects, such as twinning and stacking faults.

This work illustrates the remarkable similarities between the self-assembly and crystallization of DNA-nanoparticles and atomic crystals, with the nanoparticles in this system behaving as their true atom equivalents. It also highlights the many advantages to using DNA as a means to control not only the recognition properties and surface energy of individual particles, but also the surface energies of corresponding macroscopic particles, such that specific structures can be pre-programmed and realized experimentally. The ability to form single crystals in this way holds great promise for photonic applications as their interactions with light yields more uniform, accurate models when compared to other polycrystalline structures.

Phenomenological colloidal models of DNA-nanoparticles have captured the main features of the phase diagram for this system [34, 35] as well as the coalescence dynamics into BCC structures [36]. A refined version of the model was then constructed using the DNA functional nanoparticle coarse-grained accurate model to find the attractive pair potential between DNA-AuNPs [31]. In the model, each DNA-AuNP building block is coarse-grained into a single spherical bead interacting via an effective electrostatic repulsion and an attraction due to complementary sticky ends. The crystallization of DNA-AuNPs occurs at high salt concentration when the Debye length is smaller than the size of the particles (2R). In this regime, the surface charge density at the core radius R, which includes the radius of the gold nanoparticle and the dsDNA portion of the DNA-AuNPs, is strongly reduced by the counterions that penetrate the DNA-AuNPs. The effective charge of the particles, which is obtained from density functional theory (DFT) and confirmed by explicit ions MD simulations [37], is strongly reduced at such high salt concentrations. Therefore, a linearized mean field approach valid for the highly reduced surface charged density, known as the Derjaguin approximation, can be used to compute the effective electrostatic repulsion (electric-double layer interaction) between two charged particles (see SI in Ref. [18]). The refined colloidal model revealed the shape of grains of DNA-AuNPs assembled into BCC and FCC crystals.

8.5 Conclusions

The design of 3D crystalline assemblies via DNA hybridization has opened the possibility of a studying a new form of matter with unique functionalities. We have described here assemblies of SNAs. SNAs consist of functionalized nucleic acids covalently attached to the surface of spherical inorganic nanoparticles, forming a conjugate capable of hybridizing with neighboring nanoparticles. These DNA conjugates exhibit a synergistic combination of properties resulting from the core and nucleic acid shell and their ability to assemble in various structures leading to unique photonic, plasmonic, and catalytic properties. SNAs have attracted the interest of material scientists because they form robust equilibrium nanostructures that are responsive to external stimulus. Moreover, SNAs can be prepared from both single- and double-stranded nucleic acids, and in various DNA length to core size ratios, displaying a broad range of properties. Single stranded SNAs, for example, have been shown to respond to applied osmotic pressure [38]. In this case, the lattices of nanoparticles interconnected with DNA exhibit an isotropic transformation under compression with a remarkably strong decrease of the lattice constant, corresponding to more than 80% of the volume reduction. Moreover, nanoparticles can be introduced to generate highly functional heterogeneous superlattices of nanoparticles. Nanoparticles with all major types of functionality – plasmonic (gold), magnetic (Fe_2O_3), catalytic (palladium), and luminescent (CdSe/Te@ZnS and CdSe@ZnS) – have been incorporated into binary systems [39]. This approach demonstrates the possibility of assembling materials with novel optical and field-responsive properties.

Many of the results discussed here for SNAs can also be applied to nonspherical DNA functionalized nanoparticles. The assembly of functionalized nanoparticles is actually in its infancy. There are abundant novel structures with emergent physical properties to be designed and synthesized. Up until now, most systems synthesized typically involve binary building blocks with one pair of complementary DNA linkers. Challenges include employing three types of spherical building block (A, B, C) as well as mixtures of 1D, 2D, and 3D, anisotropic (nonspherical) nanoparticles to freely design more complex assembly with multiple types of building blocks and multiple linker interactions. Therefore, despite the exciting advances in DNA-programmable assembly achieved throughout the past two decades, there are many research opportunities ahead.

Acknowledgments

Monica Olvera de la Cruz and Ting Li acknowledge support from the Air Force Office of Scientific Research (AFOSR) Award No. FA9550-11-1-0275. R.J.M. acknowledges support from the Center for Computation and Theory of Soft Materials at Northwestern University.

References

1 Mirkin, C.A., Letsinger, R.L., Mucic, R.C. et al. (1996). A DNA-based method for rationally assembling nanoparticles into macroscopic materials. *Nature* 382 (6592): 607–609.

2 Alivisatos, A.P., Johnsson, K.P., Peng, X. et al. (1996). Organization of 'nanocrystal molecules' using DNA. *Nature* 382 (6592): 609–611.

3 Park, S.Y., Lytton-Jean, A.K.R., Lee, B. et al. (2008). DNA-programmable nanoparticle crystallization. *Nature* 451 (7178): 553–556.

4 Li, T., Sknepnek, R., Macfarlane, R.J. et al. (2012). Modeling the crystallization of spherical nucleic acid nanoparticle conjugates with molecular dynamics simulations. *Nano Letters* 12 (5): 2509–2514.

5 Jin, R.C., Wu, G., Li, Z. et al. (2003). What controls the melting properties of DNA-linked gold nanoparticle assemblies? *Journal of the American Chemical Society* 125 (6): 1643–1654.

6 Mucic, R.C., Storhoff, J.J., Mirkin, C.A. et al. (1998). DNA-directed synthesis of binary nanoparticle network materials. *Journal of the American Chemical Society* 120 (48): 12674–12675.

7 Valignat, M.P., Theodoly, O., Crocker, J.C. et al. (2005). Reversible self-assembly and directed assembly of DNA-linked micrometer-sized colloids. *Proceedings of the National Academy of Sciences of the United States of America* 102 (12): 4225–4229.

8 Maye, M.M., Nykypanchuk, D., van der Lelie, D. et al. (2006). A simple method for kinetic control of DNA-induced nanoparticle assembly. *Journal of the American Chemical Society* 128 (43): 14020–14021.

9 Geerts, N., Schmatko, T., and Eiser, E. (2008). Clustering versus percolation in the assembly of colloids coated with long DNA. *Langmuir* 24 (9): 5118–5123.
10 Biancaniello, P.L., Kim, A.J., and Crocker, J.C. (2005). Colloidal interactions and self-assembly using DNA hybridization. *Physical Review Letters* 94 (5).
11 Hill, H.D., Macfarlane, R.J., Senesi, A.J. et al. (2008). Controlling the lattice parameters of gold nanoparticle FCC crystals with duplex DNA linkers. *Nano Letters* 8 (8): 2341–2344.
12 Nykypanchuk, D., Maye, M.M., van der Lelie, D. et al. (2008). DNA-guided crystallization of colloidal nanoparticles. *Nature* 451 (7178): 549–552.
13 Maye, M.M., Kumara, M.T., Nykypanchuk, D. et al. (2010). Switching binary states of nanoparticle superlattices and dimer clusters by DNA strands. *Nature Nanotechnology* 5 (2): 116–120.
14 Cutler, J.I., Auyeung, E., and Mirkin, C.A. (2012). Spherical nucleic acids. *Journal of the American Chemical Society* 134 (3): 1376–1391.
15 Macfarlane, R.J., Jones, M.R., Senesi, A.J. et al. (2010). Establishing the design rules for DNA-mediated colloidal crystallization. *Angewandte Chemie-International Edition* 49 (27): 4589–4592.
16 Auyeung, E., Cutler, J.I., Macfarlane, R.J. et al. (2012). Synthetically programmable nanoparticle superlattices using a hollow three-dimensional spacer approach. *Nature Nanotechnology* 7 (1): 24–28.
17 Macfarlane, R.J., Lee, B., Jones, M.R. et al. (2011). Nanoparticle superlattice engineering with DNA. *Science* 334 (6053): 204–208.
18 Auyeung, E., Li, T., Senesi, A.J. et al. (2014). DNA-mediated nanoparticle crystallization into Wulff polyhedra. *Nature* 505 (7481): 73–77.
19 Macfarlane, R.J., Jones, M.R., Lee, B. et al. (2013). Topotactic interconversion of nanoparticle superlattices. *Science* 341 (6151): 1222–1225.
20 Sun, W., Boulais, E., Hakobyan, Y. et al. (2014). Casting inorganic structures with DNA molds. *Science* 346 (6210): 717–725.
21 Jones, M.R., Seeman, N.C., and Mirkin, C.A. (2015). Programmable materials and the nature of the DNA bond. *Science* 347 (6224): 840.
22 Radha, B., Senesi, A.J., O'Brien, M.N. et al. (2014). Reconstitutable nanoparticle superlattices. *Nano Letters* 14 (4): 2162–2167.
23 Park, D.J., Zhang, C., Ku, J.C. et al. (2015). Plasmonic photonic crystals realized through DNA-programmable assembly. *Proceedings of the National Academy of Sciences of the United States of America* 112 (4): 977–981.
24 Jones, M.R., Zhang, C., Ku, J.C. et al. (2010). DNA-nanoparticle superlattices formed from anisotropic building blocks. *Nature Materials* 9 (11): 913–917.
25 Padovan-Merhar, O., Lara, F.V., and Starr, F.W. (2011). Stability of DNA-linked nanoparticle crystals: effect of number of strands, core size, and rigidity of strand attachment. *Journal of Chemical Physics* 134 (24).

26 Martinez-Veracoechea, F.J., Mladek, B.M., Tkachenko, A.V. et al. (2011). Design rule for colloidal crystals of DNA-functionalized particles (vol 107, 045902, 2011). *Physical Review Letters* 107 (7).

27 Anderson, J.A., Lorenz, C.D., and Travesset, A. (2008). Micellar crystals in solution from molecular dynamics simulations. *Journal of Chemical Physics* 128 (18).

28 Allen, M.P. (2004). Introduction to molecular dynamics. In: *Computational Soft Matter: From Synthetic Polymers to Proteins*, vol. 23, 1–28.

29 Knorowski, C., Burleigh, S., and Travesset, A. (2011). Dynamics and statics of DNA-programmable nanoparticle self-assembly and crystallization. *Physical Review Letters* 106 (21): 215501.

30 Storhoff, J.J., Lazarides, A.A., Mucic, R.C. et al. (2000). What controls the optical properties of DNA-linked gold nanoparticle assemblies? *Journal of the American Chemical Society* 122 (19): 4640–4650.

31 Li, T., Sknepnek, R., and de la Cruz, M.O. (2013). Thermally active hybridization drives the crystallization of DNA-functionalized nanoparticles. *Journal of the American Chemical Society* 135 (23): 8535–8541.

32 Knorowski, C. and Travesset, A. (2012). Dynamics of DNA-programmable nanoparticle crystallization: gelation, nucleation and topological defects. *Soft Matter* 8 (48): 12053–12059.

33 Wulff, G. (1901). On the question of speed of growth and dissolution of crystal surfaces. *Zeitschrift Fur Krystallographie Und Mineralogie* 34 (5/6): 449–530.

34 Leunissen, M.E. and Frenkel, D. (2011). Numerical study of DNA-functionalized microparticles and nanoparticles: explicit pair potentials and their implications for phase behavior. *Journal of Chemical Physics* 134 (8).

35 Scarlett, R.T., Crocker, J.C., and Sinno, T. (2010). Computational analysis of binary segregation during colloidal crystallization with DNA-mediated interactions. *Journal of Chemical Physics* 132 (23).

36 Dhakal, S., Kohlstedt, K.L., Schatz, G.C. et al. (2013). Growth dynamics for DNA-guided nanoparticle crystallization. *ACS Nano* 7 (12): 10948–10959.

37 Zwanikken, J.W., Guo, P., Mirkin, C.A. et al. (2011). Local ionic environment around polyvalent nucleic acid-functionalized nanoparticles. *Journal of Physical Chemistry C* 115 (33): 16368–16373.

38 Srivastava, S., Nykypanchuk, D., Maye, M.M. et al. (2013). Super-compressible DNA nanoparticle lattices. *Soft Matter* 9 (44): 10452–10457.

39 Zhang, Y.G., Lu, F., Yager, K.G. et al. (2013). A general strategy for the DNA-mediated self-assembly of functional nanoparticles into heterogeneous systems. *Nature Nanotechnology* 8 (11): 865–872.

9

Harnessing Self-Healing Vesicles to Pick Up, Transport, and Drop Off Janus Particles

Xin Yong, Emily J. Crabb, Nicholas M. Moellers, Isaac Salib, Gerald T. McFarlin, Olga Kuksenok and Anna C. Balazs

University of Pittsburgh, Chemical Engineering Department, 3700 O'Hara Street, Pittsburgh, PA 15261, USA

9.1 Introduction

The development of nanocarriers for the targeted delivery of nanoscopic cargo (e.g. proteins and nanoparticles) can benefit such diverse areas as drug delivery, biosensing, and bottom-up self-assembly [1]. Nanoscopic lipid vesicles offer unique opportunities for encapsulating a small number of nanoparticles within their cores, and, thus, can serve as ideal carriers to controllably deliver just a few particles to specified sites [2]. Notably, nanoscopic vesicles can passively traverse the dermal layer [3] and, hence, could serve as highly effective containers for noninvasive, transdermal drug delivery. Recent advances in the synthesis of nanoscopic vesicles have led to precise control over the size distribution of these vesicles [4, 5], paving the way for realizing the above vital functions.

The imposed flow in micro- and nanofluidic environments can be harnessed to direct the self-assembly of microstructures [6], and can be used to guide targeted delivery processes. Bottom-up fabrication within fluidic devices would be greatly facilitated if we could effectively manipulate specific components on the floor of fluidic chambers by noninvasive means. Namely, it would be highly desirable if the vesicles could perform selective pick-up, transport, and deposition of specific components. In this manner, the vesicles could act as synthetic mimics of osteoclasts and osteoblasts. (In the most general sense, the osteoclasts pick up particulates from the bone surface and the osteoblasts deposit materials on the surface.)

Computational modeling has been used extensively to investigate the interactions between lipid vesicles and nanoparticles. Previous simulations revealed the shape and chemical composition of the particles as critical factors affecting encapsulation of the particles into the vesicles [7–12]. These studies provide valuable guidelines for using vesicles to controllably pick up, transport, and

Self-Assembly: From Surfactants to Nanoparticles, First Edition. Edited by Ramanathan Nagarajan.

deposit nanoparticles. Nevertheless, most of these simulations have focused primarily on microscopic vesicles. Due to the significant size difference between the particles and the vesicle, these simulations only modeled a single, flat bilayer membrane, which represents a proportion of the larger vesicle. For nanoscopic vesicles, the comparable sizes of the vesicles and particles necessitate the modeling of the curved vesicle membrane.

In our simulations, we model the interaction between a three-dimensional lipid vesicle, whose diameter is on the nanometer length scale, and spherical amphiphilic Janus nanoparticles containing hydrophobic and hydrophilic regions. Thus, the three-dimensional vesicle model allows us to capture the synergistic interactions between the particles and the deformable, curved surfaces of the nanoscopic vesicles. Using vesicles as nanocarriers, we impose flow in the fluidic chamber to drive the vesicles to move on a flat substrate, effectively pick up nanoparticles located on the surface and transport particles as cargo along the substrate. We then show that with the introduction of a "sticky" patch or a nanoscopic trench on the substrate, the vesicles can robustly drop off the cargo at desired positions in the channel. Understanding the behavior of vesicle-particle assemblies in a flow field is vital for designing effective cargo carriers in microfluidic devices, as well as gaining insight into the performance of nanocarriers in the bloodstream.

In carrying out these studies, we use dissipative particle dynamics (DPD) [13–17] to model the interaction among vesicles, nanoparticles, substrates, and solvent. DPD is a mesoscopic particle-based computational method that can effectively model physical phenomena occurring at relatively large length and time scales (compared with the length and time scales accessible with molecular dynamics [MD] simulations) within computationally reasonable time frames. One of the advantageous features of DPD is the ability to capture the hydrodynamics of complex fluids while retaining essential information about the structural properties of the components of the system [13, 16, 17].

In this study, we show that the vesicles can act as nanoscopic "vacuum cleaners", collecting and absorbing the Janus particles encountered along their paths in the flow. We demonstrate that this pick-up event is affected by the shear rate, the chemical composition of the nanoparticles, and the adhesion strength between the particles and the substrate. By exploring the parameter space, we identified a range of shear rates for which even Janus particles encompassing just a small fraction of hydrophobic region can be robustly picked up. Furthermore, a single vesicle shows the ability to carry a relatively large payload by taking up and transporting a number of these particulates. It is noteworthy that the action of these lipid vesicles roughly resembles the behavior of the osteoclasts, which can be harnessed to remove nanoscopic debris from hydrophilic surfaces in microfluidic devices. More generally, our simulations demonstrate how external flow can facilitate the incorporation of nanoparticles into nanoscale vesicles.

Prompting the vesicles to deposit the attached particles in a controllable manner is, however, considerably more challenging. We focus on two specific aspects of the deposition. First, we show that controlled deposition can be achieved on either chemically homogeneous surfaces that are topographically patterned, encompassing distinct grooves or trenches, or chemically heterogeneous surfaces, which contains a "sticky" surface patch that exhibits a preferential attraction to the particle. Hence, the vesicles also mimic osteoblasts by delivering particles to specific locations on a substrate. The results could provide additional guidelines for manipulating the assembly of nanoparticles in solution to facilitate the bottom-up fabrication. Second, we examine the deposition process in great detail. We found that the vesicle executes clean deposition of particles when the Janus particles have small hydrophobic domains. For particles with large hydrophobic domains, we observed intriguing scenarios where lipid molecules detached from the vesicle and bound to the particle as it was deposited onto the substrate. After the deposition, these lipids formed a protective coating on the particle to prevent the hydrophobic portion of particle from interacting with the hydrophilic solution in the fluidic chamber. The more striking fact is that the vesicle membrane subsequently reforms without releasing the inner fluid (i.e., lumen) of the vesicle after the particle is deposited. In effect, the adaptive vesicle is able to self-heal after the directed delivery of its cargo, and this self-healing characteristic enables the reuse of the same vesicle for subsequent targeted deliveries. Hence, our simulations show that the nanoscopic vesicles could perform selective pick up, transport, and deposition of specific components.

9.2 Methodology

The DPD method can be viewed as a coarse-grained MD approach. Similar to MD, DPD models the time evolution of a many-body system governed by Newton's equation of motion, $m\, d\mathbf{v}_i/dt = \mathbf{f}_i$. A volume of fluid is modeled with individual beads; each bead represents a cluster of molecules. Bead i experiences a force $\mathbf{f}_i$ that is the sum of three pairwise additive forces: $\mathbf{f}_i(t) = \sum(\mathbf{F}^{\mathrm{C}}_{ij} + \mathbf{F}^{\mathrm{D}}_{ij} + \mathbf{F}^{\mathrm{R}}_{ij})$, in which the sum runs over all beads j within a certain cutoff radius r_{c} from bead i. The three forces are the conservative force $\mathbf{F}^{\mathrm{C}}_{ij}$, drag or dissipative force $\mathbf{F}^{\mathrm{D}}_{ij}$, and random force $\mathbf{F}^{\mathrm{R}}_{ij}$. We describe each pairwise force below.

The conservative force is a soft, repulsive force given by $\mathbf{F}^{\mathrm{C}}_{ij} = a_{ij}(1 - r_{ij})\hat{\mathbf{r}}_{ij}$, where a_{ij} measures the maximum repulsion between beads i and j, $r_{ij} = |\mathbf{r}_i - \mathbf{r}_j|/r_{\mathrm{c}}$, and $\hat{\mathbf{r}}_{ij} = (\mathbf{r}_i - \mathbf{r}_j)/|\mathbf{r}_i - \mathbf{r}_j|$. This soft-core force leads to a degree of overlap between neighboring beads and permits the use of larger time steps than those typically used in MD simulations, which commonly involve the use of hard-core potentials (e.g. the Lennard-Jones potential). The drag force is $\mathbf{F}_{ij}^{\mathrm{D}} = -\gamma\omega_{\mathrm{D}}(r_{ij})(\hat{\mathbf{r}}_{ij}\cdot\mathbf{v}_{ij})\hat{\mathbf{r}}_{ij}$, where γ is a simulation parameter related to viscosity arising from the interactions between beads, ω_{D} is a weight function that goes to zero at r_{c}, and $\mathbf{v}_{ij} = \mathbf{v}_i - \mathbf{v}_j$. The random force is $\mathbf{F}_{ij}^{\mathrm{R}} = \sigma\,\omega_{\mathrm{R}}(r_{ij})\xi_{ij}\hat{\mathbf{r}}_{ij}$, where ξ_{ij} is a zero-mean Gaussian random variable of unit variance and $\sigma^2 = 2k_{\mathrm{B}}T\gamma$. The value of γ is chosen to ensure relatively rapid equilibration of the system temperature and the numerical stability of the simulations for the specified time step [17]. Finally, we use $\omega_{\mathrm{D}}(r_{ij}) = \omega_{\mathrm{R}}(r_{ij})^2 = (1 - r_{ij})^2$ for $r_{ij} < 1$ [17].

Each of these three pairwise forces conserves momentum locally [15, 17] and thus, DPD reproduces correct hydrodynamic behavior [13, 14, 16, 17]. The velocity-Verlet algorithm is applied to integrate the equations of motion in time. We take r_{c} as the characteristic length scale and $k_{\mathrm{B}}T$ as the characteristic energy scale in our simulations. The corresponding characteristic time scale is then defined as $\tau = \sqrt{mr_{\mathrm{c}}^2/k_{\mathrm{B}}T}$. The remaining simulation parameters are $\sigma = 3$ and $\Delta t = 0.02\tau$, with a total bead number density of $\rho = 3$.

Each vesicle in our system is generated from the self-assembly of 586 short, twin-tailed amphiphilic lipid molecules as shown in Figure 9.1a. Each lipid is composed of nine beads, with three hydrophilic beads forming the head and three hydrophobic beads in each of the two tails [8] (see Figure 9.1b). Harmonic bonds $E_{\mathrm{bond}} = K_{\mathrm{bond}}((r - b)/r_{\mathrm{c}})^2$ are assigned between neighboring lipid beads to maintain the twin-tailed structure, where $K_{\mathrm{bond}} = 64$ is the bond constant and $b = 0.5$ is the equilibrium bond length [18]. We also insert a weaker bond ($K_{\mathrm{bond}} = 16$) between the first bead on each tail (nearest to the hydrophilic head) in order to keep the tails oriented in the same direction [19]. Additionally, we stiffen the tail by introducing a three-body angle potential of the form $E_{\mathrm{angle}} = K_{\mathrm{angle}}(1 + \cos\theta)$, where θ is the angle formed by three adjacent tail beads and $K_{\mathrm{angle}} = 5$. The inclusion of the angle potential increases the stability and bending rigidity of the bilayers [18, 20].

The interior of the vesicle is filled with hydrophilic beads, modeling an aqueous solution. We measure the outer (or inner) radius of the vesicle as the average distance between the outer (or inner) lipid head beads and the center of mass of the vesicle. The obtained outer radius is 7.50 and the inner radius is 3.91.

Each Janus nanoparticle is roughly spherical and composed of 392 beads arranged in a face-centered cubic (FCC) lattice structure. The lattice constant of the FCC structure is 0.7. The particle has a radius 2.11. We assign harmonic bonds between the beads in the nanoparticle to preserve its structure. The bonds are characterized by $K_{\mathrm{bond}} = 64$ and have two equilibrium lengths $b = 0.7$ and $b = 0.7/\sqrt{2}$ (set according to the spacing between the nearest-neighbor and second nearest-neighbor beads in the FCC lattice, respectively). The composition of amphiphilic Janus particles is quantified by

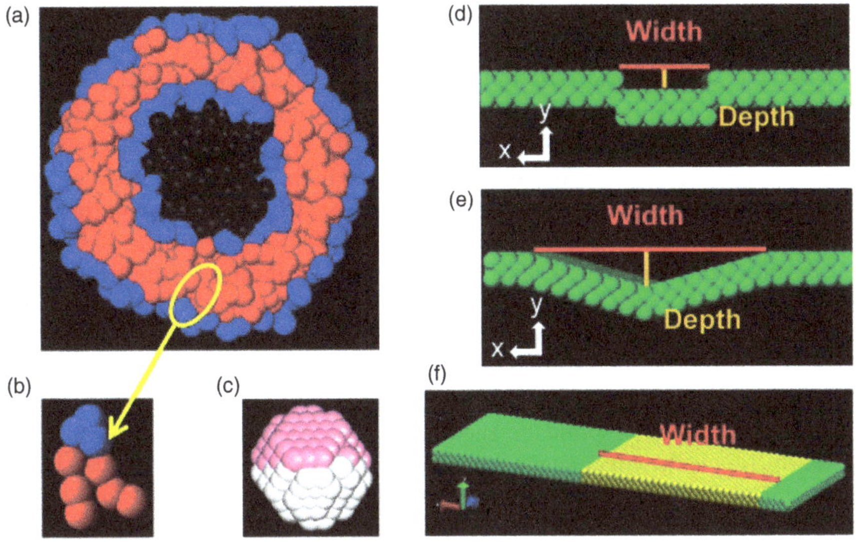

Figure 9.1 (a) Vesicle self-assembled from 586 twin-tailed lipids dispersed in a hydrophilic solvent (not displayed). Blue beads represent the hydrophilic lipid heads, red beads represent the hydrophobic lipid tails, and black beads mark the inner hydrophilic solution. (b) Coarse-grained model of twin-tailed lipid composed of three hydrophilic head beads (blue) and six hydrophobic tail beads (red). (c) Janus particle formed from 392 beads of which 196 beads are hydrophilic (white) and 196 beads are hydrophobic (pink) ($\phi = 0.500$). Beads are arranged in an FCC lattice structure with a cube side length of 0.7. (d) Step trench with width = 5.50 r_c = 3.69 nm and depth = 1.10 r_c = 0.74 nm. (e) Wedge-shaped trench with width = 16 r_c = 10.72 nm and depth = 2.20 r_c = 1.47 nm. (f) Substrate with "sticky" stripe with width = 30.25 r_c = 20.27 nm.

a parameter, ϕ, the fraction of hydrophobic beads on the particle. Figure 9.1c shows a Janus particle with $\phi = 0.5$, which is divided symmetrically into hydrophobic and hydrophilic parts. The asymmetric Janus particles with different ϕ are obtained by either replacing layers of hydrophobic beads with layers of hydrophilic beads (yielding $\phi < 0.5$) or vice versa (yielding $\phi > 0.5$), while keeping the total bead number and the structure of the particle. Here, we model a variety of asymmetric particles having $0.010 \leq \phi \leq 0.990$.

A flat substrate lies in the x–z plane at $y = 0$, which models the floor of the fluidic chamber. It consists of 5940 frozen hydrophilic beads arranged in an FCC lattice structure with a number density $\rho \approx 3$. The corresponding lattice constant is 1.3. To construct topographically patterned substrate, we create a trench aligned parallel to the z-axis by shifting the substrate beads in the negative y-direction. The trench has two distinct shapes. The step trench

(Figure 9.1d) is formed by shifting all substrate beads with x-coordinates between −2.75 and 2.75 down by 1.1. This results in a trench with a width of $w = 5.50$ and a depth of $d = 1.1$. The wedge-shaped trench has a width of $w = 16$ and a depth of $d = 2.2$ (Figure 9.1e). Its shape is characterized by the equation $y_{\text{trench}} = y_{\text{flat}} - d(1 - |x - x_c|/(w/2))$ for beads with $12 \leq x \leq 28$. Here, y_{trench} is the new y-coordinate of the bead in the trench, y_{flat} is the original y-coordinate of the bead in the flat substrate, x is the x-coordinate of the bead, and $x_c = 20$ is the center of the trench.

Furthermore, we assume that the substrate is adhesive to the nanoscopic vesicle. Namely, each hydrophilic bead in the vesicle that is in close vicinity of the surface can form bonds with the substrate beads within a critical distance (e.g. $r_{ij} < r_c$). This bond is modeled using a Hookean spring potential $F_{\text{sub/lipid}} = -K_{\text{sub/lipid}}(r_{ij} - b)$ truncated at the critical distance r_c [21, 22]. In the potential, b corresponds to the equilibrium bond length (set at $b = 0.5$) and $K_{\text{sub/lipid}}$ is the effective strength of the interaction. These bonds can break when their length exceeds the critical value (when $r_{ij} > r_c$), while new bonds can form whenever the distance between a hydrophilic vesicle bead and a surface bead is less than r_c.

In the presence of an applied shear flow, some fraction of these bonds will break. Importantly, however, the imposed flow also brings the vesicle in contact with new sites further along the surface and, hence, these components can form new bonds with the substrate. Given the number of beads that lie within a distance of $r_{ij} < r_c$ from the surface, these bonding interactions effectively create an adhesive force that prevents the vesicle from detaching from the substrate, but nonetheless allows the fluid-driven vesicle to slide along the substrate. Using this approach, we also introduce a similar adhesive interaction between the hydrophilic beads of the Janus nanoparticle and the substrate, $F_{\text{sub/Janus}}$ with effective strength $K_{\text{sub/Janus}}$.

As already mentioned, we introduce surface topographical or chemical heterogeneity to enable the fluid-driven vesicles to deliver nanoparticles to specified sites on a substrate. For the topographically patterned surface, it is taken to be chemically homogeneous in order to isolate the effect of topography. Hence, the parameters that represent chemical interactions between the surface and the components in the system, $K_{\text{sub/lipid}}$ and $K_{\text{sub/Janus}}$, are held constant. We specify the strength of the adhesion between the vesicle and the substrate at $K_{\text{sub/lipid}} = 4$, which yields an adhesion energy per unit area of approximately 10^{-2} J m^{-2}. The adhesion energy is on the same scale as that reported in other coarse-grained MD simulations for supported lipids [23, 24], as well as MD simulations with fully atomistic representations [25]. Notably, this value is also of the same order of magnitude as experimental values of the adhesion energy of ligand-receptor bonds [26]. The strength of the adhesion between the Janus particle and the surface is set in the range of $8 \leq K_{\text{sub/Janus}} \leq 12$ and uniform along the entire substrate.

In a separate series of simulations, we introduce a "sticky" stripe on the flat substrate as shown in Figure 9.1f. Compared to the rest of the substrate, the Janus particles experience a stronger adhesion on the stripe. The adhesive interaction between the patch and the particle is characterized by $K_{\text{sub/Janus}} = 12$; notably, the adhesion between the patch and the vesicle is kept the same as that for the rest of the surface (i.e. $K_{\text{patch/lipid}} = K_{\text{sub/lipid}} = 4$). Similar to the trenches, a stripe with a width of 30.25 is aligned parallel to the z-axis and perpendicular to the flow direction.

The typical size of the simulation box is $60.5 \times 40.0 \times 19.8$ with periodic boundary conditions imposed in the lateral (x- and z-) directions. The dimensions of the box in x- and z-directions are chosen to match the lattice constant of the substrate. The simulation box is filled with hydrophilic solution beads to represent an aqueous solution. The vesicles and Janus particles are driven to move along the substrate by a shear flow applied in the positive x-direction. The shear flow is imposed according to the Lees-Edwards periodic boundary conditions by displacing the images of the periodic simulation box at two boundaries, $y = 10$ and $y = -10$ [27]. The resulting flow has a velocity in the x-direction of a magnitude given by $V_{\text{sh}} = \dot{\gamma}\, y$, where $\dot{\gamma}$ is the shear rate. Below, we focus on shear rates within the range $0.015 \leq \dot{\gamma} \leq 0.035$. Above these rates, the imposed shear decreases the stability of the vesicle and prompts the rupture of the vesicle.

The hydrophilic/hydrophobic interactions between the beads are captured by specifying the interaction parameters, a_{ij}, between the components. Our choices for a_{ij} are as follows: for any two beads with the same hydrophobicity, we set the repulsion parameter to be $a_{ij} = 25$ (measured in units of $k_B T/r_c$); the interaction parameter between hydrophobic and hydrophilic moieties is specified by $a_{ij} = 100$. These values were used in a number of previous DPD simulations of lipid bilayers [18–20, 28, 29] and allow the lipids used in our model to spontaneously self-assemble into vesicles with lipid bilayer membranes.

The simulation parameters can be related to physical length and time scales through comparisons with experimental measurements of the properties of a tensionless dipalmitoylphosphatidylcholine (DPPC) membrane [18]. Typical experiments of a flat DPPC membrane in a tensionless state gives an equilibrium area per lipid of approximately 0.6 nm^2 and an in-plane diffusion constant of lipids as $D = 5\ \mu\text{m}^2/\text{s}$ [18]. These values can be used to obtain the DPD length scale of $r_c = 0.67$ nm and DPD time scale of $\tau = 7.2$ ns. Hence, a single time step in our simulations corresponds to $\Delta t = 0.02\tau = 0.14$ ns. The radius of the Janus particle and the outer radius of the lipid vesicle are approximately 1.4 and 5.0 nm, respectively. The step trench has a physical width of 3.69 nm and depth of 0.74 nm. The wedge trench is significantly larger that the step trench, which has a width of 10.72 nm and a depth of 1.47 nm. The physical width of the "sticky" stripe is 20.27 nm. Both trenches and stripe span the entire length of the substrate in the z-direction, which is 13.27 nm.

According to the physical length and time scales, the range of the shear rates in our system is $2.08 \times 10^6 \sim 4.86 \times 10^6$ s^{-1}; these shear rates are on the same order of magnitude as those used in high shear rate experiments in microfluidic devices [30]. The resulting velocity of a particle moving along the substrate is on the order of 5 mm/s, which is within the range of operating velocities in a variety of microfluidic devices [31]. (The latter value corresponds to a dimensionless velocity of 0.05 in the simulations.) We estimate the Peclet number in our system at these velocities from the in-plane diffusion coefficient of the vesicle D_v and the outer radius of the vesicle as the characteristic length scale. We obtain $D_v = 0.2$ μm^2/s, which is consistent with the experimentally measured value for the diffusion coefficient of a vesicle tethered to a lipid bilayer [32]. The Peclet number, $Pe = LU/D_v$, in our system is approximately 100, and thus, the motion of the vesicle is dominated by advection.

The above correlations between the simulation parameters and physical values provide useful guidelines for experimentally realizing the fluid-driven vesicles as nanocarriers for transporting and depositing nanoparticles. In particular, the nanoscale vesicles have been synthesized by several robust methods [4, 5], and sub-20 nanometer Janus particles can be formed through a range of experimental techniques [33–36]. The substrate could be functionalized with ligands [37, 38] that have an affinity to the vesicle surface. This adhesion is the characteristics of the ligand-receptor interactions noted above [26] and, thus, could localize the vesicles on the substrate [37, 38] until the imposed shear flow dislodges them. Through the appropriate choice of hydrophilic moieties on the Janus particles [39], the ligands on the substrate could also exhibit an affinity to the hydrophilic regions of these particles and provide adhesion between the particle and substrate. A variety of microcontact printing techniques [40] can be used to make the chemically patterned surfaces needed for depositing particles. Namely, via the latter method, the desired ligands, which have a stronger binding to the hydrophilic portions of the Janus particles compared to those on the rest of the surface, can be printed at a specific region of the surface to form the patch. On the other hand, the sub-10 nm trenches on chemically homogenous substrates can be fabricated by electron-beam lithography [41, 42].

In summary, by tailoring the chemical functionality of the nanoscopic vesicles, and Janus particles, and the chemical composition and topography of substrates, researchers would have a wide parameter space to choose from in order to experimentally match the simulation parameters. Moreover, with the different tools and techniques mentioned above, the critical components (lipid vesicles that are nanoscopic in size, Janus particles with nanometer diameters and surfaces with nanoscopic chemical or topographical patterns) for realizing this cargo-carrier system experimentally are available.

9.3 Results and Discussion

9.3.1 Selective Pick-Up of a Single Particle

9.3.1.1 Symmetric Janus Particles and Pure Hydrophilic Particles

We first determine the conditions under which a nanoscale lipid vesicle can "pick up" and transport nanoparticles along the substrate under the imposed flow. Starting with the simplest carrier-cargo system with a minimal number of components, we model the interaction between a lipid vesicle and a single symmetric Janus particle with $\phi = 0.5$ on a flat and chemically homogeneous substrate as the reference case. The interaction between the substrate and the vesicle (particle) is characterized by the bond constant of $K_{\text{sub/lipid}} = 4$ ($K_{\text{sub/Janus}} = 8$) for a shear rate of $\dot{\gamma} = 0.015$.

The symmetric Janus particle and lipid vesicle are initially placed on the substrate with the distance between the centers of the particle and the vesicle being 22.5. Hence, at early times, the vesicle is at some distance from the Janus particle and the hydrophobic portion of the particle is exposed to the hydrophilic solution (see Figure 9.2a). Once the shear flow is imposed, the vesicle is driven to move along the substrate (from right to left in Figure 9.2). When the vesicle makes contact with the particle, the hydrophobic half of the particle penetrates into and remains lodged in the bilayer lipid membrane (Figure 9.2b). This behavior can be explained from enthalpic considerations. The choice of repulsion parameter a_{ij} detailed above indicates that the hydrophobic–hydrophobic interaction between the hydrophobic particle beads and the lipid tails are more energetically favorable than the hydrophobic–hydrophilic interaction between the particle beads and the outer solution. Thus, the hydrophobic region of the particle is shielded from the hydrophilic environment when it becomes immersed in the hydrophobic inner region of the vesicle membrane.

After the particle binds to the vesicle, the shear in the positive x-direction continues to propel the vesicle along the surface. The vesicle moves over the particle, which eventually becomes localized on the right side of the carrier (see Figure 9.2c,d). The vesicle-particle assembly then moves as a unit along the surface for the remainder of the simulation.

The inset in Figure 9.2d shows the steady-state morphology of the vesicle-particle assembly. It is clear that the Janus nanoparticle is localized in the outer layer of the membrane, with the hydrophobic portion being inside the bilayer, and the hydrophilic portion being in contact with the outer solution. Notably, additional energetic input is necessary to drive the Janus particle

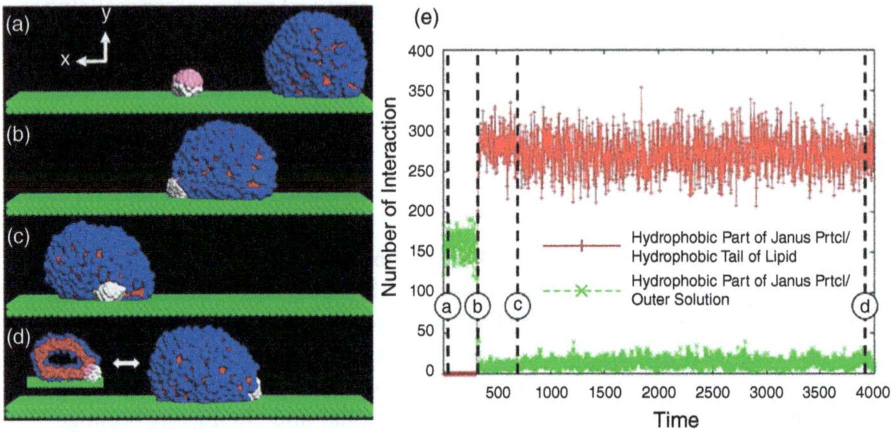

Figure 9.2 Snapshots of the system with a symmetric Janus particle (i.e. with fraction of hydrophobic beads $\phi = 0.500$) for shear rate $\dot{\gamma} = 0.015$. The substrate beads are colored as green. Here, we set $K_{\text{sub/lipid}} = 4$ and $K_{\text{sub/Janus}} = 8$ and the shear flow is in positive x-direction (from right to left). The snapshots are taken at times (a) 50, (b) 330, (c) 680, (d) 3940. The inset in (d) is the cross-sectional view of the vesicle and the attached Janus particle, which shows that the hydrophobic part of the Janus particle is absorbed into the hydrophobic part of the vesicle membrane. (e) Temporal evolution of the number of interactions between the hydrophobic beads of the Janus particle and the outer solution beans (green crosses) and between the hydrophobic beads of the Janus particle and the hydrophobic lipid tail beads (red pluses). The annotations with the letters a–d correspond to the frames (a–d), respectively.

across the interior hydrophilic layer of the membrane, and thus, promotes its complete internalization within the vesicle. When the fluid-driven vesicle and particle come into contact, the moving vesicle could impart sufficient kinetic energy to propel the particle through the entire membrane. Recently, Arai et al. performed DPD simulations of a Janus nanoparticle colliding with a nanoscopic lipid vesicle [43]. The latter studies reported the influence of the contact velocity on the last-stage morphology of the assembly. In particular, the internalization only happens for a certain range of contact velocities. If the contact velocity is too small, the kinetic energy is insufficient to drive the particle to penetrate the membrane. In our simulations, the fluid-driven vesicle and the Janus particle "collide" on the substrate with a contact velocity that is approximately equal to the velocity of vesicle. The after-collision morphology in our simulations resembles the "outer surface" state in Ref. [43]. For the shear rates considered here, the contact velocities fall in the range that yield the "outer surface" state [44] and thus, are too small to induce the internalization of the Janus nanoparticle.

To quantify the process of the vesicle membrane absorbing the hydrophobic portion of the Janus particle seen in Figure 9.2a–d, we measured the number of hydrophobic Janus particle bead/lipid tail interactions and the number of hydrophobic Janus particle bead/outer solution bead interactions as functions of time. The number of interactions is related to the number of contacts and is defined as the number of pairs of respective type of beads that are separated by a distance smaller than r_c [19]. Through the time evolutions of these two types of interactions (see Figure 9.2e), we can determine quantitatively when the vesicle picks up the particle and how much of the hydrophobic portion of the particle is immersed in the vesicle. As the Janus particle in our simulation encounters the vesicle at $t \sim 320$ (labeled as b on Figure 9.2e), the hydrophobic part of the particle gets absorbed almost instantaneously into the hydrophobic interior of the vesicle membrane. The sharp increase in the number of hydrophobic particle bead/lipid tail interactions indicates that the adsorption occurs rapidly, taking approximately 0.18 μs. Because the vesicle membrane shields the hydrophobic beads of the Janus particle from the outer hydrophilic solution, the number of hydrophobic Janus particle bead/outer solution bead interactions is significantly decreased.

Figure 9.2e also shows that the number of each type of interaction fluctuates around a constant value after the particle is adsorbed by the vesicle. This means that most of the hydrophobic half of the Janus particle remains well shielded from the outer hydrophilic solution by the vesicle's lipid membrane, as seen in the inset of Figure 9.2d. In other words, the Janus particle-vesicle assembly remains stable once it has been formed and can be successfully transported by the imposed shear flow. We find that the absorption of the symmetric Janus particle by the vesicle is robust for shear rates in the range $0.015 \leq \dot{\gamma} \leq 0.035$ and interaction strengths between the particle and the substrate of $6 \leq K_{sub/Janus} \leq 12$.

To quantitatively study the motion of the Janus particle-vesicle assembly (such as that in Figure 9.2d), we measured the speed of the attached Janus particle from the particle's trajectory aligning with the flow direction. By respectively varying the shear rate and adhesion strength between the hydrophilic particle beads and the substrate in the range of $6 \leq K_{sub/Janus} \leq 12$ and $0.015 \leq \dot{\gamma} \leq 0.035$, we observed that the speed of the particle motion increases at higher shear rates and decreases at greater adhesion strengths. In effect, increasing the magnitude of imposed flow allows the vesicle-particle to move faster under strong propulsion, while the adhesion between the particle and the substrate induces a friction force that slows down the motion of the assembly.

To examine how variations in the chemical composition of the nanoparticle affect the particle pick-up, we replaced the symmetric Janus particle with a particle composed solely of hydrophilic beads. As anticipated from the above discussion, our simulations show that the vesicle cannot pick up a purely hydrophilic particle. The hydrophilic particle is not absorbed into the bilayer lipid membrane because the hydrophilic beads prefer remaining in contact with the hydrophilic solvent and hydrophilic substrate to interacting with the hydrophobic lipid tails.

9.3.1.2 Asymmetric Janus Particles

We have demonstrated above that the symmetric Janus particle can be adsorbed and robustly transported by the moving vesicle, while the purely hydrophilic particle remained unattached. To determine the fraction of hydrophobic beads within the Janus particle, ϕ, necessary for the particle to be picked up by the fluid-driven vesicle, we systematically varied ϕ in the range $0 < \phi < 0.5$. The lowest fraction considered here, $\phi = 0.010$, corresponds to a particle with only four hydrophobic beads. We performed four independent runs for each value of ϕ.

We find that for a shear rate of $\dot{\gamma} = 0.015$ and bond constants of $K_{sub/lipid} = 4$ and $K_{sub/Janus} = 8$, all particles with the nonzero fractions of hydrophobic beads considered here (within the range $0 < \phi < 0.5$) are adsorbed and carried by the vesicle (all four runs). Even the highly asymmetric particle with $\phi = 0.010$ behaves the same as the symmetric Janus particle described above. As the vesicle encounters the particle, the hydrophobic beads penetrate into the hydrophobic interior of the lipid membrane, where they are shielded from the external hydrophilic solvent. After the particle is adsorbed, it is dragged behind the vesicle for the remainder of the simulation, with its hydrophobic beads remaining in the more energetically favorable position inside the lipid bilayer.

To better understand the pick-up process, we further explore the influences of the shear rate and the particle-substrate adhesion strength in the respective ranges of $0.015 \leq \dot{\gamma} \leq 0.035$ and $6 \leq K_{sub/Janus} \leq 12$. We generate a phase map shown in Figure 9.3 from these simulations, which presents the results of all four independent simulations for each data point. At the lowest shear rate, $\dot{\gamma} = 0.015$, the particle is picked up and transported by the vesicle quite robustly for each $K_{sub/Janus}$ and ϕ examined. Notable exceptions occur in some runs involving the $\phi = 0.010$ particle at $K_{sub/Janus} = 10$ and 12; in these cases, the highly asymmetric and predominately hydrophilic particle is either not picked up (one run at $K_{sub/Janus} = 10$) or the particle is picked up and later dropped by the vesicle (one run with $K_{sub/Janus} = 12$). In general, Figure 9.3a shows that the low shear rate facilitates the transport of particles by the moving vesicle.

As the shear rate is increased to $\dot{\gamma} = 0.025$, the particles are picked up for $K_{sub/Janus} = 8$ for all ϕ considered here and the vesicle-Janus particle assembly is robustly transported along the substrate. When the adhesion strength between the substrate and particle becomes $K_{sub/Janus} = 10$, the fate of the $\phi = 0.010$ particle is significantly affected. In one of the four runs, the particle is not picked up. In the other three runs, it is picked up, but ultimately detaches from the nanocarrier. At this combination of shear rate and adhesion strength, all other Janus particles with $\phi > 0.010$ are picked up and carried along by the moving

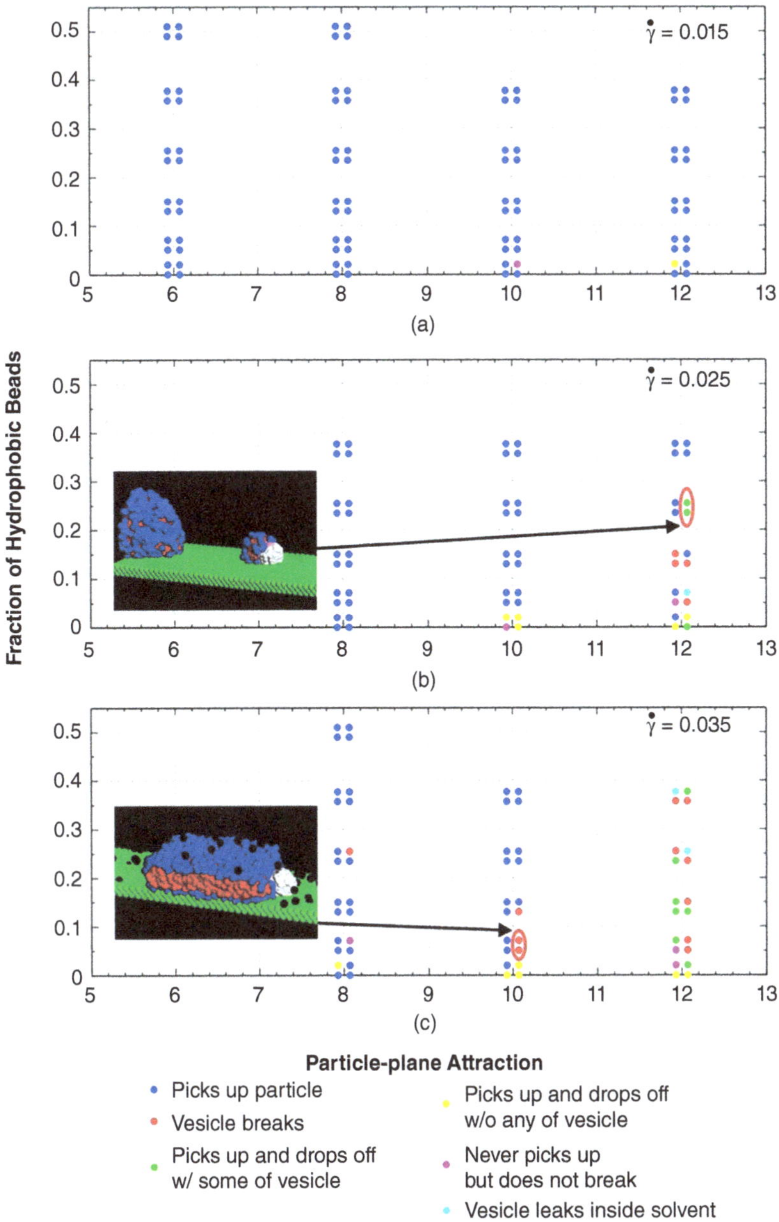

Figure 9.3 Phase diagram depicting the simulation results of a lipid vesicle picking up single Janus particle with varying fraction of hydrophobic beads ϕ and substrate adhesion $K_{sub/Janus}$ for shear rates: (a) $\dot{\gamma} = 0.015$, (b) $\dot{\gamma} = 0.025$, and (c) $\dot{\gamma} = 0.035$. Inset in (b) shows the snapshot of a simulation in which the attached particle is dropped off with torn lipids binding to its hydrophobic portion. Inset in (c) shows the complete rupture of the vesicle.

vesicle. When $K_{sub/Janus}$ is increased to 12, the robust pick up is only observed when the particle encompasses a fraction of hydrophobic beads $\phi \geq 0.365$.

The phase map also demonstrates the occurrence of possible catastrophic events in the system. Namely, the combined effects of shear and interactions with the surface can significantly distort the vesicle, and as a result, the internal fluid of the vesicle can gradually leak out through the membrane. The extreme situation is that the torn vesicle cannot self-repair and breaks entirely, releasing all internal fluid. A comparison of Figure 9.3a,b highlights the damaging effects of increasing the shear rate. As shown in the inset of Figure 9.3b, we also observed an interesting scenario where a few lipids from the membrane detached from the vesicle and bound themselves to the hydrophobic portion of the particle that was dropped off. After the drop-off, the vesicle recovered to its original closed morphology without leaking any internal fluid during the tearing of the lipids. Covered by these lipids, the hydrophobic portion of the deposited particle is protected from interacting with the hydrophilic solution. The details of this behavior are discussed in the next section.

With further increases in $\dot{\gamma}$, the catastrophic events become more prevalent. At $\dot{\gamma} = 0.035$, Figure 9.3c shows instances at $K_{sub/Janus} = 8$ where the vesicle not only drops its cargo but also undergoes catastrophic breakage. The vesicle assumes a pancake-like morphology that is caused by the vesicle-surface adhesion (see the inset of Figure 9.3c). For $K_{sub/Janus} = 10$ and 12, successful pick-up events occur only sporadically, and there is a significant probability that the vesicles are destroyed in the process of carrying its cargo.

Summarizing the data in the phase map, we find that the vesicle is prone to pick up a particle with a greater fraction of hydrophobic beads (i.e. larger ϕ) because the favorable enthalpic interaction scales with the size of the hydrophobic portion. The value of $K_{sub/Janus}$ determines the "stickiness" of the substrate to the particle. Larger $K_{sub/Janus}$ leads to greater friction between the particle and the surface and, thus, inhibits successful pick-up. Similarly, higher shear rates are also detrimental to the pick-up process. Here, the greater difference in the relative velocities between the mobile vesicle and the particle when these two species make contact can lead to their subsequent separation, even if the hydrophobic portion of the particle becomes immersed in the lipid membrane. The separation tears the vesicle and may further induce the aforementioned catastrophic events. Finally, we note that the speed of the attached particle is not greatly influenced by its fraction of hydrophobic beads because the number of interactions between the hydrophilic particle beads and the substrate, and the number of interactions between the hydrophilic lipid heads of the vesicle and the substrate do not depend significantly on ϕ.

9.3.2 Interaction Between Multiple Particles and a Lipid Vesicle

Bottom-up assembly of the nanoparticles necessitates the ability of the vesicles to controllably transport more than just one particle to a particular location. To investigate the interaction between the vesicle and a larger payload, we placed a number of Janus particles with $\phi = 0.37$ at equal distances from each other on the substrate. (Namely, the initial x- and z- coordinates of the N_{p} Janus particles on the substrate are separated by the distances $3L_x/4N_{\mathrm{p}}$ and $L_z/4$, respectively, where L_x and L_z are the box sizes in the x- and z- directions.) For the multiple particle simulations, the width of the simulation box is extended to accommodate the additional Janus particles and prevent the deformed vesicle from forming a continuous vesicular bilayer structure in the periodic z-direction. The dimensions of the simulation box in the respective x-, y-, and z-directions are now $60.5 \times 40.0 \times 29.7$, with periodic boundary conditions being applied in the x- and z-directions.

We stipulate the criteria of a successful pick up as the following: (i) "grab" the particles, (ii) transport the particles along the surface, (iii) no leakage of the inner solution, and (iv) no vesicle rupture. The event is classified as unsuccessful pick-up when any of these four requirements are not met. When all four independent simulations for a given number of particles on the substrate exhibit a successful pick-up, we consider the pick-up to be robust for this number of particles. We investigated three cases with $N_{\mathrm{p}} = 4$, 6, and 8. After 22 000 dimensionless units of time, the only scenario that displayed such robust behavior was the case involving four particles, as shown in Figure 9.4a–e. The case having six particles yielded one unsuccessful pick-up out of four runs, and all runs displayed unsuccessful pick-ups for eight particles. It can be anticipated that simulations involving an even greater number of particles would be unsuccessful in meeting all four of the criteria noted above.

While the deformation of a vesicle is visible even in the case of just one adsorbed Janus particle, it is particularly pronounced in the cases involving multiple adsorbed particles. To quantitatively characterize this morphological deformation, we calculate the relative shape anisotropy, κ^2, which is given by $\kappa^2 = \frac{(\lambda_x-\lambda_y)^2+(\lambda_y-\lambda_z)^2+(\lambda_z-\lambda_x)^2}{2R_g{}^4}$, where λ_x, λ_y, and λ_z are the eigenvalues of the gyration tensor $\mathbf{S}$, with $S_{mn} = \frac{1}{N}\sum_{i=1}^{N} r^i_m r^i_n$ [44]. Here, N is the total number of lipid beads (hydrophobic and hydrophilic) and $\mathbf{r}^i$ is the position vector of the center of the ith bead with respect to the center of mass of the vesicle. The radius of gyration, R_g, is calculated as $R_g^2 = \lambda_x + \lambda_y + \lambda_z$ [45]. The relative shape anisotropy for an ideal sphere is 0, and that for an ideal rod is 1. We note that the radius of gyration of the self-assembled vesicle (Figure 9.1a) prior to its

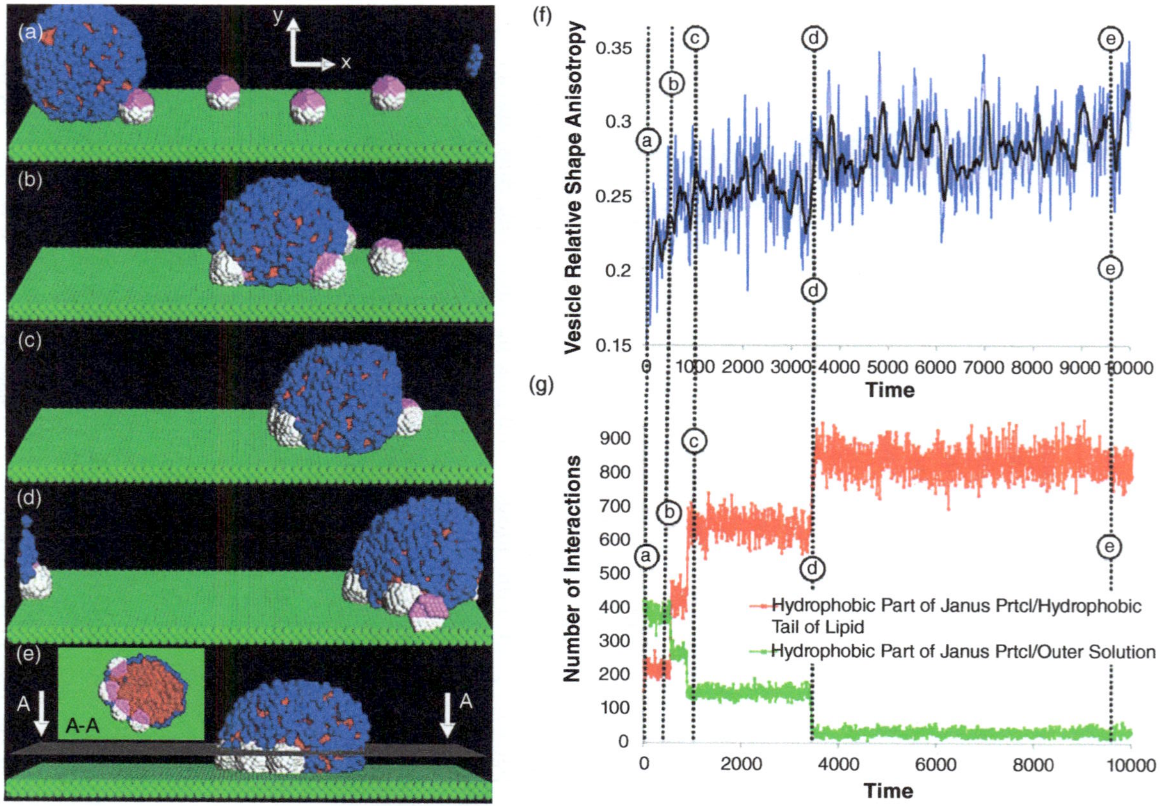

Figure 9.4 Snapshots of the system with four Janus particles with $\phi = 0.367$ taken at times (a) 0, (b) 558, (c) 894, (d) 3444, and (e) 9960. Substrate adhesion is set as $K_{sub/Janus} = 8$. Shear rate $\dot{\gamma}$ is 0.015, and shear flow is in the positive x-direction (from left to right). The inset in (e) is the cross-sectional view of the vesicle-Janus particle assembly for the cutting plane A–A. (f) Relative shape anisotropy of the vesicle with four Janus particles on the plane. (g) Number of interactions between hydrophobic beads of Janus particle and lipid tails (in red) and between hydrophobic beads of Janus particle and solvent (in green). The lines represent times (a) 0, (b) 558, (c) 894, (d) 3444, and (e) 9960, corresponding to snapshots (a–e).

interaction with the substrate is $R_g = 6.03$, which is calculated using the above formula. This value is in a good agreement with the theoretical expression for a spherical shell, $R_g^2 = \frac{3(R_o^5 - R_i^5)}{5(R_o^3 - R_i^3)}$, where $R_o = 7.50$ and $R_i = 3.91$ are the outer and inner radii of the vesicle.

To quantify the process of picking up of multiple particles for the scenario shown in Figure 9.4a–e, we examine the time evolution of the relative shape anisotropy and the total number of relevant interactions. Much as we did for the case of a single particle, we calculate the number of interactions between hydrophobic lipid tail beads and hydrophobic Janus particle beads, as well as the number of interactions between hydrophobic Janus particle beads and outer solution beads. Here, the total number of interactions involves a summation of the interactions on each Janus particle. Via these measurements, we can determine the time at which a particle is absorbed by the mobile vesicle.

The relative shape anisotropy of the vesicle, κ^2, is plotted as a function of time in Figure 9.4f; the numbers of interactions are plotted for the same time scale in Figure 9.4g. The letters on the plots correspond to the snapshots in Figure 9.4a–b. The vesicle responds to each newly picked particle with a corresponding change in κ^2. After all four particles are picked up, the structure of the whole assembly remains the same until the end of the simulation, at $t = 22\,000$. We compare the late-time value of κ^2 for a vesicle with four particles to the corresponding value for a vesicle with zero particles, calculated using the average over the last 100 frames for all four runs. Note that values of κ^2 remain relatively constant at these late times. The average final value of κ^2 for the case with four particles is 0.28 ± 0.02, as compared to 0.22 ± 0.03 in the case with zero particles. (The relatively large value of κ^2 for a vesicle without the particles is a result of the deformation induced by both the adhesion from the substrate and the external flow; when this vesicle is not bound on the surface and immersed in the solution in the absence of flow [as in Figure 9.1a], it retains a nearly spherical shape with $\kappa^2 \approx 0.075$.)

In the cases involving multiple Janus particles, these particles tend to aggregate into a large cluster on the substrate before they are adsorbed by the vesicle. By forming these clusters, some of the hydrophobic beads from multiple particles come into close contact, and thus, can be shielded from the hydrophilic solution and substrate. Notably, it is more difficult for the vesicle to pick up an entire cluster than it is to pick up a single particle. When the vesicle picks up each particle sequentially, the time separation between each pick-up allows the vesicle to relax the strain that occurred during the particle absorption, and the vesicle remains intact. But when the vesicle attempts to absorb a cluster of particles simultaneously, the sudden distortion is significant. The large instantaneous stress on the lipid membrane could readily cause either leakage or rupture of the vesicle, and it is more probable that the pick-up will be unsuccessful.

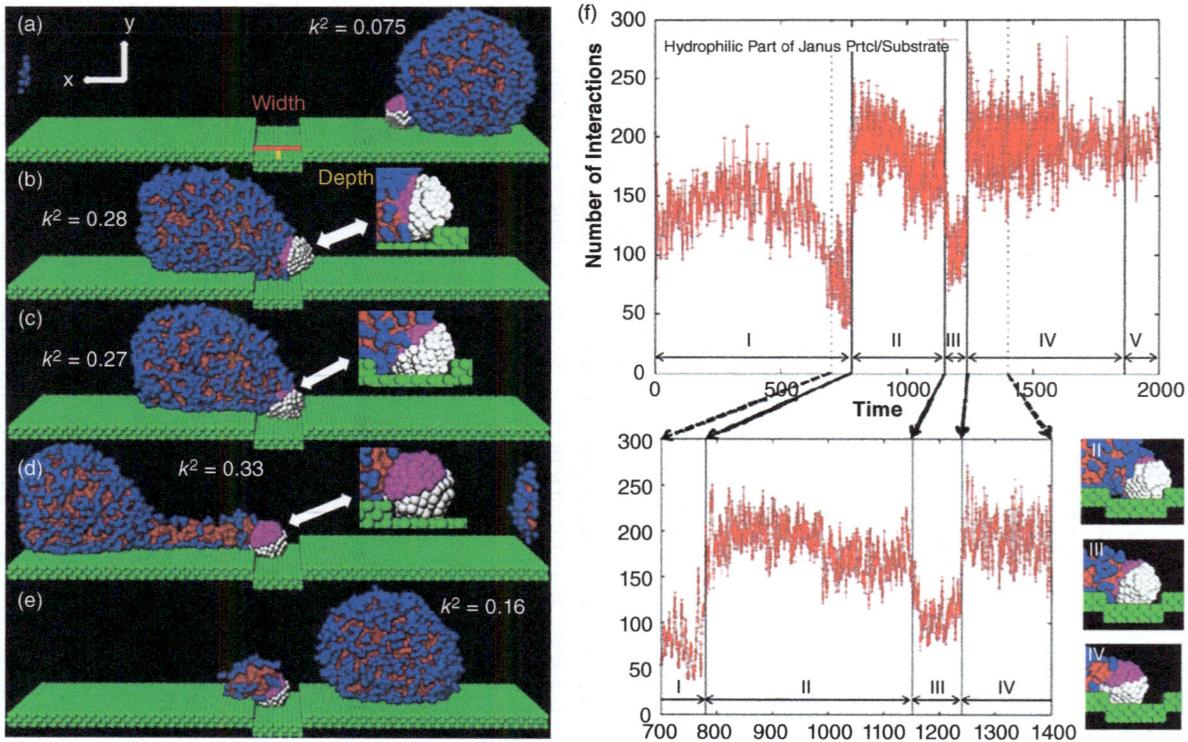

(a)
y
x
k^2 = 0.075
Width
Depth
(b)
k^2 = 0.28
(c)
k^2 = 0.27
(d)
k^2 = 0.33
(e)
k^2 = 0.16
(f)
Hydrophilic Part of Janus Prtcl/Substrate
Number of Interactions
Time
I
II
III
IV
V
0
50
100
150
200
250
300
500
1000
1500
2000
700
800
900
1000
1100
1200
1300
1400

9.3.3 Depositing Janus Particles on Patterned Surfaces

In the above sections, we pinpointed the range of shear rates, substrate adhesion strengths, and fractions of hydrophobic beads on the Janus particle for which the vesicle can pick up the particle and the intact vesicle-particle assembly is successfully transported along the surface. To fully exploit the potential of this nanoscopic cargo-carrier, it is important to achieve the controlled release or drop-off of the attached Janus particle. Potential strategies for realizing a controlled drop-off involve either tuning the shear rate or introducing substrate patterns. Our preliminary results show that, for a particular substrate adhesion, increasing the shear rate provides robust drop-offs only for the particles that encompass the lowest fraction of hydrophobic beads, $\phi = 0.010$. When the particles have a larger hydrophobic region, the vesicle is easily ruptured at high shear rates and the particle is dropped off at random positions on the surface. In other words, variations in $\dot{\gamma}$ do not provide a reliable means of achieving controllable particle drop-off. Therefore, we turn to the strategy associated with introducing patterned surfaces by changing substrate topography or adhesion. The idea of introducing heterogeneity on a surface to control the motion of microcapsules has been implemented in a few previous studies [45–47]. Both mechanically (e.g. different stiffness) and chemically (e.g. different adhesion) patterned surfaces were utilized to effectively regulate the motion.

9.3.3.1 Step Trench

Here, we first introduce a step trench and explore the behavior of the vesicle-particle assembly when it encounters the trench. Figure 9.5 displays a typical simulation involving the Janus nanoparticle, the vesicle, and the step trench. The fraction of hydrophobic beads on the Janus particle is $\phi = 0.367$, and

Figure 9.5 Snapshots of the system with one Janus particle with $\phi = 0.367$ for the step trench with $K_{sub/Janus} = 8$. Shear rate $\dot{\gamma}$ for all the simulations is set at 0.015, and shear flow is in the positive x-direction (from right to left). Substrate beads are marked in green. Particle is initially placed near the vesicle to facilitate particle pick it up. Snapshots are taken at the following times: (a) 0, (b) 840, (c) 894, (d) 1920, and (e) 2340. The κ^2 value in each snapshot is the instantaneous relative shape anisotropy of the vesicle. (f) Temporal evolution of the number of interactions between the hydrophilic beads of the Janus particle and the substrate. Regions I–V on the graph correspond to different phases of the drop-off process. In Region I ($t \approx 0$ to $t \approx 780$), the particle is being transported by the vesicle along the flat substrate. In Region II ($t \approx 780$ to $t \approx 1150$), the particle is in contact with the first edge of the trench. In Region III ($t \approx 1150$–1240), the particle has separated from the first edge of the trench and is being dragged along the flat bottom of the trench. In Region IV ($t \approx 1240$–1864), the particle is in contact with the second edge of the trench but is still connected to the vesicle. In Region V ($t \approx 1864$–2000), the particle has been deposited by the vesicle in the trench with a protective coating and remains affixed to the second edge of the trench. The three insets labeled II, III, and IV correspond to detailed view of particle deposition at the respective times $t \approx 900$, $t \approx 1160$, and $t \approx 1300$ in Regions II, III, and IV.

the particle-surface adhesion strength is $K_{sub/Janus} = 8$; the latter value is held constant for all simulations having this step trench. As the flow-driven vesicle encounters the particle (Figure 9.5a), the vesicle readily picks up the Janus particle and the vesicle-particle assembly then moves as a single unit along the surface, as was shown in the previous sections. When this assembly passes over the trench, the particle becomes lodged in the depression. While the relatively large vesicle can traverse the trough, it is highly distorted due to the trapped particle. A part of the membrane forms a narrow neck of lipids that extends from the hydrophobic portion of the Janus particle (Figure 9.5d). Ultimately, the force from the imposed flow causes this neck to break away from the particle. A small number of the membrane lipids, however, remain bound to the hydrophobic region on the Janus particle (Figure 9.5e), similar to the scenario in the inset of Figure 9.3b.

Notably, the vesicle recovers and continues moving along the substrate (Figure 9.5e). In effect, the vesicle undergoes a self-healing process. We examine the self-healing behavior by comparing the shape of the vesicle before and after the particle drop-off, again quantified by the relative shape anisotropy of the vesicle, κ^2. The instantaneous κ^2 values for the respective time frames are displayed in Figure 9.5a–e. The value of κ^2 reaches its maximum value right before the lipid "neck" becomes detached from the trapped particle (around the time shown in Figure 9.5d). Once the particle has been deposited in the trench, the relative shape anisotropy decreases significantly, as seen in Figure 9.5e, $\kappa^2 = 0.16$. According to the steady-state, time-averaged value of $\kappa^2 = 0.22 \pm 0.03$ of a shear-driven vesicle moving on a flat, adhesive surface (in the absence of particles) with $K_{sub/lipid} = 4$ and $\dot{\gamma} = 0.015$, we can conclude that the vesicle essentially resumes its normal shape after the particle deposition.

The reason why the particle becomes trapped in the trench can be understood from the changes of the number of interactions as the assembly moves over the surface. Figure 9.5f shows a sharp jump in the particle-substrate interaction count when the particle comes into contact with the first edge of the trench at $t \approx 780$ (the beginning of Region II) and the second edge of the trench at $t \approx 1240$ (start of Region IV). This increase leads to an increase in the total adhesive force acting on the particle (as calculated from $F_{sub/Janus}$). It is worth noting that in these DPD simulations, the solvent induces a depletion attraction [48–51] between the nanoparticle and walls of the trench that also contributes to the localization of the particle in the trough.

Shifting our focus to the lipids that remain bound to the deposited Janus nanoparticle, we note that these lipids help shield the hydrophobic portion of the particle from the hydrophilic solution. With the hydrophobes on the lipid tails localized on the hydrophobic sites on the particle, the hydrophilic lipid head groups extend away from the Janus sphere and thus, effectively mask the underlying hydrophobic moieties. The degree to which the particle is covered by this protective coating depends on ϕ, the fraction of hydrophobic beads on the particle ϕ. To quantify the effectiveness of this coating, we first determine the total number of hydrophobic particle beads that are separated from a lipid tail bead by one unit. The latter distance serves as the cutoff radius; beads separated by more than one unit are not considered in this count. For each value of ϕ, this total is calculated by first averaging the number of covered hydrophobic beads over several time steps for a given run and then averaging those values for all independent runs with a successful drop-off of coated particles. This average number of covered beads is then divided by the number of hydrophobic particle beads within a cutoff radius of one from the surface of the particle. Beads farther inside the particle cannot interact with beads at the particle's surface, so they are excluded from the calculations.

Figure 9.6 shows the fraction of hydrophobic sites on the Janus sphere that are coated by the lipids as a function of ϕ. For relatively low values of ϕ, the lipids cover the majority of these hydrophobes (see inset in Figure 9.6). As ϕ is increased, the coverage is less perfect and more hydrophobic beads are exposed to the outer solution. In particular, the coverage changes from 0.92 to 0.17 as ϕ is varied from 0.06 to 0.86. We find that the maximum number of lipids that are bound to particle (over the range of ϕ) is approximately equal to 25. Because

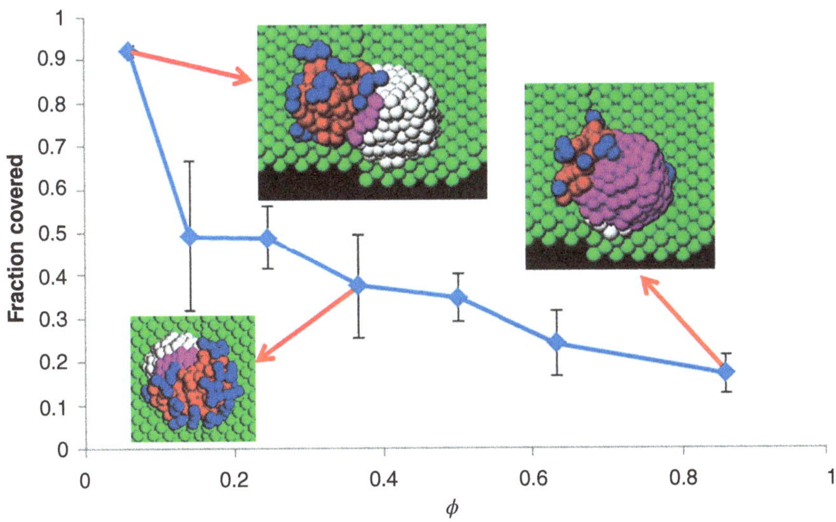

Figure 9.6 Dependence of fraction of hydrophobic particle beads covered by a lipid tail bead as a function of fraction of hydrophobic beads ϕ for the step trench with $K_{sub/Janus} = 8$ at shear rate $\dot{\gamma} = 0.015$. Data were obtained by first averaging over several time frames within each simulation with successful drop-off with a protective coating and then averaging these values for all the successful simulations with different random seeds at each value of ϕ. Snapshots show examples of particle coverage at different ratios. The error bars arise from averaging over three independent runs.

the vesicle in this simulation is self-assembled from 586 lipids, the maximum loss of vesicle mass is estimated to be less than 5%. The limited loss of vesicle mass preserves the integrity of the vesicle and allows the vesicle to be reused in successive transport events.

9.3.3.2 Wedge Trench

To generalize our findings beyond this specific topography, we also conducted simulations with a wedge-shaped trench. The major difference between two trenches is the lack of sharp vertical edges in the wedge. We first focus on a simulation run that yields successful drop-off, as shown in Figure 9.7. The particle has a chemical composition of $\phi = 0.061$, and the particle–surface adhesion strength is $K_{\text{sub/Janus}} = 10$. When the vesicle-particle assembly encounters and moves across the trench, the particle becomes localized in the center of the wedge, where the particle–surface contact area (interaction count) is maximized. This increase in the number of particle–surface interactions drives the successful drop-off of the particle. Notably, the particle deposited inside the wedge is also coated with a protective lipid layer. The latter coating is present in successful drop-offs for $\phi \geq 0.061$.

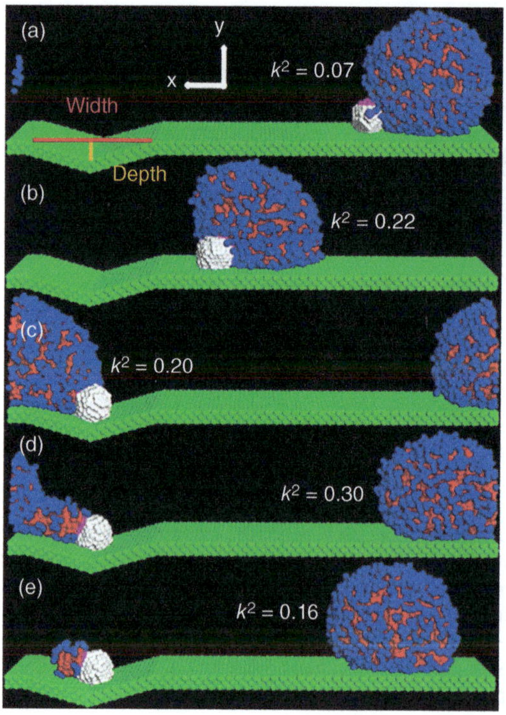

Figure 9.7 Snapshots of the system with one Janus particle with $\phi = 0.061$ for the wedge-shaped trench with $K_{\text{sub/Janus}} = 10$. The shear rate is $\dot{\gamma} = 0.015$. Snapshots are taken at the following times: (a) 0, (b) 500, (c) 1220, (d) 1400, and (e) 1800. The instantaneous relative shape anisotropy κ^2 of the vesicle is labeled in each snapshot.

As in the case of the step trench, once the lipid-coated particle becomes deposited in the trough, the torn vesicle undergoes self-healing, as evidenced by the images in Figure 9.7 and the decrease in κ^2. We note that the deformation of the carrier vesicle is generally smaller in the case of the wedge than the step trench, as can be seen by comparing Figures 9.5d and 9.7d. Importantly, particle drop-off with a protective lipid coating and vesicle self-healing were seen with both types of trenches, indicating that the behavior is robust and not dependent on the shape of the trenches considered here.

We then systematically study the single particle drop-off on the step and wedge-shaped trenches; four independent simulation runs were performed for each value of ϕ and $K_{sub/Janus}$. For the wedge-shaped trench, we performed simulations with a smaller range of ϕ values (with $0.010 \leq \phi \leq 0.5$) since Figure 9.8 indicates relatively few successful drop-off events beyond $\phi = 0.2$. We did, however, investigate higher particle-surface attraction strengths, focusing on $8 \leq K_{sub/Janus} \leq 12$. We speculate that due to weaker depletion forces between the particle and the wedge compared to the step trench, stronger particle–surface bonds are necessary to induce particle drop-off.

Notably, for $K_{sub/Janus} = 8$, the specific results depend on the trench shape. For the step trench, the particle is deposited in the trough with a protective lipid coating for every simulation with $0.010 < \phi < 0.5$ and every successful drop-off with larger ϕ. For the wedge-shaped trench, however, particle drop-off with a protective coating only occurs in two runs with $\phi = 0.061$. To induce more reliable drop-off in the wedge, it is necessary to use a higher value of the particle-substrate adhesion strength. For $K_{sub/Janus} = 10$, drop-off is much more successful, as the particle is deposited in the trench with a protective coating for all the simulations with $\varphi = 0.140$ and in three of the four runs with $\varphi = 0.061$. With the highest adhesion strength of $K_{sub/Janus} = 12$, drop-off is still more frequent but also more sporadic. Hence, the most consistent and predictable drop-off behavior for the wedge-shaped trench occurs for the particle–surface adhesion strength of $K_{sub/Janus} = 10$.

Comparing the results for the wedge-shaped trench at $K_{sub/Janus} = 10$ to those for the step trench at $K_{sub/Janus} = 8$, we find that the step trench is effective over a wider range of particle compositions, ϕ, than the wedge geometry. Namely, the step trench can induce drop-off of particles with larger ϕ and requires lower particle–surface adhesion strength to perform this function. We attribute the efficacy of the step trench to the increase in the number of particle–surface interactions (relative to a flat surface) (see Figure 9.5f) and the induced depletion attraction between the particle and the vertical wall of the trough. The depletion force acting on the particle maintains the particle affixed to these walls and effectively introduces extra friction opposing the particle's forward motion. Notably, the depletion force between the particle and an edge of the step trench could explain the occasional drop-off at the first edge encountered.

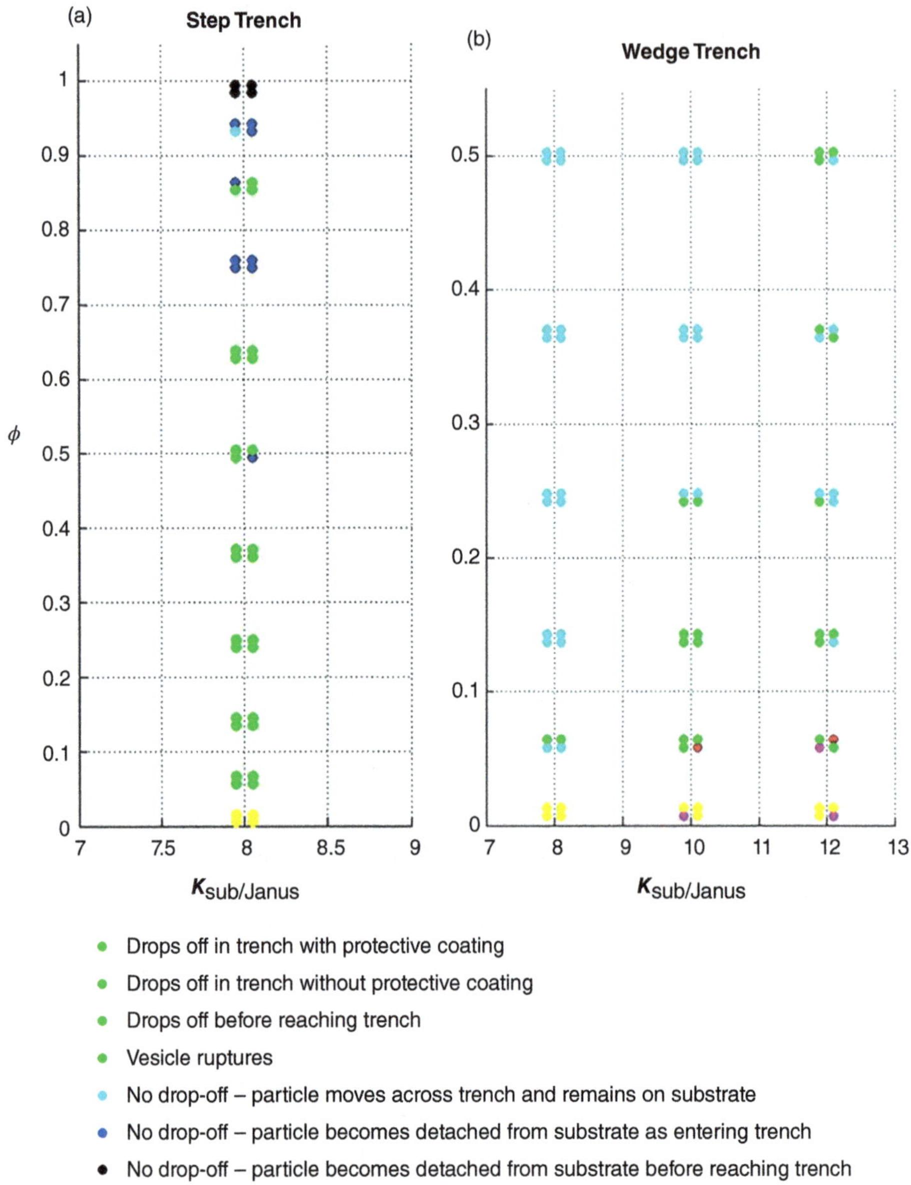

Figure 9.8 Phase diagram for drop-off of a single particle using (a) step trench and (b) wedge-shaped trench. Shear rate $\dot{\gamma}$ is held constant at 0.015. The particle–substrate adhesion $K_{sub/Janus}$ and fraction of hydrophobic beads ϕ are varied.

In contrast, for the wedge trench, drop-off occurs primarily because of the increase in the number of particle–substrate interactions (relative to the flat surface). The solvent-induced depletion attraction is less effective in the case of these slanted walls. Because the increase in the number of interactions dominates the deposition in the wedge-shaped trench, successful drop-off requires a lower fraction of hydrophobic beads ϕ and a higher particle–plane attraction strength $K_{\text{sub/Janus}}$ than those for the step trench. The wedge trench can, however, still induce drop-off much more effectively than a flat and chemically uniform substrate. Specifically, we have shown that a vesicle will not deposit a Janus particle on a flat surface at a shear rate of $\dot{\gamma} = 0.015$ when $\varphi \geq 0.061$ for any adhesion strength in the range $6 \leq K_{\text{sub/Janus}} \leq 12$.

9.3.3.3 "Sticky" Stripe

Instead of utilizing surface topography to facilitate particle deposition, we now explore the possibility of achieving robust drop-off by introducing a stripe on a flat surface on which the Janus particle experiences a stronger adhesive interaction than on the rest of the surface. At the start of these simulations, we directly introduced the vesicle-particle assembly in front of the stripe on the heterogeneous substrate, as shown in Figure 9.9a–e. When the fluid-driven vesicle-particle assembly encounters the stripe, we observe that the Janus particle is arrested at the second patch-substrate interface, as shown in Figure 9.9c, while the vesicle continues to move forward along the substrate. Similar to the cases with trenches, once the particle is localized at the patch–substrate interface, the continually moving vesicle is drastically deformed and forms a lipid neck connecting it with the hydrophobic portion of the Janus particle (see insets of Figure 9.9c,d). The neck eventually ruptures under the imposed flow and leaves a few lipids covering the hydrophobic sites of the particle.

In contrast to the cases on the topographically patterned surfaces, the drop-off on the "sticky" patch is attributed to the sharp adhesion gradient at the patch-substrate interfaces, which also corresponds to discontinuities in the potential. When the particle meets the second interface, the potential barrier induces a net enthalpic force in the direction opposite to the motion and hence, inhibits the movement of the particle. The interaction between the Janus particle and the potential barrier eventually yields the drop-off. The underlying physics of this particle deposition process also played a role in separating motile capsules as they moved along a heterogeneous surface [47].

The phase map in Figure 9.9f indicates how the fraction of hydrophobic beads, ϕ, the adhesion strength between the patch and the particle, $K_{\text{patch/Janus}}$, and the shear rate, $\dot{\gamma}$, affect the drop-off process. When $K_{\text{patch/Janus}}$ is relatively weak, the potential barrier at the second patch/substrate interface is too shallow to induce the drop-off of the particle. The whole assembly passes the yellow patch and keeps moving along the substrate. As $K_{\text{patch/Janus}}$ is increased to 11.5, making the patch more "sticky" for the particle, the potential barrier

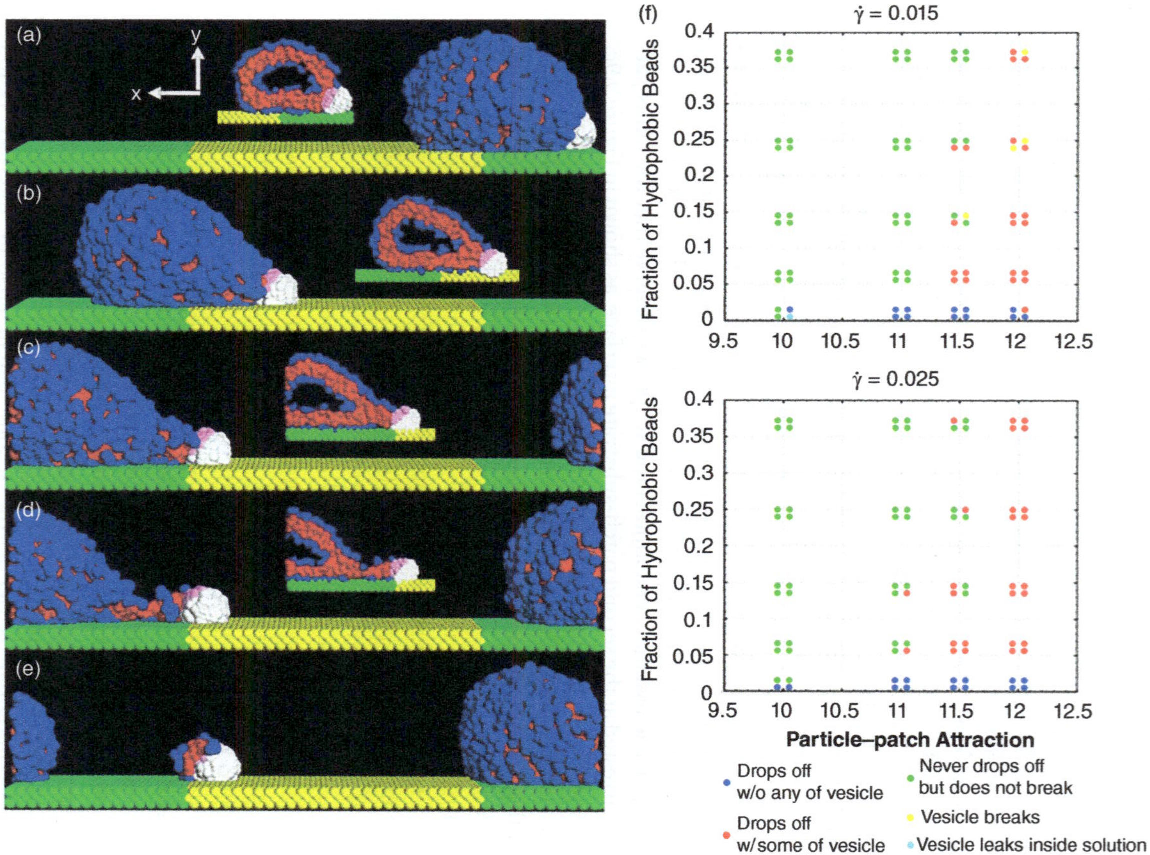

Figure 9.9 Snapshots of the system in which the vesicle drops off a Janus particle with a fraction of hydrophobic beads $\phi = 0.245$ on a "sticky" stripe for shear rate $\dot{\gamma} = 0.015$ and bond constants $K_{sub/lipid} = K_{patch/lipid} = 4$, $K_{sub/Janus} = 8$, and $K_{patch/Janus} = 12$. The snapshots are taken at times (a) 308, (b) 1000, (c) 2100, (d) 3092, (e) 3256. The insets are the cross-sectional views of the vesicle-Janus particle assembly. (f) Phase diagrams for a lipid vesicle dropping of single Janus particle on a "sticky" stripe. The shear rates are set at $\dot{\gamma} = 0.015$ and 0.025.

can impose a sufficiently strong force to arrest the particle. This force results in particles being successfully dropped off at the second edge of the stripe with a protective coating for all the simulations with $\varphi = 0.061$ and in some runs with $\varphi = 0.140$ and 0.245. Particles characterized by higher values of ϕ experience a large drag force from the vesicle that overcomes the opposing enthalpic force, and consequently, the particle passes the patch. Finally, our simulations show a relatively minor effect of varying the imposed shear rates within the range of the shear rates considered herein.

9.4 Conclusions

To provide effective guidelines for utilizing lipid vesicles as cargo carriers in micro- and nanofluidic devices, we performed computer simulations to probe the interactions between a nanoscopic vesicle and spherical Janus particles in the presence of an imposed flow. The vesicle and Janus nanoparticles were placed on a hydrophilic substrate and immersed in a hydrophilic solution. The favorable enthalpic interactions between the hydrophobic sites on the Janus particle and the hydrophobic interior of the lipid bilayer enabled the fluid-driven vesicle to successfully adsorb or "pick up" the nanoparticle. We systematically varied the chemical composition of the nanoparticles, the adhesion strength between the nanoparticle and substrate, and the shear rate, and consequently, identified the appropriate range of parameters that leads to robust carrier-cargo assembly, with the hydrophobic portion of the particle being localized within the hydrophobic domain of the bilayer. In general, this self-assembly, and hence, optimal pick-up is achieved in scenarios involving the following: Janus particles with a large hydrophobic component, low substrate adhesion, and low shear rates. We also examined the payload capacity of the nanoscale vesicle by simulating the interaction between a vesicle and multiple Janus particles, and observed that these vesicles can successfully pick up to four particles.

We then considered scenarios that would lead to controlled drop-off of the adsorbed particles, so that the vesicles could provide a means of controllably delivering nanoparticles to specified locations within the fluidic chambers. Specifically, we designed chemically homogenous substrates with nanoscale trenches and chemically heterogeneous substrates that encompass a "sticky" stripe, which has a stronger adhesive interaction with the particle. The efficacy of the trenches in trapping the particles is due primarily to the increased number of particle–substrate interactions, while the "sticky" stripe facilitates the drop-off by imposing an adhesion gradient. We isolated the range of parameters with which the vesicle can drop off the particles and remain intact as it keeps moving along the surface.

Importantly, the findings also reveal remarkable synergistic interactions between the vesicles and deposited particles that lead not only to the shielding of the particle from the outer solvent by a coating of membrane lipids but also the reforming of the disrupted vesicle. This behavior was observed to be quite robust, occurring on both topographically and chemically patterned surfaces. In effect, the vesicle provides a protective layer for the delivered particles and undergoes a self-healing process. In a generalized sense, these attributes mimic the sensory and responsive functionality exhibited by biological systems. This biomimetic behavior could have important consequences for biomedical applications, including the delivery of hydrophobic drugs, which need to be masked by a hydrophilic coating for efficient uptake.

Overall, our findings provide insight into harnessing fluid-driven vesicles to controllably transport nanoscopic cargo in nano- and microfluidic devices. The self-mending vesicle can potentially perform this function multiple times and, thus, could be used repeatedly, which would play an important role in facilitating the bottom-up assembly of nanoparticle structures. The results also suggest that the vesicles could be used to collect and remove debris from the lower surface of nano- and microfluidic devices [52].

Acknowledgments

The authors gratefully acknowledge support from the DOE, under grant number DE FG02 90ER45438. We also acknowledge helpful conversations with Prof. Todd Emrick.

References

1 Nicolas, J., Mura, S., Brambilla, D. et al. (2012). Design, functionalization strategies and biomedical applications of targeted biodegradable/biocompatible polymer-based nanocarriers for drug delivery. *Chem. Soc. Rev.* 42: 1147–1235.
2 Browne, K.P., Walker, D.A., Bishop, K.J.M. et al. (2010). Self-division of macroscopic droplets: partitioning of nanosized cargo into nanoscale micelles. *Angew. Chem. Int. Ed.* 49: 6756–6759.
3 Hood, R. R., Kendall, E. L., DeVoe, D.L. et al. (2013). Microfluidic formation of nanoscale liposomes for passive transdermal drug delivery. *Microsystems for Measurement and Instrumentation (MAMNA)*. pp. 12, 15, 14–14 May 2013
4 Jahn, A., Stavis, S.M., Hong, J.S. et al. (2010). Microfluidic mixing and the formation of nanoscale lipid vesicles. *ACS Nano* 4: 2077–2087.

5 Lee, J., Lee, M.G., Jung, C. et al. (2013). High-throughput nanoscale lipid vesicle synthesis in a semicircular contraction-expansion array microchannel. *BioChip J.* 7: 210–217.
6 Chung, S.E., Park, P., Shin, S. et al. (2008). Guided and fluidic self-assembly of microstructures using railed microfluidic channels. *Nat. Mater.* 7: 581–587.
7 Yang, K. and Ma, Y.Q. (2010). Computer simulation of the translocation of nanoparticles with different shapes across a lipid bilayer. *Nat. Nanotechnol.* 5: 579–583.
8 Dutt, M., Nayhouse, M.J., Kuksenok, O. et al. (2011). Interactions of end-functionalized nanotubes with lipid vesicles: spontaneous insertion and nanotube self-organization. *Curr. Nanosci.* 7: 699–715.
9 Vácha, R., Martinez-Veracoechea, F.J., Frenkel, D. et al. (2011). Receptor-mediated endocytosis of nanoparticles of various shapes. *Nano Lett.* 11: 5391–5395.
10 Van Lehn, R.C. and Alexander-Katz, A. (2011). Penetration of lipid bilayers by nanoparticles with environmentally-responsive surfaces: simulations and theory. *Soft Matter* 7: 11392–11404.
11 Yue, T. and Zhang, X. (2011). Molecular understanding of receptor-mediated membrane responses to ligand-coated nanoparticles. *Soft Matter* 7: 9104–9112.
12 Ding, H., Tian, W., and Ma, Y. (2012). Designing nanoparticle translocation through membranes by computer simulations. *ACS Nano* 6: 1230–1238.
13 Hoogerbrugge, P.J. and Koelman, J.M.V.A. (1992). Simulating microscopic hydrodynamic phenomena with dissipative particle dynamics. *Europhys. Lett.* 19: 155–160.
14 Koelman, P.J. and Hoogerbrugge, J.M.V.A. (1993). Dynamic simulations of hard-sphere suspensions under steady shear. *Europhys. Lett.* 21: 363–368.
15 Español, P. and Warren, P. (1995). Statistical mechanics of dissipative particle dynamics. *Europhys. Lett.* 30: 191–196.
16 Español, P. (1995). Hydrodynamics from dissipative particle dynamics. *Phys. Rev. E* 52: 1734–1742.
17 Groot, R.D. and Warren, P.B. (1997). Dissipative particle dynamics: bridging the gap between atomistic and mesoscopic simulation. *J. Chem. Phys.* 107: 4423–4435.
18 Smith, K.A., Jasnow, D., and Balazs, A.C. (2007). Designing synthetic vesicles that engulf Nanoscopic particles. *J. Chem. Phys.* 127: 084703.
19 Dutt, M., Kuksenok, O., Little, S.R. et al. (2011). Forming transmembrane channels using end-functionalized nanotubes. *Nanoscale* 3: 240–250.
20 Shillcock, J.C. and Lipowsky, R. (2002). Equilibrium structure and lateral stress distribution of amphiphilic bilayers from dissipative particle dynamics simulations. *J. Chem. Phys.* 117: 5048–5061.

21 Maresov, E.A., Kolmakov, G.V., Yashin, V.V. et al. (2012). Modeling the making and breaking of bonds as an elastic microcapsule moves over a compliant substrate. *Soft Matter* 8: 77–85.
22 Roiter, Y., Ornatska, M., Rammohan, A.R. et al. (2009). Interaction of lipid membrane with nanostructured surfaces. *Langmuir* 25: 6287–6299.
23 Hoopes, M.I., Longo, M.L., and Faller, R. (2011). Computational modeling of curvature effects in supported lipid bilayers. *Curr. Nanosci.* 7: 716–720.
24 Hoopes, M.I., Deserno, M., Longo, M.L. et al. (2008). Coarse-grained modeling of interactions of lipid bilayers with supports. *J. Chem. Phys.* 129: 175102.
25 Heine, D.R., Rammohan, A.R., Balakrishnan, J. et al. (2007). Atomistic simulations of the interaction between lipid bilayers and substrates. *Mol. Simul.* 33: 391–397.
26 Israelachvili, J.N. (2011). *Intermolecular and Surface Forces*, 3e. Elsevier.
27 Lees, A.W. and Edwards, S.F. (1972). Computer study of transport processes under extreme conditions. *J. Phys. C: Solid State Phys.* 5: 1921.
28 Alexeev, A., Upsal, W.E., Balazs, A.C. et al. (2008). Harnessing Janus nanoparticles to create controllable pores in membranes. *ACS Nano* 2: 1117–1122.
29 Dutt, M., Kuksenok, O., Nayhouse, M.J. et al. (2011). Modeling the self-assembly of lipids and nanotubes in solution: forming vesicles and bicelles with transmembrane nanotube channels. *ACS Nano* 5: 4769–4782.
30 Pipe, C.J. and McKinley, G.H. (2009). Microfluidic rheometry. *Mech. Res. Commun.* 36: 110–120.
31 Squires, T.M. and Quake, S.R. (2005). Microfluidics: fluid physics at the nanoliter scale. *Rev. Mod. Phys.* 77: 977–1026.
32 Yoshina-Isii, C., Chan, Y.-.H.M., Johnson, J.M. et al. (2006). Diffusion dynamics of vesicles tethered to a fluid supported bilayer by single-particle tracking. *Langmuir* 22: 5682–5689.
33 Wang, B., Li, B., Zhao, B. et al. (2008). Amphiphilic Janus gold nanoparticles via combining "solid-state grafting-to" and "grafting-from" methods. *J. Am. Chem. Soc.* 130: 11594–11595.
34 DeVries, G.A., Brunnbauer, M., Hu, Y. et al. (2007). Divalent metal nanoparticles. *Science* 315: 358–361.
35 Kim, B.J., Bang, J., Hawker, C.J. et al. (2007). Creating surfactant nanoparticles for block copolymer composites through surface chemistry. *Langmuir* 23: 12693–12703.
36 Lattuada, M. and Hatton, T.A. (2007). Preparation and controlled self-assembly of Janus magnetic nanoparticles. *J. Am. Chem. Soc.* 129: 12878–12889.
37 Jung, L.S., Shumaker-Parry, J.S., Campbell, C.T. et al. (2000). Quantification of tight binding to surface-immobilized phospholipid vesicles using surface

plasmon resonance: binding constant of phospholipase A_2. *J. Am. Chem. Soc.* 122: 4177–4184.

38 Voskuhl, J., Wendeln, C., Versluis, F. et al. (2012). Immobilization of liposomes and vesicles on patterned surfaces by a peptide coiled-coil binding motif. *Angew. Chem. Int. Ed.* 51: 12616–12620.

39 Lattuada, M. and Hatton, T.A. (2011). Synthesis, properties and applications of Janus nanoparticles. *Nano Today* 6: 286–308.

40 Ruiz, S.A. and Chen, C.S. (2007). Microcontact printing: a tool to pattern. *Soft Matter* 3: 168–177.

41 Xia, Y., Rogers, J.A., Paul, K.E. et al. (1999). Unconventional methods for fabricating and patterning nanostructures. *Chem. Rev.* 99: 1823–1848.

42 Manfrinato, V.R., Zhang, L., Su, D. et al. (2013). Resolution limits of electron-beam lithography toward the atomic scale. *Nano Lett.* 13: 1555–1558.

43 Arai, N., Yasuoka, K., and Zeng, X.C. (2013). A vesicle cell under collision with a Janus or homogeneous nanoparticle: translocation dynamics and late-stage morphology. *Nanoscale* 5: 9098–9100.

44 Noguchi, H. and Gompper, G. (2006). Dynamics of vesicle self-assembly and dissolution. *J. Chem. Phys.* 125: 164908.

45 Alexeev, A., Verberg, R., and Balazs, A.C. (2007). Patterned surfaces segregate compliant microcapsules. *Langmuir* 23: 983–987.

46 Usta, O.B., Alexeev, A., and Balazs, A.C. (2007). Fork in the road: patterned surfaces direct microcapsules to make a decision. *Langmuir* 23: 10887–10890.

47 Kolmakov, G.V., Yashin, V.V., Levitan, S.P. et al. (2011). Designing self-propelled microcapsules for pick-up and delivery of microscopic cargo. *Soft Matter* 7: 3168–3176.

48 Yodh, A.G., Lin, K., Crocker, J.C. et al. (2001). Entropically driven self–assembly and interaction in suspension. *Philos. Trans. R. Soc. London, Ser. A* 359: 921–937.

49 Boek, E.S. and Van Der Schoot, P. (1998). Resolution effects in dissipative particle dynamics simulations. *Int. J. Mod. Phys. C* 9: 1307–1318.

50 Whittle, M. and Dickinson, E. (2001). On simulating colloids by dissipative particle dynamics: issues and complications. *J. Colloid Interface Sci.* 242: 106–109.

51 Dinsmore, A.D., Yodh, A.G., and Pine, D.J. (1996). Entropic control of particle motion using passive surface microstructures. *Nature* 383: 239–242.

52 Kosif, I., Chang, C.C., Bai, Y. et al. (2014). Picking up nanoparticles with functional droplets. *Adv. Mater. Interfaces* 1: 1400121. doi: 10.1002/admi.201400121.

10

Solution Self-Assembly of Giant Surfactants: An Exploration on Molecular Architectures

Xue-Hui Dong[1], Yiwen Li[1], Zhiwei Lin[1], Xinfei Yu[1], Kan Yue[1], Hao Liu[1], Mingjun Huang[1], Wen-Bin Zhang[1,2], and Stephen Z. D. Cheng[1]

[1] *The University of Akron, College of Polymer Science and Polymer Engineering Goodyear Polymer Center, Department of Polymer Science, Akron, OH 44325-3909, USA*
[2] *Peking University, Center for Soft Matter Science and Engineering, College of Chemistry and Molecular Engineering, Key Laboratory of Polymer Chemistry and Physics of Ministry of Education, Beijing 100871, PR China*

10.1 Introduction

Self-assembly and self-organization processes are ubiquitous in nature and important in our daily life [1]. The resultant ordered structures construct foundations of many complex functionalities. For example, the membranes of living cells that separate and protect the interior of the cells from the surrounding environments with selective permeability are constructed by a bilayer of self-assembled phospholipids. On the other hand, the soap bubbles originate from self-assembly of surfactants that have been widely used as detergents. These small-molecule surfactants possess the same molecular characteristics as consisting of one (or more) hydrophobic tail(s) and one (or more) hydrophilic polar/ionic head(s). The self-assembly behaviors of these small-molecule surfactants have been substantially investigated for decades [2–4]. A variety of stable micellar morphologies have been experimentally observed and theoretically understood, including spheres, cylinders, and vesicles, among others [2–4]. Over the past two decades, the study has been further extended to block copolymers, which possess similar amphiphilicity but with much larger molecular sizes [5–9]. The selective partition of hydrophobic and hydrophilic components of block copolymers results in three representative equilibrium micellar structures are analogous to their small-molecule counterparts, i.e. spheres, cylinders, and vesicles, yet the aggregates of block copolymers exhibit higher stability and durability due to their physical and mechanical properties [6, 7]. The ability of amphiphilic molecules to spontaneously form ordered structures at nanometer scale give rise to extraordinary

Self-Assembly: From Surfactants to Nanoparticles, First Edition. Edited by Ramanathan Nagarajan.

properties useful for various applications in emerging nanotechnologies such as drug delivery, nanoreactors, catalytic systems, and others [10–12].

Among the self-assembly behaviors of both small-molecule surfactants and block copolymers, there are a few common fundamental principles that can qualitatively describe the experimental observations. For example, a dimensionless packing parameter, $P = V/Al$, has been utilized to characterize the self-assembled structures of small-molecule surfactants, where l and V are the length and volume of the hydrophobic tail, and A is the equilibrium interfacial area of the polar head [13]. This parameter defines a measure of the geometrical shape of surfactant molecules, and accounts for the formation of various self-assembled morphologies [13]. Generally, surfactants with P close to 1 tend to form structures with small curvature; whereas aggregates with large curvatures are usually observed for those with P significantly deviating from 1. For example, spherical micelles form when $P < 1/3$; cylindrical micelles fall in the regime $1/3 < P < 1/2$; vesicles (bilayers with a spontaneous curvature) are produced when $1/2 < P < 1$. Inverted micelle structures can be observed when P exceeds unity. Although this packing parameter is merely a geometrical description of the surfactants, the interfacial area is critically dependent on many factors, including interactions among head, initial concentrations, solvent polarities, and so forth, and thus is closely associated with the free energy of the assemblies. Despite the difference in molecular design, this concept has also been largely adopted to explain block copolymer self-assembly in solution with some modifications [7, 14].

Recently, a category of amphiphilic molecules termed "giant surfactants" has emerged [15–17]. Similar to their small-molecule analogue, they possess compact and rigid three-dimensional (3D) molecular nanoparticles (MNPs) as hydrophilic head(s), whereas the tail components include but are not limited to flexible polymer chains [16, 17]. Based on the similar criteria of small-molecule surfactants categorization, the giant surfactants can also be divided as simple giant surfactants, giant bola-form surfactants, giant gemini surfactants, giant multiheaded/multitailed surfactants, etc. [16, 17]. These giant surfactants capture the essential molecular features of small-molecule surfactants, yet with amplified sizes comparable to block copolymers. They bridge the gap between these two classes of amphiphilic materials. It has been found that giant surfactants are able to form highly diverse, thermodynamically stable/metastable micelles in solution [18–21]. They share general self-assembly behaviors as the small-molecule surfactants and block copolymers, yet enjoy their own distinctive features [18–20]. There are several comprehensive reviews and perspectives on the precise syntheses and structural formation of giant surfactants [16, 17, 22, 23]. In this chapter, we will mainly focus on the recent progress in solution self-assemblies of the giant surfactants. This chapter is organized based on the molecular architecture of the giant surfactants to highlight its importance on their phase behaviors. Self-assembly of the giant surfactants with nonpolymeric tails will be first discussed as a foundation for further discussion. The simple giant surfactants with one head and one tail will be addressed to elucidate their unique and distinctive self-assembly behaviors. The discussion will be then expanded to giant surfactants with multiple heads/tails, which illustrate the importance of molecular architectures on the self-assembly. We will also focus on the role of composition heterogeneity of tails (e.g. block polymers) during structural formation. We are going to demonstrate that, as a unique class of amphiphiles, giant surfactants open up numerous possibilities in the fabrications of well-defined nanostructures, which are not only scientifically intriguing but also technologically relevant.

10.2 Molecular Architecture of Giant Surfactants

Typical MNPs include [24] fullerenes (C_{60}) [25], polyhedral oligomeric silsesquioxanes (POSSs) [26], polyoxometalates (POMs) [27, 28], folded globular proteins (Figure 10.1a) [22], and others. With site-selective mono-/multifunctionalization or regio-selective multifunctionalization [29, 30], the surface functionality of MNPs can be finely tuned, providing remarkable opportunities for controlling the properties of the resultant giant surfactants [16, 17]. The tail part of giant surfactants is usually flexible polymer chain(s) with random coil conformation if the molecular weight is sufficiently high. When the tails are relatively short, a wormlike model may be applied. The tails can also be of various topologies (cyclic/branched/dendritic) and have heterogeneous compositions and/or stimuli-responsive features (Figure 10.1b). In some extreme cases, the tails could also be short alkyl chains or rigid particles with exact chemical compositions. The structural complexity renders the giant surfactants abundant self-assembly behaviors and tunable functional properties.

Self-assemblies of the giant surfactants are found to be highly sensitive to their primary chemical structures. Precise control of the surface functionalities of MNP heads and the composition/architecture of polymer tails is a prerequisite for a comprehensive study of the supramolecular structure formation and resultant physical properties. Because of the large size and reduced reactivity of functional groups, it is traditionally difficult to manipulate the chemical structure at the macromolecular level with molecular precision [30]. The emergence of "click" chemistry provides a highly efficient way to construct giant surfactants that would greatly control and simplify the synthesis and purification process [31–33]. We have developed sequential/orthogonal/one-pot "click" approaches as a general methodology to synthesize giant surfactants [34–37]. Detailed description and discussion can be found in our recent review and perspectives [16, 17]. Generally speaking, with the help from the

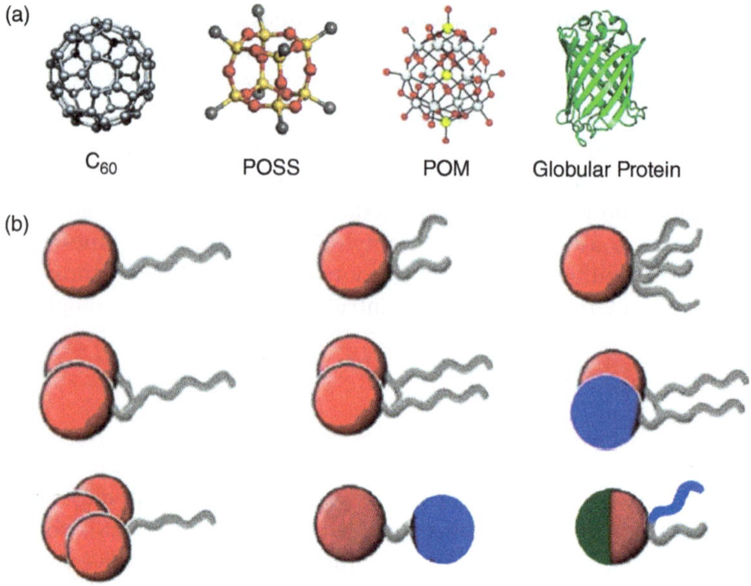

Figure 10.1 Typical MNPs (a), and giant surfactants with variable compositions and diverse topologies (b).

existing "click" toolbox, e.g. Cu(I)-catalyzed [3+2] azide–alkyne cycloaddition (CuAAC) [31, 32], strain-promoted alkyne-azide cycloaddition (SPAAC) [38], oxime ligation [39], and thiol-ene chemistry [40], and the well-developed living/controlled anionic/radical polymerization methods [41, 42], we are able to construct giant surfactants with precisely defined chemical compositions and various macromolecular architectures (Figure 10.1b) [18–21, 35, 43–49].

10.3 Giant Surfactants with Short Nonpolymeric Tails

Although deviated from classic geometry of surfactants that usually possess a compact head and a relative long tail, the conjugates composed of a hydrophilic MNP tethered by a small nonpolymeric moiety (e.g. alkyl chains or other organic ligands) can still be regarded as a member of the giant surfactant family, due to their similarity in both of molecular structures and self-assembling behaviors. Different from polymers that are polydisperse in nature, alkyl chains and organic ligands are monodisperse molecules. In this sense, the chemical composition of this class of giant surfactants is precisely defined, which provides an ideal model as the starting point for the investigation on their solution self-assembly behaviors.

One typical example is the POM-based giant surfactants consisting of hydrophilic POM(s) as head(s) and organic ligand(s) as tail(s). With site-/regio-selective mono-/multi-functionalization, organic ligands can be installed on the designated positions on POM surface. Similar to small-molecule surfactants and block copolymers, their self-assembly is determined by molecular parameters, such as surface properties of POM and molecular architectures, as well as experimental parameters including solvents, initiate concentration, etc. [50, 51]. For example, a giant surfactant composed of a Lindqvist POM head and two pyrene fluorescent probes could form vesicular structures in a mixed solvent of dimethyl sulfoxide (DMSO) and water (Figure 10.2a) [52]. Meanwhile, cylinders and spheres were observed in another type of giant surfactants with a Keggin POM and two alkylsiloxy tails (Figures 10.2b) [53]. More interestingly, POMs are highly charged MNPs and are associated with various counterions (e.g. tetrabutylammonium TBA) in solution, which provides new opportunity to manipulate their supramolecular structure formation. By substituting the counterions, different self-assembled structures can be prepared [53]. On the other hand, the interactions between POM and counterions are also sensitive to the nature of solvent. With polar solvents, the electrostatic interactions between POM head and their counterions would be screened, as evidenced by their diffusion coefficients revealed by two-dimensional (2D) nuclear magnetic resonance (NMR) spectrometry. This would enhance their hydrophilicity [52, 54]. As a result, the giant surfactants with an Anderson POM and two alkyl tails of varying lengths (Anderson-C_n) self-organized into vesicles in a mixture solvent of acetonitrile (MeCN) and water, whereas reversed vesicles were observed in a nonpolar solvent such as MeCN/toluene (Figure 10.3) [55, 56].

Recently, our group has developed another type of giant surfactants by conjugating hydrophilic and hydrophobic MNP(s), such as a carboxylic acid-functionalized hydrophilic fullerene (AC_{60}) covalently connected with one or two hydrophobic fullerene (AC_{60}–C_{60} and AC_{60}–$2C_{60}$, Figure 10.4a,b) [48]. They possess rigid and shape-persistent MNPs instead of linear flexible tails. It has been demonstrated that these rigid giant surfactants also exhibit

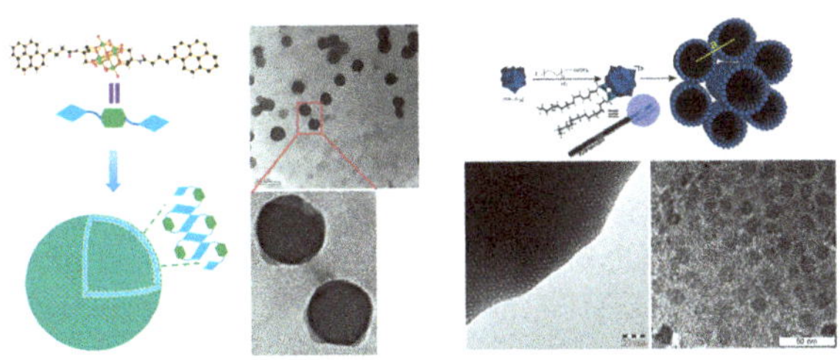

(a) (b)

Figure 10.2 Lindqvist POM-based giant surfactants with vesicular morphologies (a), and Keggin POM-based giant surfactants with cylindrical (bottom left) and spherical (bottom right) morphologies (b) [52, 53].

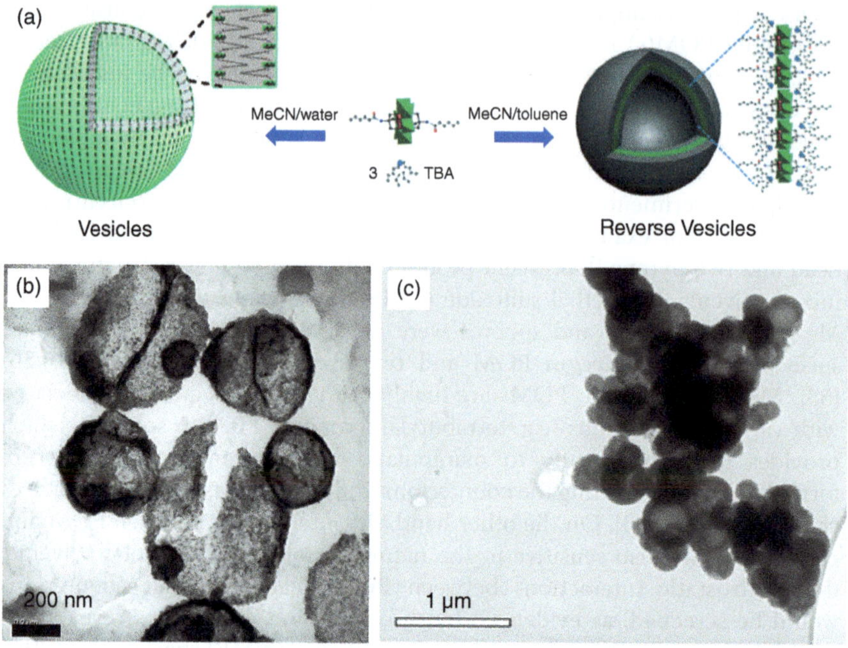

Figure 10.3 Schematic illustration (a) and corresponding bright field TEM image of Anderson-C_{16} giant surfactant with vesicular structure formed in MeCN/water (b) and reverse-vesicular structure in MeCN/toluene (c) [55, 56].

interesting self-assembly behaviors in solution. In tetrahydrofuran (THF), vesicles were observed for both AC_{60}–C_{60} and AC_{60}–$2C_{60}$, with different molecular packing (Figure 10.4c,d), as confirmed by the transmission electron microscopy (TEM) and atomic force microscopy (AFM) experimental observations. The vesicles constructed by AC_{60}–C_{60} adopt an interdigitated arrangement of hydrophobic C_{60}, while that of AC_{60}–$2C_{60}$ possess a bilayered packing. In a mixed solvent of dimethylformamide (DMF) and water, AC_{60}–C_{60} and AC_{60}–$2C_{60}$ formed spherical and cylindrical morphologies, respectively.

The architecture and solvent effects on their structural formation can be explained by the packing parameter P. In THF, ionization of carboxylic acid on AC_{60} is negligible. Hence, the P value of AC_{60}–$2C_{60}$ can be quantitatively calculated as 0.69, corresponding to a vesicular morphology ($P > 1/2$). Despite that the apparent P value of AC_{60}–C_{60} (0.35) is below this limit, the formation of vesicles becomes predictable due to the fact of the interdigitated molecular packing. In DMF/water, AC_{60} are expected to possess a larger hydrodynamic size due to the ionization of carboxylic acid groups on AC_{60}. The P value for AC_{60}–C_{60} and AC_{60}–$2C_{60}$ can be qualitatively estimated, which falls in the regime of spheres ($P < 1/3$) and cylinders ($1/3 < P < 1/2$), respectively.

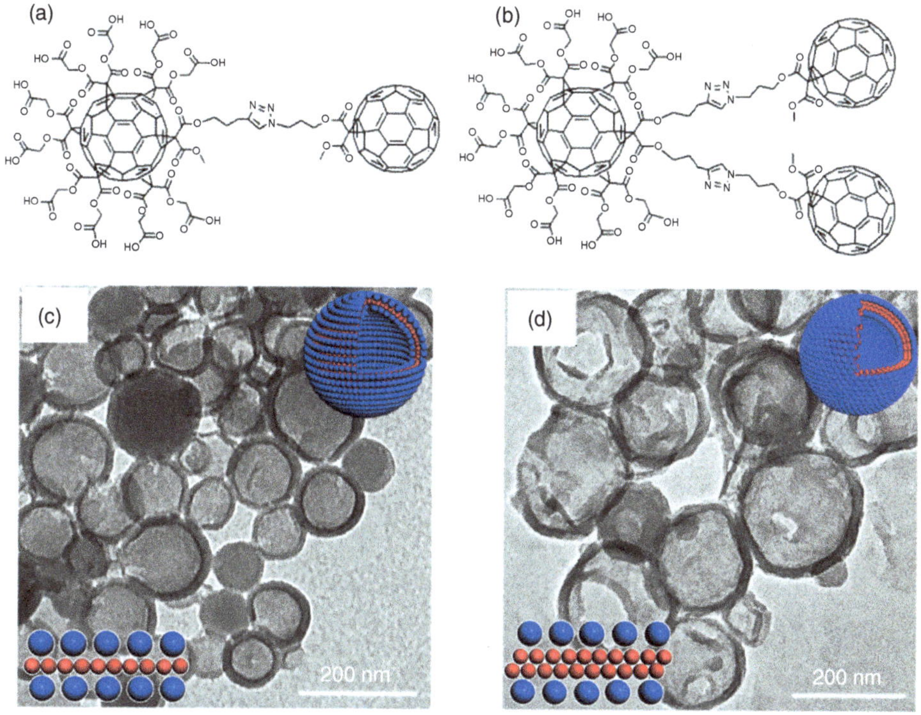

Figure 10.4 Chemical structures (a, b) and self-assembled vesicles (c, d) of AC_{60}–C_{60} and AC_{60}–$2C_{60}$. The insets are schematic illustrations of corresponding molecular packing: interdigitated (c) and bilayer (d) [48].

10.4 Giant Surfactants with a Single Head and Single Polymer Tail

Typical small molecule surfactants, such as sodium dodecyl sulfate (SDS), sodium oleate, and cetyltrimethylammonium bromide (CTAB), consist of a hydrophobic tail (usually a long hydrocarbon chain) and a hydrophilic (ionic or polar) group [57]. Correspondingly, polymer-tethered MNP with one macroionic head and one hydrophobic polymer tail represents the simplest class of giant surfactants.

The study in this area was pioneered by Nolte and co-workers using enzyme-polystyrene (PS) conjugates, which were prepared by covalently connecting a mono-maleimide-functionalized PS chain (40 repeat units) onto a lipase enzyme (Figure 10.5) [15]. Similar to small molecule surfactants, these "giant amphiphiles" could self-organize into bundles of fibers in aqueous solution, as evidenced by TEM observations (Figure 10.5). These fibers are actually the aggregates of cylindrical micelles. This study stimulates

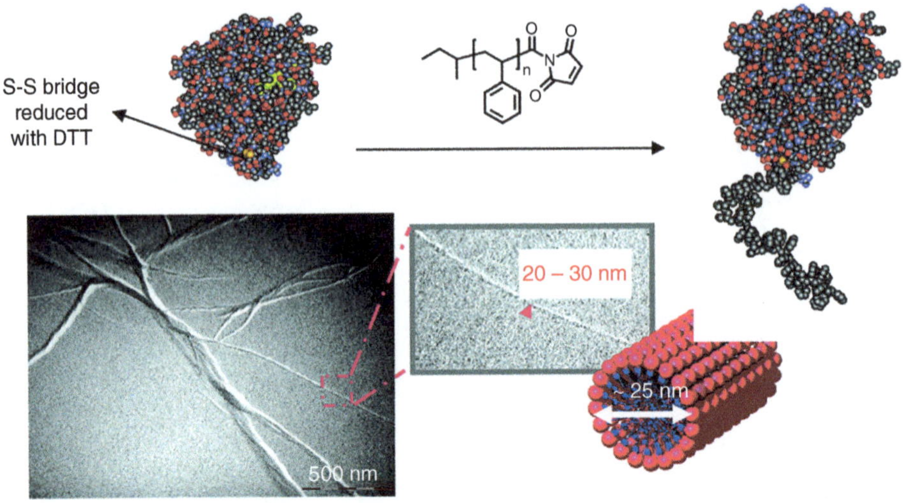

Figure 10.5 Synthesis and cylinder micelle formation of a lipase-PS giant surfactant [15].

further investigation on the molecular design, precision syntheses, and solution self-assembly of various protein-based single-head/single-tail giant surfactants. For example, a bovine serum albumin (BSA)-PS giant surfactant containing a similar tail component could self-assemble into spherical micelles with a diameter of 30–70 nm [58]. In another case, a giant surfactant based on horseradish peroxidase (HRP) with a PS tail was prepared through cofactor reconstitution method in aqueous media [59]. Vesicular micelles formed in situ during preparation. Those aggregates still retained part of catalytic activity [59]. Olsen and coworkers systematically studied the self-assembly of globular protein-based giant surfactants in concentrated solution [24, 60, 61]. Globular proteins, such as mCherry, enhanced green fluorescent protein (EGFP), were conjugated with temperature responsive polymer tails, including poly(*N*-isopropylacrylamide) (PNIPAm), poly(hydroxypropyl acrylate) (PHPA), and others, to prepare the target giant surfactants. The resulting macromolecular amphiphiles could self-assemble into various ordered structures in concentrated solution. The effects of protein–polymer interactions as well as protein surface potential on their assembly were illustrated. Phase diagram of these giant surfactants has also been constructed.

In the past a few years, extensive efforts have been devoted to the development of a new class of giant surfactants with functional POSS MNP as head due to the feasibility of precise control on surface functionalities, molecular architectures, etc. [16, 17, 35, 62, 63]. The simplest model compound, carboxylic acid-functionalized polyhedral oligomeric silsesquioxane–polystyrene conjugates (APOSS-PSs), consisting of a rigid hydrophilic APOSS head and a hydrophobic polymer tail, was prepared by a combination of living anionic polymerization, hydrosilylation reaction, and thiol-ene "click" chemistry (Figure 10.6) [18]. Morphological transitions from vesicles to wormlike cylinders and further to spheres were observed in solution with increasing degree of ionization of the carboxylic acid groups, which enhances the repulsive forces among the APOSS cages in these micellar assemblies. Considering the conformational rigidity and shape persistence of the POSS head, the repulsion force is associated with thermodynamic distance between neighboring APOSS cages, leading to micellar morphologies with different curvatures. More interestingly, it was observed that the PS tails in the micellar cores were highly stretched, which is quite similar to those of small-molecule surfactants. This phenomenon originates from the rigid conformation of the APOSS and the strong APOSS–APOSS interactions. These interactions force the PS tails to stretch in order to minimize the overall free energy. This study inspires further investigations on the rational molecular design and subsequent solution self-assembly of a library of giant surfactants in accordance to their small-molecule counterparts with various structural variations, which are going to be discussed in following sections.

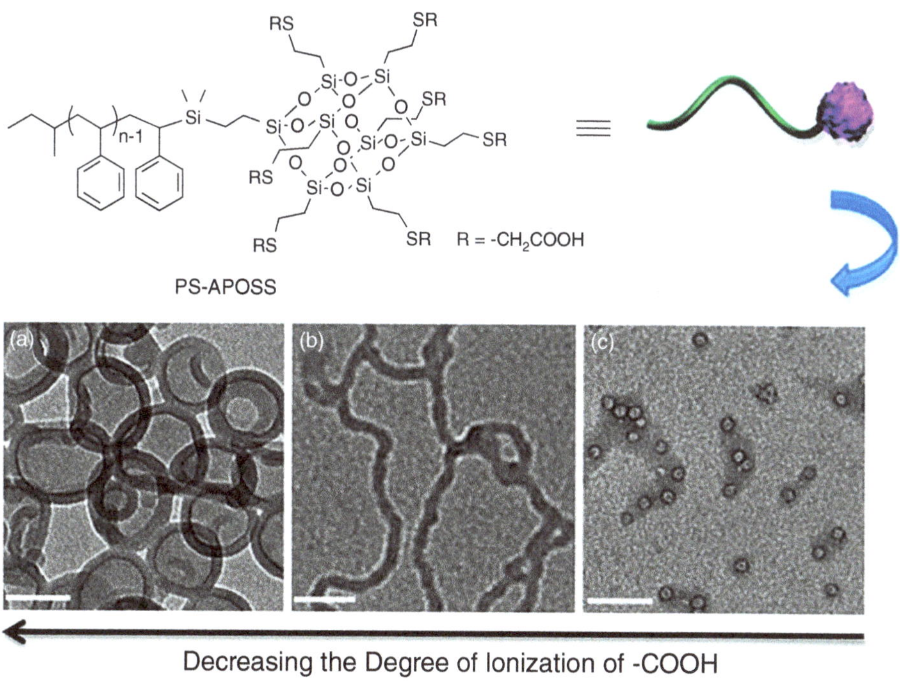

Figure 10.6 Self-assembly of APOSS-PS giant surfactant into various micellar morphologies: vesicles in 1,4-dioxane (a), cylinders in DMF (b), and spheres in DMF/NaOH (c). The scale bar is 80 nm [18].

A comprehensive study on the phase behaviors and self-assembled morphologies of giant surfactants in solution was performed by using AC_{60}–PS as model compound [19]. It was evident that the micellar structures transfer from spheres to worm-like cylinders and further to vesicles with increasing initial concentration in 1,4-dioxane/DMF/water solution, where 1,4-dioxane/DMF is a mixed common solvent and water is a selective solvent, as observed in bright-field TEM images (Figure 10.7a). On the other hand, if we keep the initial concentration being a constant (>0.5 wt%), the morphology of micelle evolves with increasing PS tail length, resulting in the different interface curvatures between the hydrophobic tail and hydrophilic head of the giant surfactants. A phase diagram was constructed based on the initial concentration and the PS tail length (Figure 10.7b). It is worth noting that the phase diagram suggests a balance between two key thermodynamic parameters during the micelle formation: the degree of head ionization and the aggregation number of macromolecules in one micelle [19].

POM-based giant surfactants were firstly prepared by Wang's group via in situ atom transfer radical polymerization (ATRP) and protonation reaction (Figure 10.8) [64]. The resultant giant surfactants were found to form vesicles during protonation due to the hydrophilic nature of protonated POM MNPs. The aggregation process from single molecules to vesicular micelles is kinetically driven [65]. Moreover, the morphological evolution from vesicles to tubular aggregates consisting of POM nanotubes covered with PS coronas could be observed after further annealing (Figure 10.8). The tubular aggregates can grow and then organize into micelle-stacked domains in a parallel manner

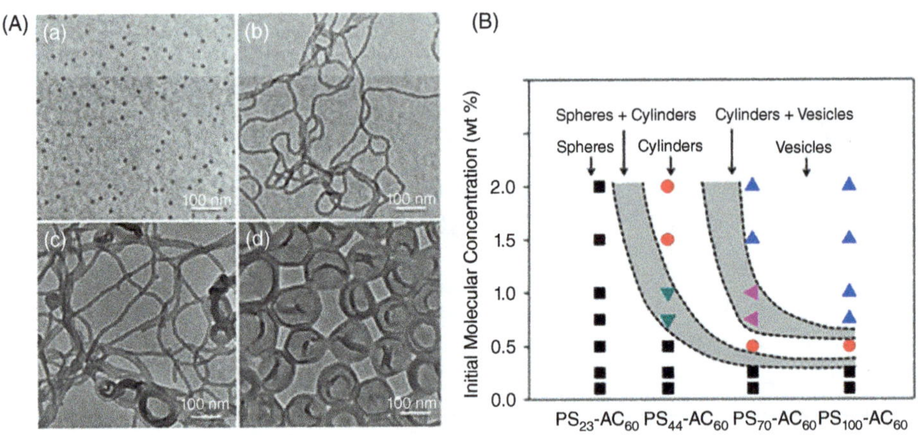

Figure 10.7 (A) Self-assembly of AC_{60}–PS_{70} giant surfactant into various micellar morphologies in the 1,4-dioxane/DMF/water with different initial concentrations: (a) 0.1 wt%; (b) 0.5 wt%; (c) 1.0 wt%; (d) 2.0 wt%; (B) Morphological phase diagram of AC_{60}–PS_n giant surfactants on the basis of the PS tail length and the initial concentration [19].

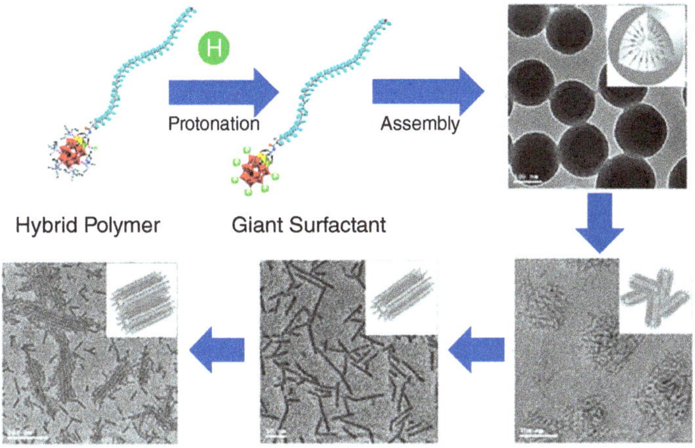

Figure 10.8 Self-assembly of POM-PS giant surfactants and the micellar morphology evolution under thermal annealing [64, 65].

(Figure 10.8). It was anticipated that such an intriguing morphological evolution involves typical step-by-step process toward the final global equilibrium state [65].

Interesting morphology evolutions of these giant surfactants with single-head/single-polymer-tail not only provide the scientific understanding of solution self-assembly behaviors, but also lead to an influx of new ideas in the bottom-up fabrication of functional nanomaterials for various potential applications.

10.5 Giant Surfactants with Multiheads and Multitails

Inspired by the intriguing properties and interesting self-assembly behaviors of single-head/single-tail giant surfactants, various topologies of giant surfactants in analogy to their small-molecule counterparts have been designed and synthesized, including giant lipids [44], symmetric and asymmetric giant gemini surfactants [20, 35, 47], giant bolaform surfactants [47], multiheaded/multitailed giant surfactants [35, 36, 47], and macrocyclic giant surfactants [66, 67]. The molecular design in giant surfactants also denotes an evolution toward complex macromolecular structures with unprecedented solution self-assembly behaviors. These well-defined macromolecules offer necessary and promising extensions of single-head/single-tail giant surfactants, and they are particularly essential for qualitative and systematical investigations on the importance of architecture effects on the self-assembly behaviors [19, 46].

It is well known that the self-assembly behaviors of giant surfactants depend on not only their primary chemical structures but also molecular topologies. In this sense, we have specifically designed and synthesized a series of topological isomers, which possess identical chemical compositions but different topologies for the tails. We have already demonstrated that the self-assembled structures is highly sensitive to the molecular topology in the solid state [46]. For example, a giant surfactant consisting of a hydroxyl-functionalized polyhedral oligomeric silsesquioxane (DPOSS) tethered with a hydrophobic PS tail (DPOSS-PS_{35}) forms a double gyroid structure, while the giant surfactant with two tails but similar composition (DPOSS-$2PS_{17}$) exhibits a hexagonally packed cylinder structure [46]. Similar topological effect has also been observed in the solution self-assembly. AC_{60}–PS_{44} with a single PS tail forms spherical micelle, while vesicles are observed for its corresponding topological isomer (AC_{60}–$2PS_{23}$) with two PS tails under the identical conditions, as observed in bright-field TEM images and dynamic light scattering studies (Figure 10.9) [19]. The packing parameter argument can be applied to explain the topological effects of AC_{60}-based giant surfactants on their self-assembly behaviors. AC_{60}–PS_{44} with a stretched single tail is expected to be in a wedge-like molecular shape, while AC_{60}–$2PS_{23}$ could be regarded as cylinder-like. The molecular shape further leads to distinct supramolecular packing and thus, different morphologies in solution. This study highlights the importance of topological effects in giant surfactants, which is less pronounced in traditional amphiphilic polymers but very common in small-molecule surfactants.

The solution self-assembly behaviors of giant gemini surfactants with two APOSS heads and two hydrophobic PS tails (PS-$(APOSS)_2$-PS) were also

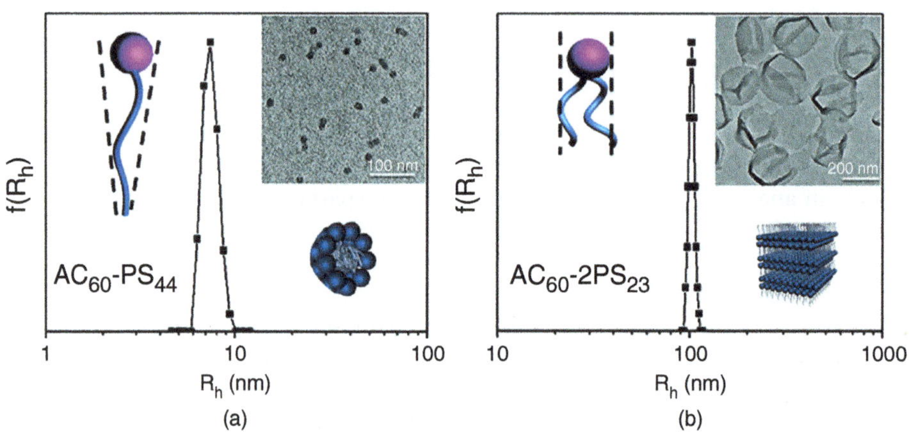

Figure 10.9 Self-assembled micellar morphologies of AC_{60}–PS_{44} (a) and AC_{60}–$2PS_{23}$ (b) in solution with a mixture of 1,4-dioxane and DMF as the common solvent and water as the selective solvent [19].

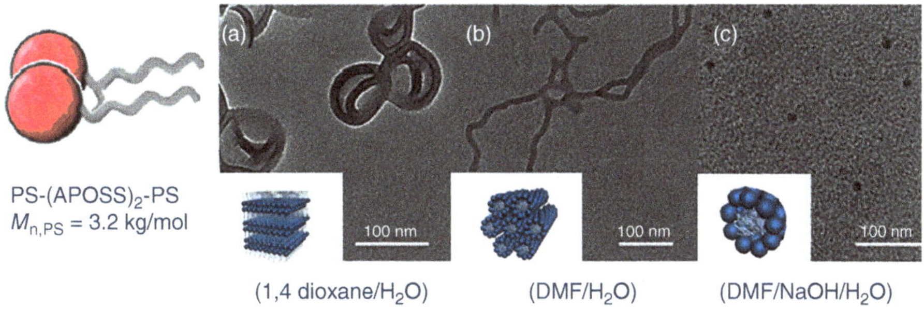

Figure 10.10 Self-assembly of giant gemini surfactants, PS-(APOSS)$_2$-PS, into various micellar morphologies in solution with an initial concentration of 2.8% and a final water content >70% using 1,4-dioxane (a), DMF (b), and DMF/NaOH (c) as common solvent [20].

studied (Figure 10.10) [20]. Similar to the corresponding single-head/single-tail counterparts (APOSS-PS) [18], the micellar morphologies transfer from spheres to cylinders and further to vesicles as the degree of ionization of the carboxylic acids on POSS heads decreases. The PS tails are less stretched in the micellar cores of these giant gemini surfactants than the corresponding APOSS-PS analogues. More importantly, it was found that the spacer length could greatly affect their self-assembly behaviors in solution. The PS tail conformation is usually less stretched in the micelles formed by the giant gemini surfactant with shorter spacer. This decreased chain stretching is due to the topological constraint imposed by the rigid spacer, which leads to an increase of local charge density, and results in redistribution of the APOSS heads on the micellar surface to minimize the overall electrostatic repulsive free energy [20].

10.6 Giant Surfactants with Block Copolymer Tails

Introducing an additional chemically distinct tail will greatly expand the scope of giant surfactants [21, 45, 49, 68]. Specifically, giant surfactants with three mutually immiscible components or blocks are expected to generate multiple interactions, resulting in the formation of diverse hierarchical assemblies [21, 49, 69]. This second tail could be attached at either end of original tail or directly on the MNPs, leading to topological isomer pairs with the same composition but different architectures [49].

Nolte and co-workers studied the solution self-assembly behaviors of giant surfactants consisting of a synthetic diblock copolymer, polystyrene-*block*-polyethylene oxide (PS-*b*-PEO), and a hemeprotein, myoglobin (Mb) or HRP [70]. Depending on the nature of the protein and the length of the PS block, these giant surfactants were found to form a large variety of micellar structures in aqueous solution, including the classical aggregates such as cylinders and vesicles, and unconventional supramolecular structures including Y junctions, octopus structures, and figure eight structures, etc. (Figure 10.11). The large number of different micellar morphologies observed in one sample may be explained by the fact that the aggregates become kinetically trapped due to the glassy nature of PS at ambient temperature. The authors also proposed another possibility that the system is under equilibrium but situated in the coexisting region between two successive morphology regimes [70].

Xia et al. demonstrated the ability to perform orthogonal conjugation reactions at the same active site on globular proteins, enabling the synthesis of protein-based giant surfactants with different architectures [71]. A site-specific conjugation among green fluorescent protein (GFP), poly(dimethylacrylamide) (PDMA), and PNIPAM results in linear and Y-shaped giant surfactants. Both of them self-assembled into spherical micelles in aqueous solution upon heating above the lower critical solution temperature (LCST) of PNIPAM. The spheres from Y-shaped conjugates possess smaller radius as compared to the linear one, consistent with the more compact molecular configuration of the branched architecture. Micellization process was completely reversible. The aggregates dissemble and conjugate monomers were restored within minutes upon cooling back to room temperature. These thermo-responsive micelle conjugates could be used as materials for enzyme encapsulation and protein therapeutics.

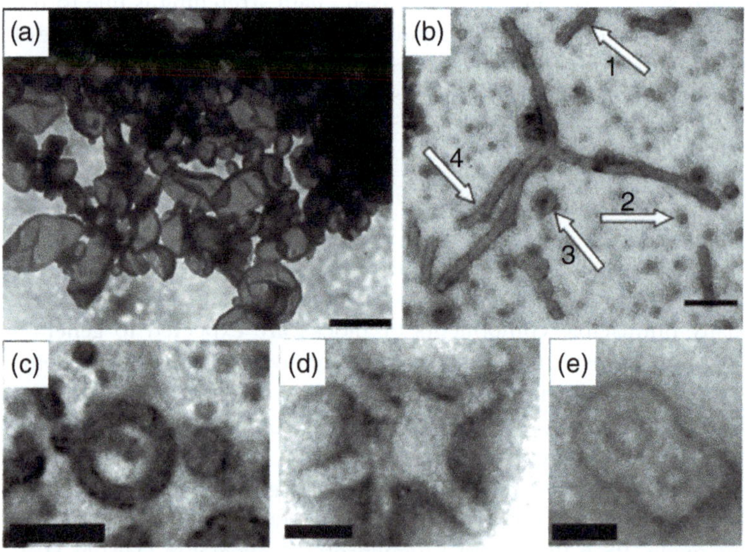

Figure 10.11 Vesicles (a) and junction structures (b) formed from HRP-*b*-PEO-*b*-PS giant surfactants, and toroids (c), octopi (d), and figure eight structure from myoglobin–*b*-PEO-*b*-PS giant surfactants [70].

Recently, our group has demonstrated the micellization of a specifically designed giant surfactant with a fluorinated polyhedral oligomeric silsesquioxane (FPOSS) head and a PS-*b*-PEO diblock copolymer tail (FPOSS-PS_{140}-*b*-PEO_{45}, the subscripts refer to the degree of polymerization) [21, 49]. It consists of three mutually immiscible components: hydrophilic PEO block, hydrophobic PS block, and omniphobic FPOSS MNP. At low water contents, this giant surfactant forms spheres and cylinders with a core-shell-corona structure (Figure 10.12). According to the relative hydrophobicity, the FPOSS aggregated as a core, which was surrounded by the PS shell and further stabilized by the PEO corona. Abundant unconventional micellar morphologies were found, including toroids, tadpoles, dumbbells, 2D hexagonally patterned colloidal nanosheets, and laterally structured vesicles with further increasing water content (Figure 10.12). Most interestingly, thermodynamically stable 2D hexagonally patterned nanosheets were observed, which can even reach tens of micrometers in lateral dimension. These nanosheets possess hexagonally patterned internal structures (Figure 10.12). Both AFM and TEM images of these nanosheets revealed that they have a higher thickness at the boundary compared to that of the interior. These nanosheets are 2D layers with hexagonally connected FPOSS domains embedded inside of a continuous PS matrix, both of which are further covered by the PEO coronas.

With a detailed calculation of free energy terms, it was concluded that these distinctive morphologies are under thermodynamic equilibrium. The

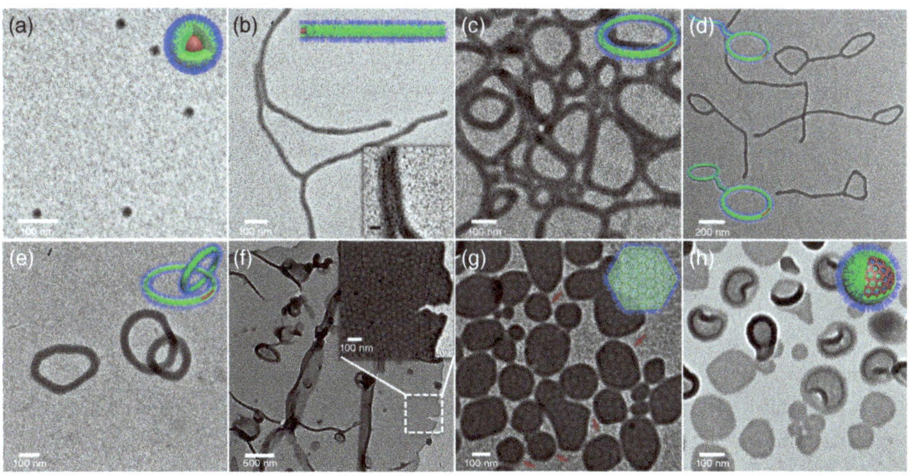

Figure 10.12 Bright-field TEM images of spheres (a), cylinders (b), toroids (c), tadpoles and dumbbells (d), interlocked toroids (e), 2D nanosheets (f, g), and laterally structured vesicles (h) formed by FPOSS-PS-*b*-PEO giant surfactants. The inset in (f) is enlarged image of the area inside the white box. The proposed models are illustrated with different colors: FPOSS (red), PS (green), and PEO (blue) [21].

formation of these morphologies is balanced by two competing free energies: the endcap/rimcap energies of cylinders/nanosheets, and the bending energy to form closed structures. Incorporating the superhydrophobic FPOSS cage enhances the endcap energy of cylinders and promotes the toroidal formation. The nanosheets are stabilized by the thicker rims to partially release the rimcap energy.

10.7 Conclusions

In this chapter, we have reviewed recent progresses in solution self-assembly behaviors of giant surfactants. With rational molecular design and high efficient coupling reactions, giant surfactants based on different functional MNPs and various polymer tails with diverse architectures can be prepared. It has been recognized that these giant surfactants could self-assemble into a large variety of ordered structures in solution, which share some general principles as small-molecule surfactants and block copolymers. Meanwhile, the giant surfactants also exhibit many unique characteristics not yet observed in their small-molecule counterparts. Packing parameter is adopted to account for micellar morphologies variations. The self-assembled structures are also sensitive to the molecular topologies/architectures, which is clearly illustrated by the comparison of the phase behaviors of topological isomers. The ultimate goal of the research in this area is to elucidate the universal principles that govern the self-assembly behaviors of giant surfactants, which could guide the design of new soft functional materials. To achieve this goal, many challenges need to be addressed. Future efforts in this direction may be focused on: (i) advances in coupling chemistry that provide higher precision in primary molecular structure control, and could bridge synthetic and biological systems (as well as other hybrid systems) without compromising the bioactivity, giving materials with controlled heterogeneity; (ii) theoretical studies and computer simulations that could provide detailed roadmaps about structural formations and transformations from various building blocks. This would shed light on the fundamental understanding of the thermodynamics and dynamics of their self-assembly; (iii) establishment of general and simple processes for biohybrids to create technologically relevant nanostructures with minimum loss of bioactivity; and (iv) integration of chemistry and physics principles for rational design and programing of hierarchical structure formation toward specific functions. Last but not least, giant surfactants would seek for applications which fit their unique properties. It is believed that these giant surfactants will stand out as a unique class of materials in addition to traditional small-molecule surfactants and block copolymers.

Acknowledgments

The authors thank the researchers whose results are cited here. This work was supported by the National Science Foundation (DMR-1408872).

References

1 Whitesides, G.M. and Grzybowski, B. (2002). Self-assembly at all scales. *Science* 295: 2418.
2 Israelachvili, J.N. (2011). *Intermolecular and Surface Forces*. Amsterdam: Elsevier.
3 Fong, C., Le, T., and Drummond, C. (2012). Lyotropic liquid crystal engineering-ordered nanostructured small molecule amphiphile self-assembly materials by design. *J. Chem. Soc. Rev.* 41: 1297.
4 May, S. and Ben-Shaul, A. (2007). Molecular packing in cylindrical micelles. In: *Giant Micelles: Properties and Applications* (ed. R. Zana and E.W. Kaler). Boca Raton: CRC Press.
5 Zhang, L. and Eisenberg, A. (1995). Multiple morphologies of "crew-cut" aggregates of polystyrene-b-poly(acrylicacid) block copolymers. *Science* 268: 1728.
6 Zhang, L. and Eisenberg, A. (1998). Formation of crew-cut aggregates of various morphologies from amphiphilic block copolymers in solution. *Polym. Adv. Technol.* 9: 677.
7 Mai, Y. and Eisenberg, A. (2012). Self-assembly of block copolymers. *Chem. Soc. Rev.* 41: 5969.
8 Price, C. (1982). Colloidal properties of block copolymers. In: *Developments in Block Copolymers*, vol. 1 (ed. I. Goodman), 39. London: Applied Science.
9 Zhulina, E.B. and Borisov, O.V. (2012). Theory of block polymer micelles: recent advances and current challenges. *Macromolecules* 45: 4429.
10 Savic, R., Luo, L., Eisenberg, A., and Maysinger, D. (2003). Micellar nanocontainers distribute to defined cytoplasmic organelles. *Science* 300: 615.
11 Park, M. (1997). Block copolymer lithography: periodic arrays of 10(11) holes in 1 square centimeter. *Science* 276: 1401.
12 Vriezema, D.M., Aragones, M.C., Elemans, J.A.A.W. et al. (2005). Self-assembled nanoreactors. *Chem. Rev.* 105: 1445.
13 Israelachvili, J.N., Mitchell, D.J., and Ninham, B.W. (1976). Theory of self-assembly of hydrocarbon amphiphiles into micelles and bilayers. *J. Chem. Soc., Faraday Trans.* 72: 1525.
14 Antonietti, M. and Forster, S. (2003). Vesicles and liposomes: a self-assembly principle beyond lipids. *Adv. Mater.* 15: 1323.

15 Velonia, K., Rowan, A.E., and Nolte, R.J.M. (2001). Lipase polystyrene giant amphiphiles. *J. Am. Chem. Soc.* 124: 4224.

16 Zhang, W.-B., Yu, X., Wang, C.-L. et al. (2014). Molecular nanoparticles are unique elements for macromolecular science: from "nanoatoms" to giant molecules. *Macromolecules* 47: 1221.

17 Yu, X., Li, Y., Dong, X.-H. et al. (2014). Giant surfactants based on molecular nanoparticles: precise synthesis and solution self-assembly. *J. Polym. Sci. B Polym. Phys.* 52: 1309.

18 Yu, X., Zhong, S., Li, X. et al. (2010). A giant surfactant of polystyrene-(carboxylic acid-functionalized polyhedral oligomeric silsesquioxane) amphiphile with highly stretched polystyrene tails in micellar assemblies. *J. Am. Chem. Soc.* 132: 16741.

19 Yu, X., Zhang, W.B., Yue, K. et al. (2012). Giant molecular shape amphiphiles based on polystyrene-hydrophilic [60]fullerene conjugates: click synthesis, solution self-assembly, and phase behavior. *J. Am. Chem. Soc.* 134: 7780.

20 Wang, Z., Li, Y., Dong, X.-H. et al. (2013). Giant gemini surfactants based on polystyrene–hydrophilic polyhedral oligomeric silsesquioxane shape amphiphiles: sequential "click" chemistry and solution self-assembly. *Chem. Sci.* 4: 1345.

21 Ni, B., Huang, M., Chen, Z. et al. (2015). Giant molecular shape amphiphiles based on polystyrene-hydrophilic [60]fullerene conjugates: click synthesis, solution self-assembly, and phase behavior. *J. Am. Chem. Soc.* 137: 1392.

22 Olsen, B.D. et al. (1659). Self-assembly of globular-protein-containing block copolymers. *Macromol. Chem. Phys.* 2013: 214.

23 Kim, M., Tang, S., Olsen, B.D. et al. (2013). Physics of engineered protein hydrogels. *J. Polym. Sci. B Polym. Phys.* 51: 587.

24 Chang, D., Lam, C.N., Tang, S. et al. (2014). Effect of polymer chemistry on globular protein–polymer block copolymer self-assembly. *Polym. Chem.* 5: 4884.

25 Hirsch, A. and Brettreich, M. (2005). *Fullerenes: Chemistry and Reactions.* Weinheim: Wiley-VCH.

26 Cordes, D.B., Lickiss, P.D., and Rataboul, F. (2010). Recent developments in the chemistry of cubic polyhedral oligosilsesquioxanes. *Chem. Rev.* 110: 2081.

27 Coronado, E. and Gomez-Garcia, C. (1998). Polyoxometalate-based molecular materials. *J. Chem. Rev.* 98: 273.

28 Dolbecq, A., Dumas, E., Mayer, C.R. et al. (2010). Hybrid organic-inorganic polyoxometalate compounds: from structural diversity to applications. *Chem. Rev.* 110: 6009.

29 Li, Y., Guo, K., Su, H. et al. (2014). Tuning "Thiol-Ene" reactions toward controlled symmetry breaking in polyhedral oligomeric silsesquioxanes. *Chem. Sci.* 5: 1046.

30 Zhang, W.-B., Tu, Y., Ranjan, R. et al. (2008). "Clicking" fullerene with polymers: synthesis of [60]fullerene end-capped polystyrene. *Macromolecules* 41: 515.
31 Kolb, H.C., Finn, M.G., and Sharpless, K.B. (2001). Click chemistry: diverse chemical function from a few good reactions. *Angew. Chem. Int. Ed.* 40: 2004.
32 Meldal, M. and Tornøe, C.W. (2008). Cu-catalyzed azide-alkyne cycloaddition. *Chem. Rev.* 108: 2953.
33 Hoyle, C.E. and Bowman, C.N. (2010). Thiol-ene click chemistry. *Angew. Chem. Int. Ed.* 49: 1540.
34 Yue, K., Liu, C., Guo, K. et al. (2012). Sequential "Click" Approach to polyhedral oligomeric silsesquioxane-based shape amphiphiles. *Macromolecules* 45: 8126.
35 Su, H., Zheng, J., Wang, Z. et al. (2013). Sequential triple "click" approach toward polyhedral oligomeric silsesquioxane-based multiheaded and multitailed giant surfactants. *ACS Macro Lett.* 2: 645.
36 Li, Y., Wang, Z., Zheng, J. et al. (2013). Cascading one-pot synthesis of single-tailed and asymmetric multitailed giant surfactants. *ACS Macro Lett.* 2: 1026.
37 Su, H., Li, Y., Yue, K. et al. (2014). Macromolecular structure evolution toward giant molecules of complex structure: tandem synthesis of asymmetric giant gemini surfactants. *Polym. Chem.* 5: 3697.
38 Agard, N.J., Prescher, J.A., Bertozzi, C.R. et al. (2004). A strain-promoted [3 + 2] azide-alkyne cycloaddition for covalent modification of biomolecules in living systems. *J. Am. Chem. Soc.* 126: 15046.
39 Heredia, K.L., Tolstyka, Z.P., and Maynard, H.D. (2007). Aminooxy end-functionalized polymers synthesized by atrp for chemoselective conjugation to proteins. *Macromolecules* 40: 4772.
40 Lowe, A.B. (2010). "Click" reactions and recent applications in polymer and materials synthesis. *Polym. Chem.* 1: 17.
41 Keddie, D.J., Moad, G., Rizzardo, E. et al. (2012). RAFT agent design and synthesis. *Macromolecules* 45: 5321.
42 Matyjaszewski, K. (2012). Atom transfer radical polymerization (ATRP): current status and future perspectives. *Macromolecules* 45: 4015.
43 Zhang, W.-B., Li, Y., Li, X. et al. (2011). Synthesis of shape amphiphiles based on functional polyhedral oligomeric silsesquioxane end-capped poly(l-lactide) with diverse head surface chemistry. *Macromolecules* 44: 2589.
44 Dong, X.-H., Zhang, W.-B., Li, Y. et al. (2012). Synthesis of fullerene-containing poly(ethylene oxide)-block-polystyrene as model shape amphiphiles with variable composition, diverse architecture, and high fullerene functionality. *Polym. Chem.* 3: 124.
45 Li, Y., Dong, X.-H., Guo, K. et al. (2012). Synthesis of shape amphiphiles based on poss tethered with two symmetric/asymmetric polymer tails via

sequential "grafting-from" and thiol–ene "click" chemistry. *ACS Macro Lett.* 1: 834.

46 Yu, X., Yue, K., Hsieh, I.F. et al. (2013). Giant surfactants provide a versatile platform for sub-10-nm nanostructure engineering. *Proc. Natl. Acad. Sci. U.S.A.* 110: 10078.

47 Yue, K., Liu, C., Guo, K. et al. (2013). Exploring shape amphiphiles beyond giant surfactants: molecular design and click synthesis. *Polym. Chem.* 4: 1056.

48 Lin, Z., Lu, P., Hsu, C.H. et al. (2014). Self-assembly of fullerene-based janus particles in solution: effects of molecular architecture and solvent. *Chem. Eur. J.* 10: 11630.

49 Ni, B., Dong, X.-H., Chen, Z. et al. (2014). "Clicking" fluorinated polyhedral oligomeric silsesquioxane onto polymers: a modular approach toward shape amphiphiles with flu- orous molecular clusters. *Polym. Chem.* 5: 3588.

50 Li, D., Yin, P., and Liu, T. (2012). Supramolecular architectures assembled from amphiphilic hybrid polyoxometalates. *Dalton Trans.* 41: 2853.

51 Yin, P., Li, D., and Liu, T. (2012). Solution behaviors and self-assembly of polyoxometalates as models of macroions and amphiphilic polyoxometalate-organic hybrids as novel surfactants. *Chem. Soc. Rev.* 41: 7368.

52 Li, D., Song, J., Yin, P. et al. (2011). Inorganic-organic hybrid vesicles with counterion- and pH-controlled fluorescent properties. *J. Am. Chem. Soc.* 133: 14010.

53 Landsmann, S., Lizandara-Pueyo, C., and Polarz, S. (2010). A new class of surfactants with multinuclear, inorganic head groups. *J. Am. Chem. Soc.* 132: 5315.

54 Yin, P., Wu, P., Xiao, Z. et al. (2011). A double-tailed fluorescent surfactant with a hexavanadate cluster as the head group. *Angew. Chem. Int. Ed.* 50: 2521.

55 Zhang, J., Song, Y.-F., Cronin, L. et al. (2010). Reverse-vesicle formation of organic-inorganic polyoxometalate-containing hybrid surfactants with tunable sizes. *Chem. Eur. J.* 16: 11320.

56 Zhang, J., Song, Y.-F., Cronin, L. et al. (2008). Self-assembly of organic-inorganic hybrid amphiphilic surfactants with large polyoxometalates as polar head groups. *J. Am. Chem. Soc.* 130: 14408.

57 Reynhout, I.C., Cornelissen, J.J.L.M., and Nolte, R.J.M. (2009). Synthesis of polymer-biohybrids: from small to giant surfactants. *Acc. Chem. Res.* 42: 681.

58 Dirks, A.J., van Berkel, S.S., Hatzakis, N.S. et al. (2005). Preparation of biohybrid amphiphiles via the copper catalysed huisgen [3 + 2] dipolar cycloaddition reaction. *Chem. Commun.* 4172.

59 Boerakker, M.J., Hannink, J.M., Bomans, P.H.H. et al. (2002). Giant amphiphiles by cofactor reconstitution. *Angew. Chem. Int. Ed.* 41: 4329.

60 Lam, C.N. and Olsen, B.D. (2013). Phase transitions in concentrated solution self-assembly of globular protein–polymer block copolymers. *Soft Matter* 9: 2393.
61 Lam, C.N., Kim, M., Thomas, C.S. et al. (2014). The nature of protein interactions governing globular protein-polymer block copolymer self-assembly. *Biomacromolecules* 15: 1248.
62 Feng, X., Zhu, S., Yue, K. et al. (2014). T10Polyhedral oligomeric silsesquioxane-based shape amphiphiles with diverse head functionalities via "click" chemistry. *ACS Macro Lett.* 3: 900.
63 Li, Y., Su, H., Feng, X. et al. (2014). Thiol-Michael "Click" chemistry: another efficient tool for head functionalization of giant surfactants. *Polym. Chem.* 5: 6151.
64 Han, Y., Xiao, Y., Zhang, Z. et al. (2009). Synthesis of polyoxometalate–polymer hybrid polymers and their hybrid vesicular assembly. *Macromolecules* 42: 6543.
65 Han, Y.-K., Zhang, Z.-J., Wang, Y.-L. et al. (2011). An intriguing morphology evolution of polyoxometalate-polystyrene hybrid amphiphiles from vesicles to tubular aggregates. *Macromol. Chem. Phys.* 212: 81.
66 Lin, Z., Lu, P., Yu, X. et al. (2014). Sequential "Click" Synthesis of "nano-diamond-ring-like" giant surfactants based on functionalized hydrophilic POSS/C60 tethered with cyclic polystyrenes. *Macromolecules* 47: 4160.
67 Li, Y., Su, H., Feng, X. et al. (2015). Precision synthesis of macrocyclic giant surfactants tethered with two different polyhedral oligomeric silsesquioxanes at distinct ring locations via four consecutive "click" reactions. *Polym. Chem.* 6: 827.
68 Bates, F.S., Hillmyer, M.A., Lodge, T.P. et al. (2012). Multiblock polymers: panacea or Pandora's box? *Science* 336: 434.
69 He, J., Yue, K., Liu, Y. et al. (2012). Fluorinated polyhedral oligomeric silsesquioxane-based shape amphiphiles: molecular design, topological variation, and facile synthesis. *Polym. Chem.* 3: 2112.
70 Reynhout, I.C., Cornelissen, J.J.L.M., and Nolte, R.J.M. (2007). Self-assembled architectures from biohybrid triblock copolymers. *J. Am. Chem. Soc.* 129: 2327.
71 Xia, Y., Tang, S., and Olsen, B.D. (2013). Site-specific conjugation of RAFT polymers to proteins via expressed protein ligation. *Chem. Commun.* 49: 2566.

Index

Self-Assembly: From Surfactants to Nanoparticles, First Edition. Edited by Ramanathan Nagarajan.